To Dottie

a colleague a [...] [...]

[...] on test equating is instrumental

to much of what ETS can do

Paul
D[...]

Test Equating

Contributors

William H. Angoff

Robert H. Berk

Henry I. Braun

William R. Cowell

Joseph Deken

John C. Flanagan

Leon Jay Gleser

David V. Hinkley

Paul W. Holland

Frederic M. Lord

Gary L. Marco

Carl N. Morris

Nancy S. Petersen

Richard F. Potthoff

Donald A. Rock

David Rogosa

Donald B. Rubin

I. Richard Savage

Elizabeth Scott

E. Elizabeth Stewart

T. W. F. Stroud

Ted H. Szatrowski

Robert L. Thorndike

Madeline M. Wallmark

Lawrence E. Wightman

Test Equating

Edited by

Paul W. Holland

Donald B. Rubin

Educational Testing Service
Princeton, New Jersey

 1982

ACADEMIC PRESS

A Subsidiary of Harcourt Brace Jovanovich, Publishers

New York   London
Paris   San Diego   San Francisco   São Paulo   Sydney   Tokyo   Toronto

ACADEMIC PRESS, INC.
111 Fifth Avenue, New York, New York 10003

United Kingdom Edition published by
ACADEMIC PRESS, INC. (LONDON) LTD.
24/28 Oval Road, London NW1 7DX

Library of Congress Cataloging in Publication Data
Main entry under title:

Test equating.

   Chiefly written papers and discussions from the
Test Equating Conference, held in Princeton, N.J.,
Apr. 2-3, 1980.
    1. Educational tests and measurements--United States
--Congresses. I. Holland, Paul W. II. Rubin, Donald B.
III. Educational Testing Service. Division of Measure-
ment, Statistics, and Data Analysis Research. IV. Test
Equating Conference (1980 : Princeton, N.J.)
LB3050.5.T47   371.2'6'0973   82-1699
ISBN 0-12-352520-9   AACR2

To Kathryn and Roberta

Contents

List of Contributors

Numbers in parentheses indicate the pages on which the authors' contributions begin.

WILLIAM H. ANGOFF (*55*), Office of Technical Development, Educational Testing Service, Princeton, New Jersey 08541

ROBERT H. BERK (*163*), Department of Statistics, Rutgers University, New Brunswick, New Jersey 08903

HENRY I. BRAUN (*9*), Division of Measurement, Statistics, and Data Analysis Research, Educational Testing Service, Princeton, New Jersey 08541

WILLIAM R. COWELL (*149*), Higher Education and Career Programs, Educational Testing Service, Princeton, New Jersey 08541

JOSEPH DEKEN (*327*), Department of General Business, University of Texas, Austin, Texas 78712

JOHN C. FLANAGAN (*243*), American Institute for Research, Palo Alto, California 94302

LEON JAY GLESER[1] (*259*), Department of Biostatistics, Harvard School of Public Health and Sidney Farber Cancer Institute, Boston, Massachusetts 02115

DAVID V. HINKLEY (*193*), Department of Applied Statistics, University of Minnesota, Minneapolis, Minnesota 55108

[1] Present address: Department of Statistics, Purdue University, Lafayette, Indiana 47907.

PAUL W. HOLLAND (*1, 9, 271*) Division of Measurement, Statistics, and Data Analysis Research, Educational Testing Service, Princeton, New Jersey 08541

FREDERIC M. LORD (*141*), Division of Measurement, Statistics, and Data Analysis Research, Educational Testing Service, Princeton, New Jersey 08541

GARY L. MARCO (*71*), College Board Statistical Analysis, Educational Testing Service, Princeton, New Jersey 08541

CARL N. MORRIS (*169*), Department of Mathematics, University of Texas, Austin, Texas 78712

NANCY S. PETERSEN (*71*), College Board Statistical Analysis, Educational Testing Service, Princeton, New Jersey 08541

RICHARD F. POTTHOFF (*201*), 803 Winview Drive, Greensboro, North Carolina 27410

DONALD A. ROCK (*247*), Division of Measurement, Statistics, and Data Analysis Research, Educational Testing Service, Princeton, New Jersey 08541

DAVID ROGOSA (*319*), School of Education, Stanford University, Stanford, California 94305

DONALD B. RUBIN (*1, 51, 301, 339*), Division of Measurement, Statistics, and Data Analysis Research, Educational Testing Service, Princeton, New Jersey 08541

I. RICHARD SAVAGE (*229, 343*), Department of Statistics, Yale University, New Haven, Connecticut 06520

ELIZABETH SCOTT (*343*), Department of Statistics, Statistical Laboratory, University of California, Berkeley, Berkeley, California 94729

E. ELIZABETH STEWART (*71*), Higher Education and Career Programs, Educational Testing Service, Princeton, New Jersey 08541

T. W. F. STROUD (*137*), Department of Mathematics and Statistics, Queen's University, Kingston, Ontario, Canada K7L 3N6

TED H. SZATROWSKI[2] (*301*), Division of Measurement, Statistics, and Data Analysis Research, Educational Testing Service, Princeton, New Jersey 08541

ROBERT L. THORNDIKE (*309*), Department of Psychology—Education, Columbia University Teachers College, New York, New York 10027

MADELINE M. WALLMARK (*343*), Higher Education and Career Programs, Educational Testing Service, Princeton, New Jersey 08541

LAWRENCE E. WIGHTMAN (*271*), Higher Education and Career Programs, Educational Testing Service, Princeton, New Jersey 08541

---

[2] Present address: Graduate School of Management, Rutgers University, Newark, New Jersey 07102.

Preface

With the advent of open testing and test disclosure legislation, the technical problems associated with test equating have received renewed interest and have been the subject of vigorous debate. The older methods of equating tests assumed various degrees of test security that are not necessarily compatible with the current legislation. Thus test equating techniques, once viewed as obscure and specialized statistical methods of interest only to testing organizations, have been thrust into the public arena for scrutiny and sharp analysis. The chapters in this book give a detailed and thorough discussion of test equating from many different points of view. It should be a valuable reference on test equating for a broad audience interested in many aspects of tests and testing—educational researchers, psychologists, educators, legislators, and others interested in the technical consequences of educational policies. We have included a bibliography on test equating and related topics to enhance the value of the book as a reference volume. The technical level of the papers varies from general discussions of issues to detailed mathematical analyses of concepts and methods. We believe that test equating is a fertile area for future research and hope that this book will help stimulate discussion.

The Program Statistics Research Project was initiated by Educational Testing Service (ETS) in the summer of 1978. This book is the first major publication to result from this project, and we are indebted to many individuals, at ETS and elsewhere, for their help in making it possible. We especially wish to thank Robert J. Solomon, Winton H. Manning, and E. Belvin Williams for their generous support and aid in initiating and continuing the Program Statistics Research Project at ETS.

Instrumental to the success of the "equating project" were our statistical consultants from outside ETS: Robert Berk, Morris DeGroot, David Hinkley, Carl Morris, Richard Potthoff, Thomas Stroud, and Ted Szatrowski. In addition to their own individual contributions in the volume and to the work that led to it, their insights and criticism have been a continual source of guidance to us. A core group of our ETS colleagues—William Angoff, Frederic Lord, Gary Marco, Ben Schrader, and Elizabeth Stewart— gave freely of their valuable time and experience to help build our understanding of ETS equating procedures. They participated with us at each stage in the development of the project and were crucial to its success. In addition, numerous individuals at ETS provided us with a wide variety of assistance in this project for which we are greatly indebted. We would especially like to thank our colleagues in the Research Statistics Group—Henry Braun, Douglas Jones, Dorothy Thayer, and Howard Wainer, as well as Ernest Anastasio, Albert Beaton, Jane Faggen, Nancy Petersen, and Larry Wightman.

We would like to extend our special thanks to Ledyard Tucker, who on numerous occasions throughout the project has helped our understanding in his own very special way.

We offer our sincerest thanks to the past and present members of our support staff for their strenuous efforts on behalf of this project. Jill Callahan planned and flawlessly ran the conference on which this book is based. Vera House coordinated all of the activities after the conference necessary to produce this volume. Charity Hicks, Linda DeLauro, and especially Kathleen Fairall typed many of the papers that constitute this book and otherwise aided in its preparation in ways too numerous to list. We have, indeed, been very fortunate to have had such a tireless and outstanding staff.

Introduction: Research on Test Equating Sponsored by Educational Testing Service, 1978–1980

Paul W. Holland

Donald B. Rubin

"Virtue can only flourish amongst equals"
Wollstonecraft, 1790

Although this quotation was obviously written before there was any interest in equating large-scale examinations, we feel that when properly interpreted, it describes very well the reasons Educational Testing Service (ETS) has expended substantial effort on the problem of equating tests: Only when tests are equated can it be fair to give them to different people and treat the scores as if based on the same test.

ETS develops and administers a wide variety of tests and examinations each year. Many of the testing programs [for example, the Scholastic Aptitude Test (SAT), the Graduate Record Examination (GRE), and the Graduate Management Admissions Test (GMAT)] use a new version or *form* of the test at each major administration. Although the different test forms for a given test are very similar in format and in the types and ranges of difficulty of the questions asked, the actual questions used are completely

1

different in each new form. Of course, the reason for having different versions of the tests is to maintain fairness and security; if the same questions were used at each administration of the test in one year, the people taking the exam later in the year would have an obvious advantage over those who took it earlier.

Therefore, when an admissions officer at a college sees two applicants with, as an example, identical SAT scores of 680, it is quite common for these two scores to be based on two different forms of the SAT. Yet for ease of comparison, the admissions officer implicitly assumes the two test scores mean the same thing, as if they were based on the same test.

The process by which ETS insures that this implicit assumption is accurate is called test equating. Equating tests involves both techniques for carefully constructing tests and methods for making statistical adjustments to the scores obtained on different test forms to compensate for differences in their relative difficulty. The papers in this book focus exclusively on the statistical aspects of test equating rather than on the problems of test construction.

ETS has been equating tests since its inception and has pioneered the development of several test-equating methods. However, the testing environment has changed over the years. For example, the number of examinees (and, consequently, of exams) has increased and the importance of the exams for admissions has increased in some areas. In addition, testing has come under greater external scrutiny; the New York State LaValle Bill mandates disclosure of tests as well as individual answer sheets.

In 1977, we initiated a study of test equating in the Research Statistics Group at ETS. The objective was to review the methods currently used at ETS as well as to investigate new possibilities and formulate the mathematical and statistical foundations of these procedures. The Program Statistics Research Project was funded by ETS with one of its two primary goals being the study of test equating; the other goal was the investigation of statistical methods for evaluating the validity of tests.

In addition to the statisticians at ETS, we enlisted the help of several external statistical consultants, since we felt that a fresh look at the problem would be beneficial. Three small conferences were held with ETS staff and the outside consultants: December 12, 1978; March 13, 1979; and June 4–5, 1979. The culmination of these meetings was the two-day Test Equating Conference on April 2–3, 1980, held at ETS; the list of attendees is presented in Display 1. Although our activities on test equating started nearly three years ago, the current testing environment emphasizes the importance of our interests. Open testing, involving disclosure of test items, creates a need for new approaches to equating, and the work of the Program Statistics Research Project has been instrumental in the development of appropriate new equating methods.

This book presents the written papers and discussions from the April 1980 test equating conference. The contributions that appear here are finished versions of the verbal presentations, although in the case of some discussions, the contributions were made in writing with the benefit of the written paper at hand. With one exception, all the papers in this volume either were prepared for and presented at the test equating conference or are a direct outgrowth of material presented at the conference. The single exception is William Angoff's "Summary and Derivation of Equating Methods Used at ETS." This piece is a slightly modified version of a paper that Angoff has used for pedagogical purposes within ETS for several years. We felt its inclusion would enhance the value of the book and we are grateful to him for allowing us to include it.

The structure of the book represents an attempt to organize the papers by common themes. Part I, Analyses of Equating Procedures Used at ETS, concentrates on a better understanding of the current equating techniques in use at ETS. There are five pieces in Part I. The Braun and Holland paper, "Observed-Score Test Equating: A Mathematical Analysis of Some ETS Test Equating Procedures," lays down a mathematical framework for studying the kinds

DISPLAY 1

Conference Participants from ETS

Ann Angell	Frederic M. Lord
William H. Angoff	Gary L. Marco
Isaac Bejar	Christopher C. Modu
John Bianchini	Ruth Mroczka
Edwin Blew	Nancy Petersen
James Braswell	Barbara Pitcher
Henry I. Braun	Charlotte Rentz
Marston Case	Donald B. Rubin
Linda Cook	Janice Scheuneman
William Cowell	William B. Schrader
Neil Dorans	June Stern
James Douglass	E. Elizabeth Stewart
Daniel Eignor	Martha Stocking
Jane Faggen	Spencer Swinton
Alice Gerb	James B. Sympson
Harold Gulliksen	Dorothy T. Thayer
Marilyn Hicks	Lawrence A. Thibodeau
Paul W. Holland	Howard Wainer
Douglas H. Jones	Relinda C. Walker
Neal Kingston	Madeline Wallmark
Janet Levy	Victor E. Wichert
Samuel Livingston	Lawrence E. Wightman

Conference Participants from Outside ETS

of equating procedures currently used at ETS. The discussion of this paper, by Rubin, emphasizes the relative ease with which procedures that are tied to observables may be understood and the need for more flexible statistical tools to estimate equating functions. The third piece, "Summary and Derivation of Equating Methods Used at ETS" by Angoff, considers linear and equipercentile equating methods as applied to several commonly occurring data collection procedures. "A Test of the Adequacy of Linear Score Equating Models," by Petersen, Marco, and Stewart, presents the application of linear and equipercentile equating methods in 11 different equating situations; item characteristic curve equating is used to establish the criterion against which the other equatings are compared. The discussion of the paper, by Stroud, summarizes the results and makes suggestions for further research.

Part II of this book, Item Response Theory Equating Methods, deals with the theory and practice of test equating using a stochastic model that characterizes each test item by parameters, and then sets the probability of a correct response on an item as a function of the item parameters and of individuals' ability parameters. Such an approach to tests is referred to here as item response theory (IRT). The first piece in this part, "Item Response Theory and Equating—A Technical Summary," is a concise statement of the underlying theoretical framework written by one of the earliest and most creative contributors to this area of research, Frederic Lord. "Item Response Theory Pre-Equating in the TOEFL Testing Program," by Cowell, provides a description of the successful application of IRT three-parameter equating methods to the Test of English as a Foreign Language Program. The discussion of this paper, by Berk, presents a mathematical consideration of (a) alternative forms of model specification in IRT in order to study robustness and (b) the problem of deciding how well two tests can be equated. "On the Foundations of Test Equating," by Morris, considers the conceptual underpinnings of IRT test equating using a priori defined equivalence classes; this treatment leads to formal definitions of "strong equating" and "weak equating" and thereby to alternative ways to equate tests using IRT. The last piece in Part II, the discussion of Morris's paper by Hinkley, presents ideas useful for assessing the consistency of various equating methods.

Part III of this book, New Methods for Test Equating, discusses alternatives not currently being used. The first piece in Part III, "Some Issues in Test Equating," by Potthoff, is an expansive treatment of test equating consisting of a statement of general perspectives, mathematics for linear test equating, the derivation and illustration of a new method for determining scores, and the consideration of anomolies in definitions of fairness. The discussion of this paper, by Flanagan, focuses on the advantages of using equipercentile equating rather than linear equating, but also provides some personal historical perspectives on the problem. Rock's contribution, "Equating Using the Confirmatory Factor Analysis Model," proposes the use of maximum likelihood factor analysis as a method of estimating parameters and testing assumptions in some common test-equating designs. Gleser's discussion of Rock's paper primarily deals with three main topics: the adequacy of true score models for subtests and the need for item design, the adequacy of the computer program COFAMM and the need to allow for inequality constraints on factor loadings, and the possible advantages of using weighted sums of subtest scores as summary scores for tests. The next paper, "Section Pre-Equating: A Preliminary Investigation," by Holland and Wightman, is an in-depth presentation of initial theory and results for a potentially powerful new method, section pre-equating. This method capitalizes on current technology for estimating parameters from "missing data" and presents the results of an initial application. The first discussion, by Savage, suggests using a Bayesian approach to the problem of imputing correlations that arises in section pre-equating; he also makes several suggestions for further work. Added discussion, by Rubin and Szatrowski, extends the current computational theory to handle section pre-equating data with special covariance patterns.

The final part of this book, Part IV, Related Topics and Discussion, contains papers touching on a variety of topics in testing practice that are important to equating tests. The first, "Item and Score Conversion by Pooled Judgment" by Thorndike, explains the use of pooled judgment of item difficulty as a procedure for equating new forms of a test; evidence suggests that although four or five judges can provide fairly reliable difficulty measurements, the correspondence between judged and empirically established difficulty is far from perfect. The discussion of this paper by Rogosa considers a variety of statistical issues related to the analysis of such data. Deken's "Partial Orders and Partial Exchangeability in Test Theory" demonstrates that equating, and more fundamentally, scoring tests, requires the creation of a full ordering from the natural partial ordering of right-wrong responses to a test. Rubin's discussion emphasizes the need to rely on underlying theoretical considerations as well as statistical properties in moving from a partial to a full ordering. The final piece in Part IV is a collection of brief general comments by Savage, Scott, and Wallmark about the conference. The volume ends with a bibliography on test equating and related topics.

We expect that this collection will be of interest to a broader audience than those concerned with the purely technical aspects of test construction and scoring. For example, it should be invaluable to those interested in studying equating as an important issue arising from open testing and item disclosure In addition, it should be of interest to educational psychologists, especially those with a psychometric background but interested in applications. Furthermore, as statisticians, we expect that the papers will be of interest to both applied and mathematical statisticians. Test equating is a rich and fertile area for statistical research with immediate and important applications. This volume is only a beginning, and we look forward to its serving as a stimulus for further work on test equating and related research.

Part I

ANALYSES OF EQUATING PROCEDURES USED AT ETS

Observed-Score Test Equating: A Mathematical Analysis of Some ETS Equating Procedures*

Henry I. Braun

Paul W. Holland

1. Introduction

Our goal in this paper is to provide a single mathematical framework in which several different test-equating procedures used at ETS can be examined. In particular, we adopt a single definition of observed-score test equating (Definition 2 in Section 2) and exploit it in the analysis of several test-equating procedures. This approach results in an emphasis on the assumptions underlying equating procedures. It also allows us to identify procedures that do *not* satisfy the definition, and hence whose claims to legitimacy must be justified on other grounds. Both possibilities are illustrated in this paper.

We focus on what is called "observed-score" test equating because it forms the foundation and is the underlying theory for most of the equating procedures in current use at ETS. The alternatives to observed-score equating—

* This research was a collaboration in every respect and the order of the authorship is alphabetical. It was supported by the Program Statistics Research Project at ETS.

9

true score equating (Levine, 1955) and item response theory equating (Lord, 1981)—will not be considered here. These two alternatives to observed-score equating are based on assumptions about the psychological mechanisms that generate an individual's performance on tests. Observed-score test equating does not make psychological assumptions; rather, it makes statistical assumptions, which we discuss in detail.

Because we are interested in a mathematical theory that allows us to discuss all observed-score-equating procedures within a single framework, we emphasize *population* quantities rather than *sample* estimates of these quantities. Sampling and sampling variability will be mentioned but will not play a central role in our analysis—except in a brief discussion of the "standard error of equating" in Section 5.

In the remainder of this introduction we consider why test equating is a necessary part of many ETS testing programs. Since the general topic of test equating includes a much wider set of problems than we can cover here, this will help to focus our discussion on test-equating problems that have direct relevance to ETS and the testing programs of its clients.

A feature of many testing programs is that there is no single *form* or version of the test. Instead, there is a continual process in which a test form is developed, then used at one or more administrations, and finally retired from use and replaced by new forms of the test. While such a sequence of tests may exhibit an evolutionary trend, we are more interested in the case in which tests are homogeneous in form and content. In this sense, our discussion follows Levine (1955) in assuming that all of the tests to be equated are "parallel" in structure, timing, item types, and formats, as well as subject matter. They will differ, however, in their *relative difficulty*.

It is inevitable that test forms will vary somewhat in difficulty because the process of test construction must involve numerous human decisions that are based (at best) on pretest information about test items that does not perfectly predict the characteristics of final assembled tests. It is this form-to-form variation in test difficulty that makes *test equating* necessary. When we equate tests we are making numerical adjustments to the scores obtained on each form of the test to compensate for the form-to-form variation in difficulty. In simple terms, the points earned on a harder test are "worth more" than those earned on an easier form of the test. Angoff (1971) discusses the circumstances of test use under which it is especially important that the scores obtained on different forms of the test be as comparable as possible.

In Fig. 1 we have schematically represented the sequence of test forms and have indicated the relationship of several important concepts to this sequence.

(a) *Item-level responses:* This is the vector of coded responses to each item (or question) on the test form. It is the basic information that each test-taker provides by taking the test. In most ETS testing programs, responses to items are coded as *right, wrong,* or *omitted.* Each test-taker's responses to the questions on Form-X are represented by a vector **x**.

(b) *Raw stores:* These are scalar summaries of the vectors of item-level responses. We denote a raw score on a form by the same capital letter as the form (e.g., X for Form-X). The raw scores on Form-X are computed by the function $r_X(\mathbf{x})$. There are two common r functions in use at ETS:

$r_X(\mathbf{x})$ = number of correct answers indicated by **x**

and

$r_X(\mathbf{x})$ = number of correct answers minus a fraction of the number of incorrect answers indicated by **x**. (The fraction depends on the number of options on the multiple-choice items.)

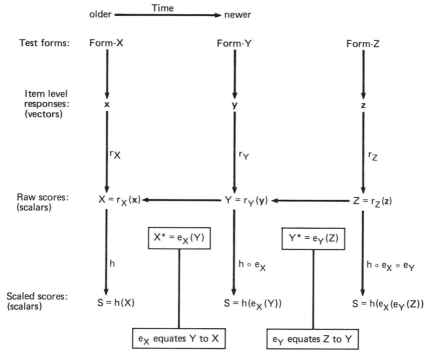

Fig. 1 A schematic representation of a system of tests in which equating occurs. Here r_X, r_Y, r_Z are functions that reduce item-level responses to raw scores (e.g., $r_X(x)$ = the total number of correct answers indicated by **x**); e_X, e_Y are the equating functions; and h is the scaling function.

In the scheme in Fig. 1, r_Y could differ from r_X, but normally this does not occur.

(c) *Equating functions:* The process of test equating results in the selection of equating functions that map the raw scores obtained from a newer test form into raw scores obtained from an older test form. In Fig. 1 we have denoted by e_X the equating function that takes Y raw scores into X raw scores. The result of applying e_X to a raw score Y will be denoted by X^* to remind us that it is like an X raw score but in fact was computed from the item-level responses on a different test form.

(d) *The scaling function:* This is a more-or-less arbitrary final transformation that is applied to the raw scores of X to express the test results in terms of a convenient set of units. An example is the well-known 200–800 SAT scale. For simplicity h is often taken to be a linear function.

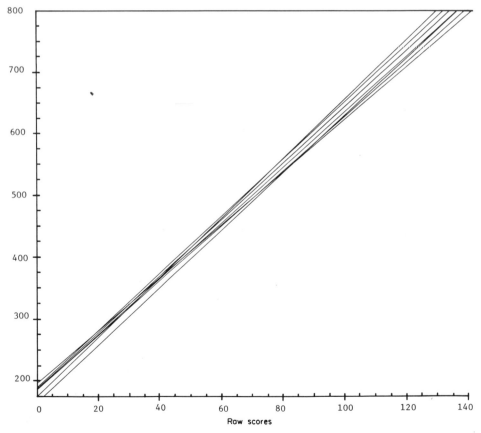

Fig. 2   Conversion lines for seven forms of the same test.

(e) *The conversion functions:* These functions transform the raw scores of a test form into the scaled scores. For Form-X, the conversion function and the scaling function are identical. For Form-Y, the conversion function is the composition of h and e_X, i.e.,

$$S = h(e_X(Y)).$$

For Form-Z, the conversion function is the composition of the conversion function for Form-Y and e_Y, i.e.,

$$S = h(e_X(e_Y(Z))).$$

For test forms that are temporally quite distant from the initial form, the conversion function may involve the composition of h with a large number of equating functions.

Figure 2 shows the conversion functions for seven recent forms of one ETS graduate-level admissions testing program. A raw score of 100 converts to values ranging from 622 to 654 across these seven test forms. A raw score of 120 converts to values ranging from 709 to 750. The variation in these seven conversion functions is a direct result of the fact that these seven tests vary somewhat in their relative difficulty. Testing programs differ in the degree to which their conversion functions differ from form to form.

In order to ascertain whether one test form is more or less difficult than another, the following conceptual framework is useful. Let P denote a population of test-takers for which the test forms in question are appropriate. For each person in P we could administer either Form-X or Form-Y, for example. Thus prior to any actual testing there are two *potential* raw scores associated with each person in P—an X and Y raw score. Let $F(x)$ and $G(y)$ be two distribution functions defined in the following way:

$$F(x) = \text{proportion of people in } P \text{ for which } X \text{ is } \leq x,$$
$$G(y) = \text{proportion of people in } P \text{ for which } Y \text{ is } \leq y. \tag{1}$$

F and G are population quantities rather than sample estimates. F and G may be *estimated* by methods that we shall discuss shortly. Suppose that we are able to estimate F and G and discover that

$$F(x) \leq G(x) \qquad \text{for all} \quad x. \tag{2}$$

We will denote this by $F \leq G$. When $F(x) \leq G(x)$ it means that fewer people in P score lower than x on X than score lower than x on Y. Since this holds for all x, it is intuitively plausible to say that if $F \leq G$, then X is easier than Y for the population P. Since F and G are both distributions of scores for the same population P, we know that if F and G differ, it is because the tests differ and not because the people taking the tests differ in ability. If X is easier than Y, then more people in P will get high scores on X than on Y, so that $F \leq G$.

In order to estimate F and G so that their relative difficulty for a given population P can be ascertained, the following method is commonly employed. A random sample P_1 is drawn from P and tested with Form-X. This yields an estimate of F. A second random sample P_2 is drawn from P and tested with Form-Y. This yields an estimate of G. In practice, at ETS, P_1 and P_2 are obtained simultaneously by alternating (or "spiraling") forms X and Y when they are packaged in boxes for shipping so that when they are handed out to the test-takers approximate random halves of the population of test-takers take each form.

2. Observed-Score Test Equating

There is some disagreement over what test equating *is* and the proper method for doing it. But there is little disagreement about how to tell when an attempt to equate two test forms has failed. This is, we believe, a fundamental fact and forms the entire justification for several methods of test equating that are generally referred to under the rubric of "observed-score test equating."

2.1. How to Tell if e_X Fails to Equate Y to X

We continue with the notation developed in Section 1. Assume that there is a population P that may be tested by Form-X or by Form-Y and we let $F(x)$ and $G(y)$ denote the distribution functions of the X and Y raw scores in P, respectively. Furthermore, we assume that an equating function e_X has been selected according to some definition of test equating so that for each potentially observable raw score Y we can compute the corresponding $X^* = e_X(Y)$. We let $F^*(x)$ be the distribution function of values of X^* over P, i.e.,

$$F^*(x) = \text{proportion of people in } P \text{ for which } X^* \leq x$$

$$= G(e_X^{-1}(x)), \tag{3}$$

where $e_X^{-1}(x)$ is the *inverse* of the equating function e_X. If we knew F and F^*, then we could compare them. If we discovered that $F \neq F^*$, it would mean that the distribution of X and X^* was not the same even though the people taking the tests were the same. We would then have to conclude that e_X had failed to equate Y to X on P. We shall elevate this observation to the level of a formal definition.

Definition 1.   *If there exists a population P for which* $F \neq F^*$, *then Form-X and Form-Y are not equated by* e_X *on P.*

Thus our definition leaves open the possibility that Form-X and Form-Y are equated by e_X on one population but not on another.

Because all distributions involving test scores are discrete, it is usually impossible to achieve $F = F^*$ exactly, so that it would probably be useful to weaken Definition 1 slightly and require that F and F^* differ, in some metric for distribution functions, by no more than a given amount. Furthermore, our greatest concern about the quality of the equating would occur if $F \leq F^*$ or $F^* \leq F$ so that the proper metric ought to give this type of discrepancy a large weight. We will not worry about this detail in this paper but the discreteness of the distributions F, G, and F^* will arise from time to time in our discussion.

In order to apply Definition 1 we may estimate F and F^* as described in Section 1 for F and G.

2.2.   Definition of Observed-Score Equating

In this paper we shall adopt the view that all of the so-called observed-score equating methods are based on the attempt to turn the negative aspects of Definition 1 into a positive definition of test equating. We shall state this positive version as

Definition 2.   *Form-X and Form-Y are equated on P by* e_X *if* $F = F^*$.

Definitions 1 and 2 involve dependence on the population P. We believe that it is very important to be explicit about this dependence on P, because one of the basic assumptions of observed-score test equating is that the choice of the underlying population has a negligible effect on the equating function. In our view, Definition 1 *always* provides a vehicle for ascertaining whether or not an equating function for one choice of P also works for another population. The influence of the underlying population on the selection of an appropriate equating function is always an empirical question that can, in principle, be tested with data.

2.3.   Equipercentile Equating

We may use Definition 2 to obtain a formula for the equating function that equates Y to X on P. This is the so-called equipercentile equating function (Angoff, 1971). It requires us to define the inverse function of F, F^{-1}, which is only problematic if the number of possible values of the raw scores X and Y is small (again the discreteness arises). $F^{-1}(p)$ is that value of x for which $F(x) = p$. We state the result as a theorem.

Theorem 1. *The requirement that $F = F^*$ implies that*

$$e_X(Y) = F^{-1}(G(Y)).$$

The proof is omitted since it is well known when F and G are continuous. When the discreteness of F and G must be taken into account, the proper statement of the theorem is more complicated than we wish to be in this paper. In view of Theorem 1 we shall call $F^{-1}(G(Y))$ the *equipercentile equating function* for P. It follows from Theorem 1 and Definition 2 that the equipercentile equating function for P always equates X and Y on the given population P. It does not follow that it equates X and Y on any other population. Nor does it follow that the sample estimate of $F^{-1}(G(Y))$, found by estimating F and G as described earlier, is a useful equating function to use in practice. Angoff (1971, and his paper in this volume) describes estimates of the equipercentile equating function that are based on *continuous* approximations to the *discrete* distribution functions F and G. All that can be said for using the equipercentile equating function for P is that it is then impossible to use that population and Definition 1 to show that X and Y are *not equated* by e_X. For some choices of P and some tests, this may be a satisfactory definition of equating.

It is important to remember that Definition 2 and the equipercentile equating function that results from it via Theorem 1 are completely *atheoretical* in the sense that they are totally free of any conception (or misconception) of the subject matter of the two tests X and Y. They would apply as mathematical definitions whether or not the content of X and Y is the same. We are only prevented from applying Definition 2 to equate a verbal test to a mathematical test by common sense. This is an inherent problem with observed-score equating.

We point out, however, that the intuitive force of using the distribution functions $F(x)$ and $G(y)$ to assess the relative difficulty of X and Y is weakened if the tests differ in a structural characteristic like length (i.e., number of items). In this case, $F(x)$ and $G(y)$ refer to raw scores that are based on tests of different lengths and that are therefore incomparable unless more is known about the internal structure of the two tests. Strictly speaking, the theory of test equating examined here assumes tests of equal length. Of course, in practice *small* variations in test length are usually ignored in the applications of these methods.

2.4. Linear Equating

To estimate the equipercentile equating function for a given population P requires estimates of F and of G. If one proceeds nonparametrically, these estimates in turn require the estimation of a large number of parameters. Generally speaking, it is better to estimate fewer rather than more parameters, so we are led to look for compromises, i.e., smooth forms for $e_X(Y)$. The smoothest form of e_X we might consider is linear, i.e.,

$$e_X(Y) = aY + b, \tag{4}$$

where a and b are parameters to be estimated.

Strictly speaking, (4) coincides with the equipercentile equating function if and only if

$$F^{-1}(G(Y)) = aY + b \tag{5}$$

or, equivalently, if

$$G(Y) = F(aY + b). \tag{6}$$

Equation (6) is equivalent to the assumption that F and G are of the form

$$F(x) = H[(x - \mu_X)/\sigma_X] \tag{7}$$

and

$$G(y) = H[(y - \mu_Y)/\sigma_Y] \tag{8}$$

for some distribution function H. In (7) and (8) μ_X, μ_Y, σ_X^2, σ_Y^2 are the means and variances, respectively, of X and Y over P. If we apply (7) and (8) to (5), we obtain the following formulas for a and b in (5):

$$a = \sigma_X/\sigma_Y, \tag{9}$$

$$b = \mu_X - (\sigma_X/\sigma_Y)\mu_Y. \tag{10}$$

Applying these to (4) we obtain the usual formula for the linear equating function, i.e.,

$$X^* = e_X(Y) = \mu_X + (\sigma_X/\sigma_Y)(Y - \mu_Y). \tag{11}$$

Although these formulas may be derived in numerous ways, we prefer the present development because it emphasizes the fact that a linear equating function may be viewed as a low-parameter approximation to the equipercentile equating function. Again the discreteness of the distributions causes a small problem. Equation (6) cannot hold exactly if F and G are discrete. Only if (6), (7), and (8) are reasonable approximations can (11) be a reasonable approximation to the equipercentile equating function. As usual, we will ignore the problems caused by discreteness.

Another criterion for choosing a and b in (5) is to find a "best linear approximation" to $F^{-1}(G(Y))$. One such approach is to choose a and b to minimize

$$E(F^{-1}(G(Y)) - aY - b)^2 \tag{12}$$

where the expectation is over the distribution of Y, i.e., G. It is well known that this criterion leads to selecting $aY + b$ to be the linear regression function of $X^* = F^{-1}(G(Y))$ in Y, i.e.,

$$e_X(Y) = \mu_{X^*} + \rho_{X^*Y}(\sigma_{X^*}/\sigma_Y)(Y - \mu_Y), \tag{13}$$

where ρ_{X^*Y} is the correlation of Y with $X^* = F^{-1}(G(Y))$. Since $F^{-1}(G(Y)) = X^*$ has the same distribution as X (disregarding discreteness again), it follows that

$$\mu_{X^*} = \mu_X \qquad \text{and} \qquad \sigma_{X^*} = \sigma_X$$

so that (13) may be rewritten as

$$e_X(Y) = \mu_X + \rho_{X^*Y}(\sigma_X/\sigma_Y)(Y - \mu_Y). \tag{14}$$

Except for the factor ρ_{X^*Y}, (14) is identical to (11). In general, $\rho_{X^*Y} \neq \rho_{XY}$, and so (14) does not commit the fundamental statistical *faux pas* of equating —i.e., confusing (11) with the *regression* of X on Y. In many applications ρ_{X^*Y} will be very high, far higher than ρ_{XY}, and thus substituting the approximation $\rho_{X^*Y} = 1$ into (14) (which leads directly to the usual linear equating formulas (9), (10), and (11)) may *not* be unreasonable. Note that ρ_{XY} is ordinarily not estimable from the data at hand, since a sample of individuals taking both tests X and Y would be required. On the other hand, ρ_{X^*Y} can be estimated once reasonable estimates of F and G are obtained.

In cases where ρ_{X^*Y} is substantially smaller than 1 it is advisable to use a nonlinear approximation to $F^{-1}(G(Y))$. Unfortunately, this idea does not seem to have been explored extensively. Instead, there is a tendency to jump from the low-parameter linear equating function directly to the high-parameter nonparametric estimate of $F^{-1}(G(Y))$ rather than trying to develop an approximation of intermediate complexity. We believe this is a place where modern smoothing methods prove useful (Good and Gaskins, 1980; Mallows, 1980).

3.   Equating Experiments

3.1.   Introduction

Until this point we have been concerned with the formal definition of equating and the population parameters that are involved when we equate Y to X on P. Throughout, our discussion has been in terms of population quantities. In practice these quantities, whether they be entire distribution functions or parameters of a linear transformation, must be estimated from real data obtained from an equating experiment. We shall discuss two such experiments that are commonly employed at ETS, random-groups and common-item equating. The validity of the second method requires more assumptions to be satisfied, but this is compensated for by its increased versatility.

3.2. The Random-Groups Equating Experiment

In a *random-groups equating experiment* a group of test-takers is divided in half randomly and two different tests are administered to each half. There are several ways in which the initial group of test-takers is selected. For example, it may be all persons taking the test on a given day, or it may be a random or nonrandom sample of this larger population. In both cases, there is a population P that could take either test; therefore prior to testing there are two potentially observable raw scores, X or Y, associated with each person in P. When X is given to one random group, we obtain X raw scores on a random sample from P. When Y is given to another random group, we obtain Y raw scores on a second random sample from P. Thus we end up with two random samples, one from F and one from G, and can use these data to estimate the parameters of these distributions and carry out whichever equating procedure has been selected.

Thus a random-groups equating experiment is a very direct, assumption-free, way of obtaining the data that are necessary for doing observed-score equating on a given population. It is not the only way but, conceptually, it is, the simplest. If one of the forms has been used before, the random-groups equating experiment requires reusing the older form. When the security of test forms is hard to insure after they have been used once, the random-groups equating experiment is not a practical way to equate tests. We note, however, that the alternative of common-item equating, to be discussed below, cannot always be implemented because of the structure of the test form. In such cases, a new equating method called *section pre-equating* may be employed. It is discussed in papers by Holland and Wightman and Rubin and Szatrowski of this volume.

3.3. Common-Item Equating

In this section we shall discuss a method of observed-score test equating that is indirect and that requires certain untestable assumptions to hold but does not require reusing older test forms. This method makes use of an *anchor test* V and involves *two* populations of test-takers P and Q. It is called *common-item equating* (CIE).

3.3.1. THE DATA USED IN CIE

Suppose that a random sample of people from a population P are tested with Form-X and another random sample of people from a population Q

are tested with Form-Y. Suppose, in addition, that both samples also take another, shorter test V, which is similar to X and Y in format and content. In Fig. 3a we schematically represent the data that is collected in this situation and contrast it to the information obtained in a random-groups equating experiment in Fig. 3b. The essential difference between the two situations described in Fig. 3 is that in 3b P_1 and P_2 are random samples from the same population P, whereas in 3a P_1 and Q_2 are random samples from two different populations P and Q. The anchor test V is given to both samples P_1 and Q_2 and thus constitutes the items "common" to both tests. Common-item equating (CIE) makes heavy use of V to adjust for the fact that P and Q are different populations.

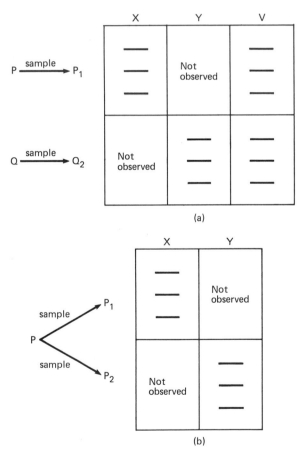

Fig. 3 Schematic representation of the data collected in two equating methods. (a) Data collected in a common-item equating experiment. (b) Data collected in a random-groups equating experiment.

The anchor test may be a separately timed section or it may consist of a number of items embedded in the operational test. However, the following formulas assume that the score on V is not counted in with the score on X or Y. The following calculations also implicitly make the assumption that the presence of the anchor test does not differentially effect the scores of P_1 and Q_2 on X and Y.

3.3.2. THE SYNTHETIC POPULATION

The equating population used in CIE is a synthetic population made from both P and Q. This is done by conceptualizing a larger population that has P and Q as two mutually exclusive and exhaustive strata. We let

$$T = fP + (1 - f)Q \tag{15}$$

denote the synthetic population in which the strata P and Q are weighted by f and $1 - f$, respectively. For a specific choice of f, T is the synthetic population on which the equating will be done. Distributions are computed for T by first computing conditional distributions for each of the two strata, P and Q, and then forming weighted averages of these conditional distributions using f and $1 - f$ as the weights. The usual choice of f is

$$f = N_1/(N_1 + N_2), \tag{16}$$

where N_1 and N_2 represent the sample sizes of P and Q, respectively. This choice of f represents an averaging or compromise between P and Q.

3.3.3. COMMON-ITEM EQUIPERCENTILE EQUATING (FREQUENCY ESTIMATION)

This is a form of common-item equating that makes the fewest assumptions and is therefore the easiest to describe. It is the analog of equipercentile equating described earlier. The objective is to estimate $F_T(x)$ and $G_T(y)$, which are the distributions of X and Y over the synthetic population T. We also need to define the following quantities:

$F_P(x|v)$ = conditional distribution function of X given $V = v$ in P,

$F_Q(x|v)$ = conditional distribution function of X given $V = v$ in Q,

$G_P(y|v)$ = conditional distribution function of Y given $V = v$ in P,

$G_Q(y|v)$ = conditional distribution function of Y given $V = v$ in Q,

$K_P(v)$ = distribution function of V in P,

$K_Q(v)$ = distribution function of V in Q,

$K_T(v)$ = distribution function of V in T.

Then we have

$$K_T(v) = f K_P(v) + (1 - f) K_Q(v) \qquad (17)$$

by definition of the synthetic population. We note that $F_Q(x|v)$ and $G_P(y|v)$ can not be estimated from the data collected in CIE and so must be specified by untestable assumptions. The following two equations are of fundamental importance:

$$F_T(x) = \int F_P(x|v) \, dK_P(v) f + \int F_Q(x|v) \, dK_Q(v) \, (1 - f), \qquad (18)$$

$$G_T(y) = \int G_P(y|v) \, dK_P(v) f + \int G_Q(y|v) \, dK_Q(v) \, (1 - f). \qquad (19)$$

The integrals in (18) and (19) are really sums because all the distributions are discrete, but it is more compact to use the preceding notation. A major simplification occurs if we make the assumption that the distribution of X given $V = v$ is the same in the two populations P and Q, and that the same is true for Y given $V = v$. This simplification is stated in Theorem 2.

Theorem 2. *If $F_P(x|v) = F_Q(x|v)$ and if $G_P(y|v) = G_Q(y|v)$, then*

$$F_T(x) = \int F_P(x|v) \, dK_T(v) \qquad (20)$$

and

$$G_T(y) = \int G_Q(y|v) \, dK_T(v). \qquad (21)$$

The proof of Theorem 2 is straightforward and is omitted. Equations (20) and (21) are important because they express the distributions of interest, namely, $F_T(x)$ and $G_T(y)$, in terms of quantities that can, in principle, be estimated from the data collected in CIE. In fact, this method requires great amounts of data before it is satisfactory because it involves estimating $F_P(x|v)$ and $G_Q(y|v)$, which are both functions of two variables. For this reason great care must be exercised in using this method of estimating $F_T(x)$ and $G_T(y)$ in practice. To our knowledge, little has been done to make this approach practical, and, because of its unstable sampling characteristics, it is generally not preferred when other methods are feasible. Modern methods of data smoothing should have important contributions to make here.

It is also possible to criticize the assumptions of Theorem 2 and it may be possible to use information external to that usually collected in CIE to put forward other values for $F_Q(x|v)$ and $G_P(y|v)$. If this is possible, then formulas

(18) and (19), rather than (20) and (21), are the correct ones to use in computing $F_T(x)$ and $G_T(y)$,

Once F_T and G_T have been estimated (by $\hat{F}_T$ and $\hat{G}_T$, respectively), the estimated equating function is

$$\hat{e}_X(Y) = \hat{F}_T^{-1}(\hat{G}_T(Y)). \tag{22}$$

3.3.4. COMMON-ITEM LINEAR EQUATING (TUCKER EQUATING)

Just as there is an analog to equipercentile equating in CIE, so too is there an analog to linear equating. It is called Tucker equating at ETS in honor of Dr. Ledyard Tucker, who pioneered its early use. It begins by assuming that the equating function is linear and of the form

$$e_X(Y) = \mu_X + \frac{\sigma_X}{\sigma_Y}(Y - \mu_Y), \tag{23}$$

where μ_X, σ_X, μ_Y, and σ_Y are now all to be computed over the synthetic population T. In order to compute these quantities, the mean and variance formulas analogous to (19) or (20) are used. The formula for the mean of X over T is

$$\mu_X = \int E_P(X|v)\, dK_P(v) f + \int E_Q(X|v)\, dK_Q(v)\,(1 - f), \tag{24}$$

whereas that for the variance of X over T is

$$\sigma_X^2 = \int \mathrm{Var}_P(X|v)\, dK_P(v) f + \int \mathrm{Var}_Q(X|v)\, dK_Q(v)\,(1 - f)$$

$$+ \int E_P^2(X|v)\, dK_P(v) f + \int E_Q^2(X|v)\, dK_Q(v)\,(1 - f) - \mu_X^2. \tag{25}$$

The formulas for μ_Y and σ_Y^2 are the same with X replaced by Y everywhere. In formulas (24) and (25),

$$E_P(X|v) = \int x\, dF_P(x|v),$$

$$\mathrm{Var}_P(X|v) = \int [x - E_P(X|v)]^2\, dF_P(x|v),$$

with analogous definitions for $E_Q(X|v)$ and $\mathrm{Var}_Q(X|v)$.

In order to simplify formulas (24) and (25) two different types of assumptions are made. The first type of assumption is analogous to that made earlier in Theorem 2. We assume that the conditional expectation and conditional variance of X given V is the same in P and Q and that of Y given V is also the same in P and Q. These are untestable assumptions. The result is summarized in Theorem 3.

Theorem 3. *If* $E_P(X|v) = E_Q(X|v)$ *and if* $\text{Var}_P(X|v) = \text{Var}_Q(X|v)$, *then*

$$\mu_X = \int E_P(X|v)\, dK_T(v), \tag{26}$$

$$\sigma_X^2 = \int \text{Var}_P(X|v)\, dK_T(v) + \int E_P^2(X|v)\, dK_T(v) - \mu_X^2. \tag{27}$$

Analogous results hold for μ_Y *and* σ_Y^2 *with* X *replaced by* Y *and* P *by* Q.

The second type of assumption might be called a *model* assumption because it assumes a simplified form for $E_P(X|v)$, $E_Q(X|v)$, $\text{Var}_P(X|v)$ and $\text{Var}_Q(X|v)$. These are the typical assumptions made in regression analysis, namely, that the conditional expectations are linear and the conditional variances are constants, i.e.,

$$E_P(X|v) = (a_{X|V,P})v + b_{X|V,P}, \tag{28}$$

$$E_Q(Y|v) = (a_{Y|V,Q})v + b_{Y|V,Q}, \tag{29}$$

$$\text{Var}_P(X|v) = \sigma_{X|V,P}^2, \tag{30}$$

$$\text{Var}_Q(Y|v) = \sigma_{Y|V,Q}^2. \tag{31}$$

The consequences of these two sets of assumptions are summarized in Theorem 4, which gives the usual formulas for Tucker equating.

Theorem 4. *If* $E_P(X|v) = E_Q(X|v) = (a_{X|V,P})v + b_{X|V,P}$ *and if* $\text{Var}_P(X|v) = \text{Var}_Q(X|v) = \sigma_{X|V,P}^2$, *then*

$$\mu_X = (a_{X|V,P})\mu_{V,T} + b_{X|V,P}, \tag{32}$$

$$\sigma_X^2 = \sigma_{X|V,P}^2 + (a_{X|V,P})^2\sigma_{V,T}^2. \tag{33}$$

Analogous results hold for μ_Y *and* σ_Y^2 *with* X *replaced by* Y *and* P *by* Q.

In the usual version of Tucker equating the assumptions of Theorem 4 are made for both X and Y and the sample data are used to obtain least-squares estimates of $a_{X|V,P}$, $b_{X|V,P}$, $\sigma_{X|V,P}^2$, $a_{Y|V,Q}$, $b_{Y|V,Q}$, and $\sigma_{Y|V,Q}^2$. All the data are pooled to estimate $\mu_{V,T}$ and $\sigma_{V,T}^2$ [which assumes that f is given by (17)]. Equations (32) and (33) and their analogs for Y are used to estimate μ_X, μ_Y, σ_X^2, σ_Y^2. Finally, Eq. (23), with sample estimates replacing population quantities, is used to obtain the linear equating function of Y to X.

4. The Statistical Assumptions Underlying Equating

4.1. Introduction

The justification of any statistical procedure rests on certain assumptions about the problem and the nature of the data available. The validity of some of these assumptions can be checked by recourse to the data, but for others this is impossible. Occasionally, special studies may be carried out to provide appropriate data for examining the validity of otherwise untestable assumptions. In our view it is essential that the output of an equating program include diagnostic checks of the various aspects of the procedure.

Common-item equating depends on a number of assumptions that are listed in Table 1. In Section 4.2 we undertake a theoretical analysis of one untestable assumption and develop some understanding of the circumstances under which this assumption may fail to hold. Section 4.3 presents a check of some testable assumptions and carries out a numerical example that indicates to what extent deviations from the assumptions affect the final answer. We concentrate on the assumptions underlying Tucker equating in this paper because of its widespread use. However, the assumptions underlying any equating method should be clearly formulated in statistical terms and checked whenever possible.

4.2. Checking the Untestable Assumptions Made in CIE

The assumption that the conditional distribution of X given V is the same in P (where it can be estimated) and in Q (where it can not be estimated) is inherently untestable by a direct, assumption-free approach. However, if we

TABLE 1

Assumptions Made by Tucker Equating

Assumptions	Possible checks
(1) $T = fP + (1 - f)Q$ with $f = N_1/(N_2 + N_3)$ is adequate	Recompute $\mu_{Y,T}$ and $\sigma^2_{Y,T}$ for several choices of f to see sensitivity of final equating line
(2) $F^{-1}(G(Y))$ is linear	Compare with equipercentile function estimated using a smooth version of frequency estimation
(3) Conditional distributions are the same in P and Q	Untestable unless more data are collected. Possible to make an occasional experimental check
(4) Linearity of regression functions	Standard regression diagnostics
(5) Homogeneity of residual variances	Standard regression diagnostics

are willing to push these assumptions back onto *other* untestable assumptions, we can make some headway.

For simplicity, we shall adopt the framework of Tucker equating even though the ideas we discuss can be extended easily to frequency estimation. Let us assume that the conditional expectations of X given V in P and Q are linear, i.e.,

$$E_P(X|V) = (a_{X|V,P})v + b_{X|V,P}, \tag{34}$$

$$E_Q(X|V) = (a_{X|V,Q})v + b_{X|V,Q}. \tag{35}$$

In the usual form of Tucker equating we assume

$$a_{X|V,P} = a_{X|V,Q} \tag{36}$$

and

$$b_{X|V,P} = b_{X|V,Q}. \tag{37}$$

However, suppose we allowed the possibility that (36) and (37) did not hold? Why might these regression coefficients be different? One reason is that $a_{X|V}$ and $b_{X|V}$ may depend on the distributions of other *measured* variables in P and Q. Let W denote another variable that is measured on all test-takers (examples are sex and, possibly, student background variables) and suppose that the conditional expectation of X given V and W is linear and the *same* in both P and Q, i.e.,

$$E(X|v, w) = av + b + cw. \tag{38}$$

The expectation of X given V in P is

$$E_P(X|v) = E_P(E(X|V, W)|v) = E_P(aV + b + cW|v)$$
$$= av + b + cE_P(W|v). \tag{39}$$

If we assume that the expectation of W given V is also linear, i.e.,

$$E_P(W|v) = a^*v + b^*, \tag{40}$$

then we have

$$E_P(X|v) = (a + ca^*)v + (b + cb^*). \tag{41}$$

In Q the expected value of X given V is

$$E_Q(X|v) = (a + ca^{**})v + (b + cb^{**}), \tag{42}$$

where we have assumed that

$$E_Q(W|v) = a^{**}v + b^{**}. \tag{43}$$

Thus the regression of X on V can differ in P and Q merely because of the distribution of another variable W. If W is measured on all test-takers, then

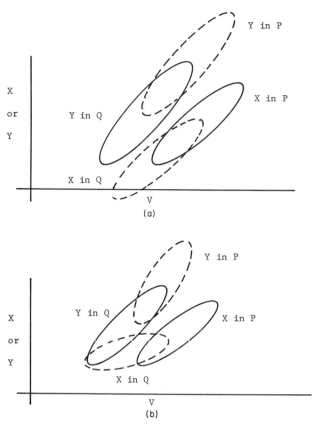

Fig. 4 (a) Schematic scatter plot of X and Y against V for P and Q. Solid ellipses represent observed data. Dotted ellipses represent assumptions made in Tucker equating. (b) An example of how the Tucker equating assumption might be violated.

we can estimate the magnitude of this difference. Figure 4 shows (a) what the untestable assumptions of Tucker equating look like in schematic form and (b) how they might be violated in the presence of "floor" effects.

The preceding analysis suggests that it is useful to maintain a file of descriptive statistics for each equating population so that equating anomalies can be understood in terms of unexpected demographic shifts and the like.

4.3. Checks of Statistical Assumptions

As is pointed out in Table 1, the validity of some of the assumptions underlying linear, common-item equating can be checked by the application of standard regression diagnostics. Among these assumptions are the linearity of the regression of the operational test score on the anchor test score and the homogeneity of the residual variance about this regression.

In Figs. 5 and 7, the sample conditional mean of an operational test score given the anchor test score has been plotted against the anchor test score for two recent administrations of an SAT-V and SAT-M. Although the regressions are substantially linear near the center of the anchor score scale, there are deviations at the extremes. Typically, approximately 1% of the population falls in each of the extreme regions where the assumption of linearity is invalid. The sequence of sample conditional means is a nonparametric estimator of the regression function. The difference between this estimator and the one based on the assumption of linearity of the regression is plotted against anchor test score in Figs. 6 and 8. These plots highlight the apparent departures from linearity found in the data.

Of course, it is of more than peripheral interest to determine the effect of the departure from the assumptions on the final equating. One approach is to replace the quantities $E_P(X|v)$ and $\mathrm{Var}_P(X|v)$ appearing in Eqs. (26) and (27) by their sample estimates based on the subset of the population having exact anchor score v. When these unsmoothed estimates are employed in deriving the means and variances of the synthetic population

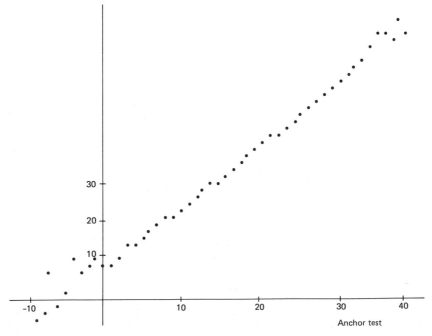

Fig. 5   Linear regression and mean residuals versus anchor test score of SAT-V.

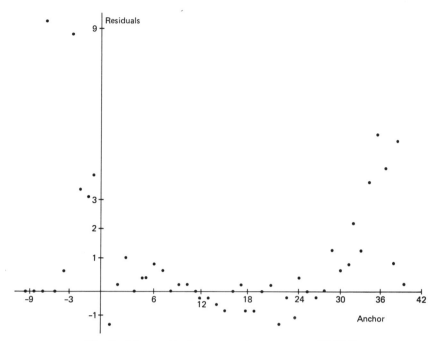

Fig. 6 Mean residuals versus anchor test score of SAT-V.

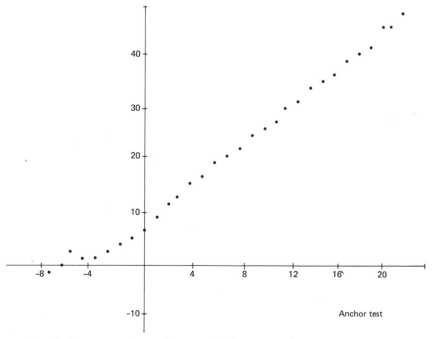

Fig. 7 Linear regression and mean residuals versus anchor test score of SAT-M.

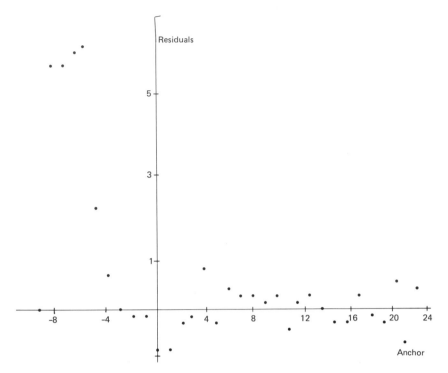

Fig. 8   Mean of residuals versus anchor test score of SAT-M.

$fP + (1 - f)Q = T$ on the two forms, the results may then be compared with those derived on the basis of the two assumptions being questioned. The results are presented in Tables 2 and 3.

A comparison of the conversion of two scores on the new form into scaled scores reveals that slight differences in the estimated means and standard deviations of the synthetic population on each form can have a nonnegligible effect on the final transformation. Although the two equatings of the SAT-V are remarkably similar, the equatings of the SAT-M produce a difference of 17 points at a raw score of 15. The existence of such a discrepancy does not prove, of course, that the assumption of linearity is incorrect and one would be loathe to use the nonparametric regression estimate unquestioningly. A smoothed version of this regression function should yield equating results that are more like those produced by the standard method. Nonetheless, our example points out the need for such diagnostics in everyday operation. They permit constant monitoring of the data as well as tracking steady departures from the assumptions.

TABLE 2

Alternative Equatings of SAT-V

	Equating based on assumption of linearity of regression	Equating based on nonparametric regression
Estimates of parameters in synthetic population		
Old form		
Mean	35.66	36.05
S.D.	15.11	15.25
New form		
Mean	34.48	34.74
S.D.	15.89	16.05
Conversion formula: New scaled score = a (new raw score) + b		
a	0.95	0.95
b	2.87	3.03

Typical conversions		
Raw score	*Scaled score*	*Scaled score*
25	374	375
50	544	545
85	781	782

TABLE 3

Alternative Equatings of SAT-M

	Equating based on assumption of linearity of regression	Equating based on nonparametric regression
Estimates of parameters in synthetic population		
Old form		
Mean	24.80	24.60
S.D.	13.72	13.58
New form		
Mean	26.08	24.60
S.D.	12.82	12.81
Conversion Formula: New scaled score = a (new raw score) + b		
a	1.07	1.06
b	−3.10	−2.91

Typical conversions		
Raw score	*Scaled score*	*Scaled score*
15	394	377
30	520	511
60	774	780

5.   Sources of Bias and Variability
that Affect Equating Results

The analysis made in this paper provides us with a framework for classify-ing several types of errors that may accumulate in a chain of equatings like the one outlined in Section 1. This section discusses such errors in terms of the statistical concepts of sampling variability and bias. A complete system for assessing "equating error" should bring all of these sources of variation into the picture. We believe that an important area of research is to provide a workable way in which all of these sources of variation can be reasonably assessed. It will then be possible to know when equating should be done and, contrariwise, when it will merely add noise to the test scores.

5.1.   Sampling Variability and the Standard
Error of Equating

All the methods we have considered here are based on sampling in-dividuals from the population or populations concerned. Consequently, all measurements are subject to sampling variability. Standard statistical tech-niques may be used to derive estimated sampling errors for all the equating methods discussed so far. These sampling errors are based on the assumption that the samples are random. The *standard error of equating* (SEE) is a measure of this type of variability.

5.1.1.   THE STANDARD ERROR OF EQUATING
FOR RANDOM GROUPS

We now derive the SEE for the *linear* equating function in the simplest equating experiment—two random groups. Our purpose is to illustrate the calculations and to give a formula for the SEE in a form that is more general than that given in Lord (1950) or Angoff (1971).

The setting is as follows. There are n_X examinees randomly sampled* from population P and tested with test X and there are n_Y examinees randomly sampled from P and tested with test Y. We will allow n_X and n_Y to be unequal. The score for the ith X-tested examinee is X_i and the score for the jth Y-tested examinee is Y_j. Thus random sampling allows us to assume that the $\{X_i\}$ are

---

* The samples that are used operationally at ETS are, with few exceptions, not strictly random. Instead they are obtained by "spiraling" or alternating test booklets containing the different test forms in the packages that are shipped to test centers. This results in samples that both experience and theory suggest will be slightly less variable than strictly random samples. Thus the formulas for the SEE derived from random sampling theory will tend slightly to overestimate the true value of the SEE.

independent and identically distributed (iid) with a common distribution function $F(x)$ and the $\{Y_j\}$ are iid with distribution function $G(y)$. Furthermore, the X and Y samples are independent. We shall need a notation for certain moments of the X and Y distributions. Let

$$\mu_X = E(X), \qquad \mu_Y = E(Y),$$
$$\sigma_X^2 = \text{Var}(X), \qquad \sigma_Y^2 = \text{Var}(Y).$$

The linear equating function for equating Y to X is given by

$$X^*(y) = \mu_X + \frac{\sigma_X}{\sigma_Y}(y - \mu_Y), \tag{44}$$

where y varies over all Y raw scores. The usual *estimate* of the population quantity $X^*(y)$ is

$$\hat{X}^*(y) = \bar{X} + \frac{s_X}{s_Y}(y - \bar{Y}), \tag{45}$$

where $\bar{X}$, s_X^2 are the sample mean and variance of the $\{X_i\}$, etc. The following theorem gives the asymptotic variance of $\hat{X}^*(y)$. The SEE is the square root of this variance.

Theorem 5

$$\text{Var}(\hat{X}^*(y)) = \sigma_X^2 \left\{ \frac{1}{n_X} + \frac{1}{n_Y} + \left(\frac{\text{skew}(X)}{n_X} + \frac{\text{skew}(Y)}{n_Y} \right) Z(y) \right.$$
$$\left. + \left(\frac{2 + \text{kurt}(X)}{4n_X} + \frac{2 + \text{kurt}(Y)}{4n_Y} \right) Z^2(y) \right\}$$
$$+ \ (\textit{higher order terms}),$$

where

$$\text{skew}(X) = \mu_{3X}/\sigma_X^3, \qquad \text{skew}(Y) = \mu_{3Y}/\sigma_Y^3 \qquad \textit{(skewness coefficients)},$$
$$\text{kurt}(X) = (\mu_{4X}/\sigma_X^4) - 3, \qquad \text{kurt}(Y) = (\mu_{4Y}/\sigma_Y^4) - 3 \quad \textit{(kurtosis coefficients)},$$
$$\mu_{3X} = E(X - \mu_X)^3, \qquad \mu_{3Y} = E(Y - \mu_Y)^3 \quad \textit{(3rd central moment)},$$
$$\mu_{4X} = E(X - \mu_X)^4, \qquad \mu_{4Y} = E(Y - \mu_Y)^4 \quad \textit{(4th central moment)},$$

and

$$Z(y) = (y - \mu_Y)/\sigma_Y.$$

The "higher order terms" referred to in Theorem 5 are positive powers of n_X^{-1} and n_Y^{-1}. The distribution functions $F(x)$ and $G(y)$ are concentrated on finite sets of possible values, consequently, we will ignore the details necessary to insure the validity of the approximation as they are always satisfied by such distributions.

Before proving the theorem we shall state two corollaries of interest.

Corollary 1. If X and Y have Normal distributions, then skew$(X) = 0$, skew$(Y) = 0$, and kurt$(X) = $ kurt$(Y) = 0$ so that

$$\text{Var}(\hat{X}^*(y)) = (\sigma_X^2/n_h)(2 + Z^2(y)) + \text{(higher order terms)},$$

where $n_h = [\frac{1}{2}(n_X^{-1} + n_Y^{-1})]^{-1}$ is the *harmonic* mean of n_X and n_Y.

Corollary 2. If X and Y have Normal distributions and $n_X = n_Y = N/2$, then

$$\text{Var}(\hat{X}^*(y)) = \frac{2\sigma_X^2}{N}(2 + Z^2(y)) + \text{(higher order terms)}.$$

Corollary 2 is the result stated in Lord (1950), whereas Corollary 1 is a version of Lord's results when n_X and n_Y are unequal. Corollary 1 is a more appropriate formula for the SEE than the one given in Angoff (1971) for the case of unequal n_X and n_Y; however, unless n_X and n_Y are very different, the two formulas give very similar results. For example, if $n_Y/n_X = 2$, the corresponding ratio of the values from the two SEE formulas is 1.06, whereas $n_Y/n_X = 4$ implies that this ratio is 1.25. Because of the well-known fact that the harmonic mean is always less than the arithmetic mean, it follows that the formula for the SEE given in Corollary 1 gives values that are always *larger* than the values given by the corresponding formula given in Corollary 2 and in Angoff (1971). We now turn to a proof of Theorem 5.

Proof of Theorem 5. From Kendall and Stuart (1958, p. 233) we have the basic result that if $\{X_i\}$ are iid then $(\overline{X}, s_X)$ has an approximate bivariate Normal distribution with mean vector (μ_X, σ_X) and covariance matrix $n_X^{-1}\Sigma_X$, where

$$\Sigma_X = \begin{bmatrix} \sigma_X^2 & \mu_{3X}/2\sigma_X \\ \mu_{3X}/2\sigma_X & (\mu_{4X} - \sigma_X^4)/4\sigma_X^2 \end{bmatrix} = \sigma_X^2 \begin{bmatrix} 1 & \frac{1}{2}\text{skew}(X) \\ \frac{1}{2}\text{skew}(X) & \frac{1}{4}(2 + \text{kurt}(X)) \end{bmatrix}.$$

Similarly, $(\overline{Y}, s_Y)$ is approximately Normal with mean vector (μ_Y, σ_Y) and covariance matrix $n_Y^{-1}\Sigma_Y$, where

$$\Sigma_Y = \sigma_Y^2 \begin{bmatrix} 1 & \frac{1}{2}\text{skew}(Y) \\ \frac{1}{2}\text{skew}(Y) & \frac{1}{4}(2 + \text{kurt}(Y)) \end{bmatrix}.$$

In addition, $(\overline{X}, s_X)$ is independent of $(\overline{Y}, s_Y)$ since the samples are random. We observe that for a fixed value of y, $\hat{X}^*(y) = f(\overline{X}, s_X, \overline{Y}, s_Y)$, where $f(a, b, c, d) = a + (b/d)(y - c)$.

We use the Taylor expansion of $f(\overline{X}, s_X, \overline{Y}, s_Y)$ about the point $(\mu_X, \sigma_X, \mu_Y, \sigma_Y)$ to compute the approximate variance of $\hat{X}^*(y)$. This requires the partial derivatives of f, which are

$$\frac{\partial f}{\partial a} = 1; \quad \frac{\partial f}{\partial b} = \frac{y - c}{d}; \quad \frac{\partial f}{\partial c} = -\frac{b}{d}; \quad \frac{\partial f}{\partial d} = -\frac{b}{d}\frac{y - c}{d}.$$

From the general theory of the δ *method* or *Taylor expansion method* for calculating asymptotic variances (Kendall and Stuart, 1958, p. 232), we have

$$\mathrm{Var}(\hat{X}^*(y)) = (\partial f)\Sigma(\partial f)^T + \text{(higher order terms)},$$

where (∂f) denotes the (row) vector of partial derivatives of f evaluated at the point $(\mu_X, \sigma_X, \mu_Y, \sigma_Y)$ and Σ is the matrix

$$\Sigma = \begin{bmatrix} n_X^{-1}\Sigma_X & 0 \\ 0 & n_Y^{-1}\Sigma_Y \end{bmatrix}.$$

The vector (∂f) equals $(1, Z(y), -\sigma_X/\sigma_Y, -(\sigma_X/\sigma_Y)Z(y))$. Thus

$$\mathrm{Var}(\hat{X}^*(y)) = \frac{1}{n_X}(1, Z(y))\Sigma_X \begin{pmatrix} 1 \\ Z(y) \end{pmatrix} + \frac{1}{n_Y}\frac{\sigma_X^2}{\sigma_Y^2}(-1, -Z(y))\Sigma_Y \begin{pmatrix} -1 \\ -Z(y) \end{pmatrix}$$

$$+ \text{(higher order terms)}$$

$$= \frac{\sigma_X^2}{n_X}\left\{ 1 + \mathrm{skew}(X)Z(y) + \left(\frac{2 + \mathrm{kurt}(X)}{4}\right)Z^2(y) \right\}$$

$$+ \frac{\sigma_X^2}{n_Y}\left\{ 1 + \mathrm{skew}(Y)Z(y) + \left(\frac{2 + \mathrm{kurt}(Y)}{4}\right)Z^2(y) \right\}$$

$$+ \text{(higher order terms)},$$

which is easily seen to simplify to the stated result.   Q.E.D.

The assumptions of Corollary 1 are not always true for data from real tests. In particular, distributions of test scores are often skewed (i.e., $\mathrm{skew}(X) < 0$ if X is an easy test and $\mathrm{skew}(X) > 0$ if X is a hard test) and also exhibit nonzero kurtosis values. Thus using the formula given in Corollary 1 rather than the more general result given in Theorem 5 will result in a *biased* estimate of the SEE. On the other hand, sample estimates of $\mathrm{skew}(X)$ and $\mathrm{kurt}(X)$ are not necessarily well behaved and care must be exercised in their use to evaluate the magnitude of this bias.

The reader is referred to Angoff (1971) for formulas that may be used to compute the SEE for other types of equating experiments. The formulas given there resemble Corollary 1 and 2 in that they assume negligible values of skewness and kurtosis.

5.1.2. The Standard Error of Equating for Chains of Equatings

It is sometimes of interest to know how the standard error of equating (SEE) propagates through a chain of equatings. We now examine this question and give a formula that can be used to answer it in certain cases.

Suppose $\hat{X}^*(y)$ is the estimated equating function for equating Y to X and that $\hat{Y}^*(z)$ is the estimated equating function for equating Z to Y. Suppose further that we are interested in the composed function

$$\hat{g}(z) = \hat{X}^*(\hat{Y}^*(z)). \tag{46}$$

The standard errors of equating for $\hat{X}^*(y)$ and $\hat{Y}^*(z)$ are defined by

$$(SEE_X(y))^2 = Var(\hat{X}^*(y)) \tag{47}$$

and

$$(SEE_Y(z))^2 = Var(\hat{Y}^*(z)). \tag{48}$$

We shall also assume that $\hat{X}^*(y)$ and $\hat{Y}^*(z)$ are unbiased estimates of $X^*(y)$ and $Y^*(z)$, respectively. Finally, we assume that the equating experiments performed to obtain $\hat{X}^*$ and $\hat{Y}^*$ are statistically independent. Under these conditions the following theorem gives the SEE for $\hat{X}^*(\hat{Y}^*(z))$.

Theorem 6

$$Var\{\hat{X}^*(\hat{Y}(z))\} = (SEE_X(Y^*(z)))^2 + \left(\frac{dX^*}{dy}(Y^*(z))\right)^2 (SEE_Y(z))^2$$

$$+ (higher\ order\ terms).$$

The proof is similar to the Taylor expansion argument used in Theorem 5 and is omitted. The validity of this result again depends on the validity of a first-order, Taylor series approximation to the function of interest.

Since the slope of $X^*(y)$ is generally close to 1, it follows that the SEE for a chain of independent equatings accumulate according to the usual rules for adding variances.

5.2. Sources of Bias in Equating

We restrict our use of the term *bias* to *statistical bias*, i.e., the difference between the average value of an estimator over repeated samplings from the same population and the value of the parameter being estimated. The term *bias* is sometimes used to refer to the difference between the estimated equating function and the true value of the equating function, i.e., the one that would be calculated if the population distributions of the test scores were known exactly. Although the notion underlying the second usage of the term *bias* is important, we shall not consider it further here, since it is concerned with the consequences of using an estimated equating function rather than the statistical properties of the estimate.

There are two major sources of bias in equating: population variability and model errors. We shall discuss each of these in turn.

5.2.1. POPULATION VARIABILITY

Generally speaking, in a system like that illustrated in Fig. 1 it is impossible to restrict all equatings to a single population. Instead, each equating will take place on a different population. This introduces the possibility that an equating function that works for one population may not work for another. For example, if $X_P^*(Y)$ equates Y to X on P and $Y_Q^*(Z)$ equates Z to Y on Q, it may not be the case that the composite functions $X_P^*(Y_Q^*(Z))$ equates Z to X on P or on Q. In common-item equating (CIE) the point can be made even more forcefully. The population used in CIE is the synthetic one composed of a weighted combination of P and Q (Section 3.3.2). There is no guarantee that the resulting equating is satisfactory for equating Y to X on either P or Q separately. This potential dependence on the population used for equating introduces what may be called *population variability* into a chain of equatings.

The statistical theory of population variability, as opposed to sampling variability, has not been extensively studied in the literature on test equating. We shall give an indication of the types of results that can be obtained.

Suppose there are two populations P and Q and three tests X, Y, and Z. Let the distribution functions of test scores for X, Y, and Z be F, G, and H, respectively. We will put a subscript P or Q on the distribution functions to denote the population over which each is computed. The equipercentile equating function for equating Y to X on P is

$$X_P^*(y) = F_P^{-1}(G_P(y)), \tag{49}$$

and the corresponding function for equating Z to Y on Q is

$$Y_Q^*(z) = G_Q^{-1}(H_Q(z)). \tag{50}$$

The composed function

$$g(z) = X_P^*(Y_Q^*(z)) \tag{51}$$

might be viewed as equating Z to X. However, on what population would this be true? Theorem 7 shows that when $P = Q$, $g(z)$ does equate Z to X on P as we would expect.

Theorem 7 *If* $P = Q$, *then*

$$X_P^*(Y_Q^*(z)) = F_P^{-1}(H_P(z))$$

so that $g(z)$ from (51) is the equipercentile equating function for equating Z to X on P.

Proof

$$X_P^*(Y_Q^*(z)) = X_P^*(Y_P^*(z)) \text{ (since } P = Q)$$
$$= F_P^{-1}(G_P(G_P^{-1}(H_P(z))))$$
$$= F_P^{-1}H_P(z) \text{ (since } G_P(G_P^{-1}(y)) = y). \quad \text{Q.E.D.}$$

We interpret Theorem 7 by saying that when $P = Q$ there is no bias in using $X_P^*(Y_Q^*(z))$ to equate Z to X. However, when $P \neq Q$ the question of on what population, if any, does $g(z)$ equate Z to X is not so easily answered. The bias in using $g(z)$ to equate Z to X on P is given by

$$\text{bias}(z) = X_P^*(Y_P^*(z)) - X_P^*(Y_Q^*(z)). \tag{52}$$

When this bias is small, we can approximate it by expanding X_P^* in a Taylor series, i.e.,

$$\text{bias}(z) \simeq (dX_P^*/dy)(Y_P^*(z))\{Y_P^*(z) - Y_Q^*(z)\}. \tag{53}$$

To better understand the import of this formula, suppose that the equipercentile equating function is, in fact, linear. Then

$$X_P^*(y) = \mu_{X,P} + (\sigma_{X,P}/\sigma_{Y,P})(y - \mu_{Y,P}),$$
$$Y_Q^*(z) = \mu_{Y,Q} + (\sigma_{Y,Q}/\sigma_{Z,Q})(z - \mu_{Z,Q}) = A_Q z + B_Q,$$
$$Y_P^*(z) = \mu_{Y,P} + (\sigma_{Y,P}/\sigma_{Z,P})(z - \mu_{Z,P}) = A_P z + B_P.$$

Hence

$$dX_P^*/dy = \sigma_{X,P}/\sigma_{Y,P}$$

and

$$Y_P^*(z) - Y_Q^*(z) = (A_P - A_Q)z + (B_P - B_Q). \tag{54}$$

In many applications we would expect $\sigma_{X,P}$ and $\sigma_{Y,P}$ to be nearly equal so that from (53) and (54) we see that bias(z) is approximately the difference, at $Z = z$, between the two equating lines for equating Z to Y on P and Q, respectively.

5.2.2. MODEL ERRORS

By a model error we mean an error made because a simplifying assumption in the form of a distribution is false. We can distinguish two types of such errors. *Testable model errors* are those, like linearity of regression functions, that with sufficient data, can be shown to be adequate or not. *Nontestable model errors* are those where there are no data available at all to prove or disprove the assumption. The most notable assumptions that may be subject to nontestable model errors are those in frequency estimation and in Tucker equating which assert that the conditional distribution of X given V in P is the same as that of X given V in Q. In Section 4 we discussed model errors in great detail. The effect of model errors is to introduce bias in the estimation of the equating function.

6. Equating Two Tests through a Third

In this section we discuss a procedure similar to common-item equating. The setting is the same as that of Section 4—test X is given to population P, test Y to population Q, and a common test V to both P and Q. Suppose we first equate Y to V on Q and then we equate V to X on P. This results in a two-stage transformation of Y scores into X scores. We shall say that this transformation equates Y to X *through* V. Angoff (1971) refers to this procedure as Design V. Lord (1950) analyzes the sampling behavior of this procedure assuming that $P = Q$ and that the data are generated by random sampling from P. Our interest centers on the *relationship* between (a) equating Y to X through V and (b) equating Y to X on either P or Q, in the sense of Definition 2. In Theorem 8 we give a condition under which equating Y to X through V gives the same result as equating Y to X on P or Q using V as a covariate (cf. Section 4.3). We shall not assume that P and Q are the same, and for simplicity we shall focus our discussion on equipercentile equating. Similar results can be given for linear equating.

As one might guess, when P and Q are different, equating Y to X through V usually does not give the same results as equating Y to X on P or Q using frequency estimation. Our results show that this is not merely due to sampling variability but is true even at the population level.

Employing the notation of Section 3.3, let F, G, and K denote the distribution functions of X, Y, and V, respectively. The subscripts P and Q indicate the population over which each distribution function is computed. For convenience we shall assume that all distribution functions encountered have densities, which we denote by the corresponding lowercase letters, f, g, and k.

We begin by computing the transformation of Y into X that results when they are equated through V. When Y is equated to V on Q, we obtain the equating function

$$V_Q(y) = K_Q^{-1}(G_Q(y)). \tag{55}$$

When V is equated to X on P, we obtain the equating function

$$X_P(v) = F_P^{-1}(K_p(v)). \tag{56}$$

Composing $X_P(v)$ with $V_Q(y)$ yields the desired transformation of y to x:

$$X_{PQ}^*(y) = F_P^{-1}(K_P(K_Q^{-1}(G_Q(y)))). \tag{57}$$

When $P = Q$, we have

$$K_P = K_Q \quad \text{and} \quad G_Q = G_P$$

so that $X_{PQ}^*(y)$ reduces to the usual equipercentile equating function (i.e., $X_{PP}^*(y) = F_P^{-1}G_P(y)$). When $P = Q$, the use of V in this manner is quite different from common-item equating where V was employed as a covariate to correct for minor differences arising from sampling fluctuations. In fact, Lord showed that when $P = Q$ the sampling variability of equating Y to X through V is usually *larger* than that of the random-groups method.

We now consider the conditions under which $X_{PQ}^*(y)$ properly equates Y to X on P *and* on Q. Suppose first that $X_{PQ}^*(y)$ does equate Y to X on Q. Then $X_{PQ}^*(y)$ must satisfy the equation

$$X_{PQ}^*(y) = F_Q^{-1}G_Q(y) \tag{58}$$

or

$$F_P^{-1}K_PK_Q^{-1}G_Q = F_Q^{-1}G_Q. \tag{59}$$

This is equivalent to

$$F_Q(x) = K_Q(K_P^{-1}(F_P(x))). \tag{60}$$

Similarly, $X_{PQ}^*(y)$ equates Y to X on P if and only if

$$G_P(y) = K_P(K_Q^{-1}(G_Q(y))). \tag{61}$$

Thus we see that $X_{PQ}^*(y)$ is the equipercentile equating function for equating Y to X on both P and Q if and only if the two distribution functions, F_Q and G_P, for which we have no *direct* empirical estimates are related to those for which direct estimates exist as specified by Eqs. (60) and (61).

The method of equating referred to in Section 3.3.3 as *frequency estimation* also implies a specific functional form for the unobserved distributions F_Q and G_P. Carrying out the calculation in terms of the densities rather than the distribution functions, the required form of f_Q may be derived under the usual assumptions:

$$f_Q(x) = \int f_Q(x|v)k_Q(v)\,dv$$

$$= f_P(x) \int \left[\frac{f_P(x|v)}{f_P(x)}\right]k_P(v)\left[\frac{k_Q(v)}{k_P(v)}\right]dv$$

$$= f_P(x) \int \left[\frac{k_Q(v)}{k_P(v)}\right]k_P(v|x)\,dv$$

or

$$f_Q(x) = f_P(x)E_P(k_Q(V)/k_P(V)|X = x). \qquad (62)$$

Similarly,

$$g_P(y) = g_Q(y)E_Q(k_P(V)/k_Q(V)|Y = y). \qquad (63)$$

We now determine where (62) and (60) or (63) and (61) give identical values for f_Q and g_P. This is most easily answered by differentiating (60) and (61) to yield equivalent formulas in terms of the density functions. Differentiating (60) yields

$$f_Q(x) = f_P(x)\frac{k_Q(K_P^{-1}(F_P(x)))}{k_P(K_P^{-1}(F_P(x)))} \qquad (64)$$

and (61) yields

$$g_P(y) = g_Q(y)\frac{k_P(K_Q^{-1}(G_Q(y)))}{k_Q(K_Q^{-1}(G_Q(y)))}. \qquad (65)$$

If we let

$$V_P(x) = K_P^{-1}F_P(x)$$

and

$$V_Q(y) = K_Q^{-1}G_Q(y),$$

then (64) and (65) reduce to

$$f_Q(x) = f_P(x)\frac{k_Q(V_P(x))}{k_P(V_P(x))} \qquad (66)$$

and

$$g_P(y) = g_Q(y)\frac{k_P(V_Q(y))}{k_Q(V_Q(y))}. \qquad (67)$$

Thus each "equating" method assumes that the ratio f_Q/f_P (equally, g_Q/g_P) has a particular form and the two forms are remarkably similar, although it is unlikely that they will yield exactly the same results. From (62), the ratio is the regression function in P of $k_Q(V)/k_P(V)$ on X evaluated at $X = x$; from (66) the ratio is $k_Q(V_P(x))/k_P(V_P(x))$.

We summarize this discussion in Theorem 8.

Theorem 8. *Equating Y to X through V is equivalent to* (a) *equipercentile equating of Y to X on P using frequency estimation if and only if for every y,*

$$k_P(V_Q(y))/k_Q(V_Q(y)) = E_Q(k_P(V)/k_Q(V)|Y = y)$$

or (b) *equipercentile equating of Y to X on Q using frequency estimation if and only if for every x,*

$$k_Q(V_P(x))/k_P(V_P(x)) = E_P(k_Q(V)/k_P(V)|X = x).$$

The mean value theorem for expectations asserts the existence of a value $V_P^*(x)$ for every x such that

$$E_P(k_Q(V)/k_P(V)|X = x) = k_Q(V_P^*(x))/k_P(V_P^*(x)).$$

Thus the two methods will agree if for every x, $V_P(x) = V_P^*(x)$. Although it is possible to construct a number of degenerate situations for which this relation holds, it cannot be true in general. One need only note that $V_P(x)$ is determined by the marginal distributions of X and V in P, while $V_P^*(x)$ is determined by the conditional distribution of V given X in P.

7. Double Equating and the Dogleg

There are a number of problems that tend to arise when equating procedures are actually carried out over a number of years. In this section we discuss two such issues: the accumulation of errors through a chain of equatings and the inability, under certain circumstances, of linear equating to maintain the desired score scale range.

7.1. Double Equating

All the sources of equating error in Table 1 accumulate over a chain of equatings and can affect the stability of the score scale in a variety of ways. A technique known as "double part-score equating" or, as we shall refer to it, "double equating," has long been used at ETS to counteract these potential sources of instability. In double equating a test Z is equated to two earlier tests, X and Y, and the two resulting conversion functions are averaged in a certain way to yield a final conversion function. Figure 9 shows the results of (a) a chain of single equatings and (b) a chain of double equatings.

We see in Fig. 9a that for every form there is only *one* path that leads from it to the initial form, i.e., form-a. Furthermore there are no cycles of forms in Fig. 9a as there are in Fig. 9b. A *cycle* of forms is a sequence of forms such as n, k and, g in 9b in which n and k have been equated, k and g have been equated, and n and g have been equated. Suppose for the moment that n, k, and g had all been equated on the same population P. Let the conversion function of form g be h_g and let the equating functions be the equipercentile equating functions given by

$$n \text{ to } k: F_k^{-1} \circ F_n,$$

$$k \text{ to } g: F_g^{-1} \circ F_k, \tag{68}$$

$$n \text{ to } g: F_g^{-1} \circ F_n.$$

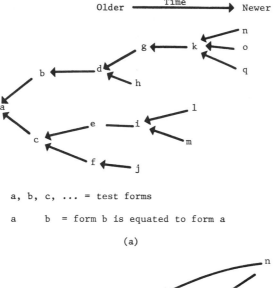

a, b, c, ... = test forms

a   b   = form b is equated to form a

(a)

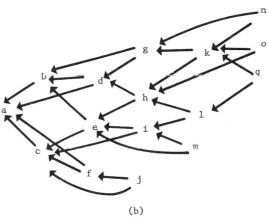

(b)

Fig. 9 (a) A chain of single equatings in which each new form is equated back to a *single* old form. (Two new forms may be equated back to the same old form.) (b) A chain of double equating in which each new form is equated back to two old forms.

The conversion function of k is the composite function

$$h_g \circ F_g^{-1} \circ F_k. \tag{69}$$

The conversion function of n if we go through k is

$$h_g \circ (F_g^{-1} \circ F_k) \circ (F_k^{-1} \circ F_n), \tag{70}$$

while the conversion function of n if we go to g directly is

$$h_g \circ F_g^{-1} \circ F_n. \tag{71}$$

However, because we are assuming that all these equatings are on the same population, it follows that $F_k \circ F_k^{-1}$ is the identity function (ignoring discreteness again), and we see that (71) and (70) are equal. This is a general fact. If all the equatings in Fig. 9b are on the same population and if we had used the correct population values for the distribution functions to compute the equipercentile equating functions, then we would get the *same* conversion functions no matter which old form we equated each new form to. The same is true of linear equating on a common population using Eq. (11). Thus if we assume that the populations on which the equatings are done are all the same and that the two conversion lines that result from double equating differ only because of sampling error, it is possible to apply standard methods to obtain limits within which we would predict the two conversion lines to fall. This provides a check on the assumption that the effect of the equating population on the final equating results is small. These limits will depend on the length of the shortest chain back to the common "ancestor" form.

However, we may find that the two conversion lines differ by more than sampling variability. This will then tell us that the other sources of variation listed in Table 1 have been operating and that either model error or population dependence or both is operating to produce two *different* conversion lines. Thus an important role of double equating is to diagnose when there is a problem with equating a form. It does not tell you what the problem is.

When the two conversion functions differ, they are averaged in a certain way to yield a single conversion function that is then used to scale the scores from the new form. How should this averaging be done? Normally, the "bisector" of the two lines is used. The bisector goes through the point where the two lines intersect and bisects the angle they form. If the two lines are close together over the range of interest, it probably does not matter what type of average is used. However, it is instructive to consider what a proper averaging of the two lines would be if the only factor operating was sampling variability. Suppose the true conversion function is linear, i.e.,

$$x' = ay + b \tag{72}$$

and that the two estimated conversion lines only differ from (72) by sampling variability, i.e.,

$$\hat{x}' = \hat{a}y + \hat{b}, \qquad \overset{*}{x}' = \overset{*}{a}y + \overset{*}{b}, \tag{73}$$

where

$$(\hat{a}, \hat{b}) \sim N((a, b), \Sigma_1)$$

and
$$\tag{74}$$

$$(\overset{*}{a}, \overset{*}{b}) \sim N((a, b), \Sigma_2).$$

The covariance matrices in (74) are small because $(\hat{a}, \hat{b})$ and $(\overset{*}{a}, \overset{*}{b})$ differ from (a, b) by a small random amount. The question of averaging the lines in (73) may be viewed as forming optimal estimates of $ay + b$ for each fixed value of y. A standard way of combining independent estimates of the same quantity is to weight $\hat{x}' = \hat{a}y + \hat{b}$ and $\overset{*}{x}' = \overset{*}{a}y + \overset{*}{b}$ by the inverses of their variances, i.e.,

$$\hat{\sigma}^2(y) = \text{Var}(\hat{x}') = y^2\sigma_{\hat{a}}^2 + \sigma_{\hat{b}}^2 + 2y\sigma_{\hat{a}\hat{b}}, \tag{75}$$

$$\overset{*}{\sigma}^2(y) = \text{Var}(\overset{*}{x}') = y^2\sigma_{\overset{*}{a}}^2 + \sigma_{\overset{*}{b}}^2 + 2y\sigma_{\overset{*}{a}\overset{*}{b}}. \tag{76}$$

Let

$$w(y) = \frac{1/\hat{\sigma}^2(y)}{(1/\hat{\sigma}^2(y)) + (1/\overset{*}{\sigma}^2(y))}. \tag{77}$$

This way of combining $\hat{x}'$ and $\overset{*}{x}'$ yields

$$\begin{aligned}\tilde{x}' &= \hat{w}(y)\hat{x} + (1 + \hat{w}(y))\overset{*}{x}' \\ &= (\hat{w}(y)\hat{a} + (1 - \hat{w}(y)\overset{*}{a})y + \hat{w}(y)\hat{b} + (1 - \hat{w}(y))\overset{*}{b}, \end{aligned} \tag{78}$$

which is not the bisector but the weighted average of the two conversion lines, which, in general, is not even a straight line. Since the parameter of interest in (72) is a linear function of a and b and since we are usually going to be dealing with estimates that are approximately normal, it is our conjecture that the bisector cannot be justified as an optimal estimator. It may possibly be justified as a reasonable compromise when sampling error is *not* the cause of the discrepancies between the two conversion functions. Figure 10 shows two intersecting conversion lines, the bisector and the simple average line ($\hat{w} = \frac{1}{2}$).

7.2. The "Dogleg"

Among the many practical problems that arise in test equating, we mention only one—the "*Dogleg*" problem.

Imagine that an acceptable linear conversion line, of the form $y = ax + b$, has been derived by standard methods. Hence y denotes the scaled score and x the raw score; a and b represent the slope and intercept of the conversion function. Let X_M denote the maximum possible raw score and y_M the corresponding maximum scaled score. It may be that y_M does not exceed Y_M, the desired maximum achievable scaled score. For the SAT, for example,

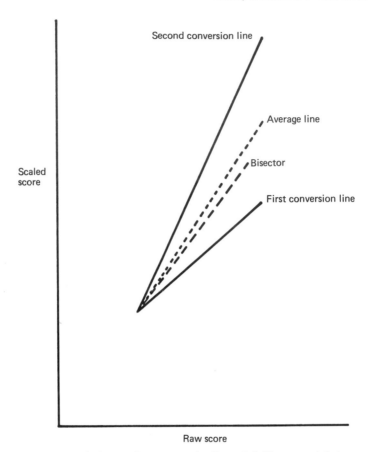

Fig. 10 Exaggerated picture of two conversion lines, their bisector and their average.

$Y_M = 800$. This situation, with $y_M < Y_M$, is illustrated in Fig. 11. The goal is to modify the equating line so that the maximum possible scaled score does equal the desired value Y_M. We note that when equipercentile equating is used, it is usually possible to arrange that $y_M = Y_M$. Thus one interpretation of the problem is that it arises when the linear equating function is not a good approximation to the equipercentile equating function. Note also that a similar problem may arise at the lower end of the scale.

One common solution is to bend the line at some point, so that the new segment passes through the point (x_M, Y_M). This is illustrated in Fig. 12 and suggests the origin of the name "dogleg." The equation of the dogleg segment is

$$y = Y_M + (a + c)(x - x_M) \quad \text{for} \quad x > x(c),$$

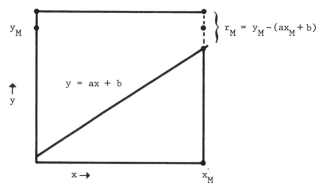

Fig. 11 Examples in which the conversion function does not reach the desired maximum scaled score.

where c represents the amount by which the slope of the dogleg segment exceeds that of the correct conversion line and $x(c)$ is the abscissa of the point of intersection of the two lines.

The appropriate choice of c, and consequently $x(c)$, is somewhat arbitrary. Our own proposal would be to specify a loss function that would incorporate the various costs associated with the use of the dogleg. A dogleg can then be chosen from among a specified set of possibilities to minimize the loss function. Although the specification of the loss function is itself somewhat arbitrary, it has the virtue of forcing the user to consider more clearly the consequences of using the dogleg.

We believe that the loss function should include two components: one related to the number of people whose scores would be affected by the dogleg, the second related to how different from the true line the dogleg appears. The two components are counterbalancing in the sense that minimizing the first tends to force $x(c)$ toward X_M, while minimizing the second tends to drive $x(c)$ away from X_M. In addition to determining the form of the two

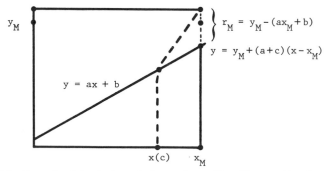

Fig. 12 Example of the dogleg solution to the problem illustrated in Figure 11.

losses, the user must also specify their relative weights before solving the optimization problem.

Generalizations to more complicated loss functions or broader families from which to choose the dogleg (e.g., curvilinear rather than simply linear) are certainly possible. Our purpose in introducing this topic, however, is to illustrate how certain difficulties arising in the context of equating need not, and probably should not, be considered as pure equating problems. As a case in point, the dogleg can be considered as one solution to the problem of modifying a proper conversion line in order that an external constraint on the system be satisfied.

8.   Discussion

There are approaches to test equating other than the one described here. Two notable alternatives are "true score equating" and "item response theory (IRT) equating." True-score equating is discussed in this volume in Part I, and IRT equating is discussed in Part II. Morris's discussion (this volume) of strong and weak equating proceeds from the IRT perspective but carries out the calculations in a way that differs somewhat from the other methods.

Our emphasis on observed-score equating stems from the fact that it is the approach favored for many years by most ETS testing programs. Furthermore, since the underlying motivation for observed-score equating is quite simple, it seems a good place to begin a mathematical analysis. In fact a new observed-score equating method, section pre-equating, has been developed as a direct result of the analysis given here. Section pre-equating is described in Holland and Wightman (this volume).

We would like to emphasize that we are most comfortable with the application of the theory given in this paper when the tests being equated are similar in format, content, difficulty, reliability, etc. Attempts to equate tests that are very different in some or all of these aspects should be considered very carefully. In such circumstances, equating may not be the appropriate procedure, for it may give the false impression of "making equal that which is not," e.g., attempting to equate a mathematics and a verbal test. "Vertical equating," that is, putting the scores of tests of widely different difficulty on the same scale, also raises new issues that we have not dealt with here.

ACKNOWLEDGMENTS

We would like to thank Gary Marco and Linda Cook for making available the data employed in Section 4, and Roger Lustig for carrying out the associated computations.

REFERENCES

Angoff, W. H. (1971). Scales, norms, and equivalent scores. In R. L. Thorndike (Ed.), *Educational Measurement* (2nd ed.). Washington, D.C., American Council on Education, 508–600.

Good, I. J., and Gaskins, R. A. (1980). Density estimation and bump-hunting for the penalized likelihood method exemplified by scattering and meteorite data. *Journal of the American Statistical Association, 75*, 42–73.

Kendall, M. G., and Stuart, A. (1958). *The Advanced Theory of Statistics.* New York: Hafner.

Levine, R. S. (1955). *Equating the Score Scales of Alternate Forms Administered to Samples of Different Ability* (RB-55-23). Princeton, N. J.: Educational Testing Service.

Lord, F. M. (1950). *Notes on Comparable Scales for Test Scores* (RB-50-48). Princeton, N. J.: Educational Testing Service.

Lord, F. M. (1981). *Applications of Item Response Theory to Practical Testing Problems.* Hillsdale, N. J.: Lawrence Erlbaum Associates.

Mallows, C. L. (1980). Some theory of non-linear smoothers. *Annals of Statistics, 8*, 695–715.

Discussion of "Observed-Score Test Equating: A Mathematical Analysis of Some ETS Equating Procedures"

Donald B. Rubin

The Braun–Holland paper is a particularly valuable and appropriate first contribution for this volume on test equating because it precisely states and motivates much of current practice at ETS. I think that a great virtue of the discussion is that the methods are stated and motivated without reference to hypothetical underlying stochastic models. In test equating, this emphasis on observable quantities is commonly referred to as *observed-score* test equating.

In other contexts (surveys and inference for causal effects), I've called such emphasis "phenomenological." When issues are defined phenomenologically, that is, in terms of observable quantities, there is no ambiguity about the real-world meaning of assumptions and definitions. In contrast, when issues are defined nonphenomenologically, that is, in terms of hypothetical underlying models and hypothetical unknown parameters, assumptions are not always easy to relate to real-world observations.

An example of a nonphenomenological concept often used in discussions of test equating is a person's *true score* on a test. Although *true score* is a useful designation for a parameter in a specific family of mathematical models for test-taking behavior, it does not have any direct real-world meaning. More distressing, the use of the term *true score* can imply, especially to the uninitiated reader, the measurement of an unchanging ability (such as *IQ*, whatever that means). Terms like *true score* and *standard error of measurement of a test* have their place, but I believe that this place is within the context of technical discussion of particular kinds of models, and not within the context of discussion for a general audience.

As an illustration of the potential problems created by the use of the terms *true score* and *standard error* as if they were phenomenologically based, consider a group of test-takers who received 300 on an SAT. This group might be told that 300 is not necessarily the true score for any member of the group, but that because the error of measurement is about 30 points, 95% of their true scores on the test lie between 240 and 360. One phenomenological reading of this statement results in a prediction for what would happen if the test-takers decided to retake the SAT, i.e., that 95% of their scores the next time would lie between 240 and 360 with an average score of 300. But this reading is probably false: Empirical evidence suggests that if such a group retook the SAT, the average score would be higher than the first score of 300; a fairer indication for the average score on the second test might be 340. I believe that test-takers should receive empirical facts and not easily misunderstood theoretical statements, even if the theoretical statements are technically correct within the context of a family of models.

Another phenomenological reading of *true score/standard error* concepts appeared in *The New York Times* (December 28, 1980, p. 14E):

> Robert C. Seaver, Vice President of the College Board, also observed last week that the gains cited by Mr. Messick fall within the 30-point "standard error of measurement" for the S.A.T. He meant that scoring this and other, similar tests is imprecise. There is only a two-thirds chance that the score a student receives is within 30 points, plus or minus, of how well that student actually did.

This statement is particularly unfortunate because it suggests that there are errors in the *scoring process*—i.e., that if the same answer sheet were passed through the scoring machine again, a different answer would result!

To summarize this point: Nonphenomenologically based concepts can be difficult to tie to real-world processes; the Braun and Holland presentation, although mathematically precise, is easy to link to testing practice because of its phenomenological focus.

Another nice feature of the Braun–Holland paper is the clear distinction between the conceptual/mathematical issues that define the need for test equating and the statistical issues that arise when estimating equating functions using samples from populations. The fact that statistical methods of inference play so slight a role in the paper reflects the lack of influence modern statistical methods have so far had on current methods for test equating.

The crucial statistical problem is stated clearly, however, when comparing linear and equipercentile equating methods:

> . . . there is a tendency to jump from the low-parameter linear equating function directly to the high-parameter nonparametric estimate of $F^{-1}(g(Y))$ rather than trying to develop an approximation of intermediate complexity. We believe this is a place where modern smoothing methods may prove useful.

The key statistical issue is stated again when considering nonparametric methods for common-item equating:

> . . . great care must be exercised in using this method of estimating $F_T(x)$ and $G_T(y)$ (the distributions of the two tests in the asymptotic population T) in practice. To our knowledge, little has been done to make this approach practicable and it is generally in disrepute because of its unstable sampling characteristics. Modern methods of data smoothing should have important contributions to make here.

Analogous statistical problems occur whenever we try to estimate the distribution of a random variable from a simple random sample. The center of the distribution is typically relatively well estimated by the empirical distribution function (the nonparametric estimator), since the variabilities of the intermediate order-statistics are limited by the high density of points near the center of distribution. In contrast, the tails of the distribution are typically poorly estimated by the empirical distribution function, since usually there are few observations in the tails. An *ad hoc* rule trying to emulate the consequences of empirical Bayes rules would combine the nonparametric estimator (with its small variance for central values and large variance for extreme values) with a parametric estimator that uses all data points to estimate the density at all values (e.g., under the normal *linear* model, global estimates of the mean and variance provide estimates of the entire distribution function). Such an approach would allow the estimated distribution function (or equating function) to follow the slow undulations of the nonparametric estimator in the middle, where the nonparametric estimator is to be trusted, yet the approach would smooth the nonparametric distribution function toward the parametric estimator (or equipercentile equating function toward the linear equating function) in the tails where the nonparametric estimator is not as trustworthy.

Although it is easy to create *ad hoc* rules for such smoothing, some thought seems to be required to develop a family of such estimators based on sound statistical principles, like those evolving from a likelihood analysis under a complete model specification. The creation of estimators of this type not only will be of interest to ETS in its efforts to improve the equating of tests, but will, more generally, be of great interest to statisticians trying to go beyond the limitations of the first generation of robust statistical procedures.

Summary and Derivation of Equating Methods Used at ETS*

William H. Angoff

1. Introduction

This summary of equating methods has had several antecedents. The first was an internal research bulletin by Lord (1950), in which he described five methods of linear equating and provided standard error formulas for them. A more recent antecedent is a chapter by Angoff (1971) in R. L. Thorndike's *Educational Measurement*, in which a detailed treatment of several linear methods, as well as equipercentile analogs of those methods, is presented, along with a description of the contexts in which they are appropriately used.

One might infer from the title that the paper contains a summary of all the methods of score equating used at ETS at the present time. However, as is often the case, the title is not fully descriptive. There are additional procedures at ETS, some still in development, yet to become operational. There are others that have recently become operational but are still being modified as the need for modification becomes clear. Finally, as would be expected, detailed variations and modifications of the general procedures

* This paper was not presented at the conference. Earlier drafts of it have been used for instructional purposes at Educational Testing Service.

55

described here are often employed at ETS as special conditions warrant. For reasons implicit in the intent of a summary, variations of this sort are not described here.

On the other hand, not all the procedures outlined in this paper are in use at ETS in precisely the form described. The equipercentile methods, for example, are now applied not by hand, but by computer. It seemed, however, that it would be more informative and instructive to the reader if they were described in terms of the steps normally carried out by hand. Other methods and formulas also appear here, even though they have rarely, if ever, been used operationally at ETS. They are included here because they are available to us as part of a larger methodological development (e.g., the major axis equations for equating unequally reliable tests, administered to random halves of a group) or because they may be needed in future applications.

The following notation is used:

1. *Variables*: x, y = raw scores on test forms to be equated. Form x is new; Form y is the old form to which Form x is to be equated. v = raw score on equating test, Form v, common to the administration of both Form x and Form y. S = standard score scale.

2. *Groups*: α, β = groups taking Forms x and y, respectively. $t = \alpha + \beta$.

3. *Statistics*: A_{gh}, B_{gh} = slope and intercept parameters, respectively, of the linear conversion from the scale of h to the scale of g.

Equating is the process of developing a conversion from the system of units of one form of a test to the system of units of another form so that scores derived from the two forms *after conversion* will be equivalent and interchangeable. Only when the two forms are truly measures of the same psychological function will the conversion be unique and applicable to all types of individuals and groups.

Two general models of equating are considered here: equipercentile and linear equating. Equipercentile equating is based on the following definition: Two scores, one on Form x and the other on Form y—where x and y are equally reliable and parallel measures, in the sense described above—may be considered equivalent if their respective percentile ranks in any given group are equal. Thus equipercentile equating of Form x to Form y operates to match all moments (i.e., all characteristics of the shape, in addition to the mean and standard derivation) of the distribution of Form x scores to the respective moments of the distribution of Form y scores.

Linear equating is based on the following definition: Two scores, one on Form x and the other on Form y—again, where x and y are equally reliable and parallel measures—may be considered equivalent if their respective standard score deviates in any given group are equal:

$$(y - M_y)/s_y = (x - M_x)/s_x. \tag{1}$$

When these terms are rearranged, (1) may take the form

$$y = A_{yx}x + B_{yx}, \tag{2}$$

where $A_{yx} = s_y/s_x$ and $B_{yx} = M_y - A_{yx}M_x$, the slope and intercept, respectively, of the conversion equation. Linear equating is carried out on the assumption that all moments of the distributions of Forms x and y above the second moment are essentially equivalent and that all that is needed to equate all moments is an adjustment in mean and standard deviation (the first and second moments).

When the shapes of the distributions of Forms x and y are identical, equipercentile equating yields a straight line, identical to the linear equating line described earlier. When the test forms are unequally difficult and have different distribution shapes, equipercentile equating will result in a curve concave to the upper left if Form x, plotted on the abscissa, is the easier test; and concave to the lower right if Form x is the more difficult test.

2. Design I—Random Groups, One Form Administered to Each Group

Select a large heterogeneous group, divide at random into two half-groups. Administer Form x to the α group and Form y to the β group.

2.1. Linear Procedure for Equally Reliable Tests

Following Eq. (1),

$$(y - M_{y_\beta})/s_{y_\beta} = (x - M_{x_\alpha})/s_{x_\alpha}, \tag{3}$$

which, on rearrangement, takes the form

$$y = A_{yx}x + B_{yx} \tag{4}$$

in which $A_{yx} = s_{y_\beta}/s_{x_\alpha}$ and $B_{yx} = M_{y_\beta} - A_{yx}M_{x_\alpha}$. If there already exists an equation

$$S = A_{Sy}y + B_{Sy} \tag{5}$$

for converting y scores to scaled scores, the substitution of (4) in (5) will result in the following conversion from scores on the scale of Form y to scaled scores:

$$S = A_{Sx}x + B_{Sx}, \qquad (6)$$

where $A_{Sx} = A_{Sy}A_{yx}$ and $B_{Sy} = A_{Sy}B_{yx} + B_{Sy}$.

2.2. Equipercentile Analog—Equally Reliable Tests

Form two distributions, one of Form x for group α, the other of Form y for group β. Compute and plot midpercentiles for each distribution. Read about 30 corresponding percentiles from each distribution, and plot and smooth the paired percentiles on arithmetic graph paper. Read off y scores corresponding to given x scores. If there already exists a conversion table, graph, or equating relating y to S, then the operation of "putting x on the scale" will involve converting x scores to their equivalent y scores, and finding the S scores corresponding to those y scores.

2.3. Linear Procedure—Unequally Reliable Tests

If Forms x and y differ significantly in reliability, there is no way to develop a transformation of scores that will make them equivalent. Nevertheless, situations arise in practice that call for such "equating." In such instances, it is appropriate to take the reliabilities of the two forms into account, as in the equation

$$(y - M_{y_\beta})/s_{\tilde{y}_\beta} = (x - M_{x_\alpha})/s_{\tilde{x}_\alpha}, \qquad (7)$$

where $s_{\tilde{x}} = s_x\sqrt{r_{xx'}}$ and $s_{\tilde{y}}\sqrt{r_{yy'}}$. The equation for converting scores from the scale of Form x to the scale of Form y becomes

$$y = (s_{\tilde{y}_\beta}/s_{\tilde{x}_\alpha})x + M_{y_\beta} - (s_{\tilde{y}_\beta}/s_{\tilde{x}_\alpha})M_{x_\alpha}, \qquad (8)$$

also of the form $y = A_{yx}x + B_{yx}$, where

$$A_{yx} = s_{\tilde{y}_\beta}/s_{\tilde{x}_\alpha} \qquad \text{and} \qquad B_{yx} = M_{y_\beta} - A_{yx}M_{x_\alpha}.$$

3. Design II—Random Groups, Both Forms Administered to Each Group, Counterbalanced

Again, select a large homogeneous group and divide into two random halves, group α taking Form x followed by Form y, group β taking Form y followed by Form x.

3.1. Linear Procedure for Equally Reliable Tests

It is assumed that the standardized practice effect of Form x on Form y is the same as the standardized practice effect of Form y on Form x: $K_x/s_x = K_y/s_y = H$; and that the best estimate of H is the average of the mean difference on Form x and the mean difference on Form y when each is expressed in standard deviation units:

$$H = \tfrac{1}{2}[(M_{x_\beta} - M_{x_\alpha})/s_x + (M_{y_\alpha} - M_{y_\beta})/s_y]. \tag{9}$$

In determining the values to be used to substitute in Eq. (1), $(y - M_y)/s_y = (x - M_x)/s_x$,

$$M_x = \tfrac{1}{2}(M_{x_\alpha} + M_{x_\beta} - K_x), \tag{10}$$

$$M_y = \tfrac{1}{2}(M_{y_\alpha} + M_{y_\beta} - K_y), \tag{11}$$

$$s_x^2 = \tfrac{1}{2}(s_{x_\alpha}^2 + s_{x_\beta}^2), \tag{12}$$

$$s_y^2 = \tfrac{1}{2}(s_{y_\alpha}^2 + s_{y_\beta}^2). \tag{13}$$

When these values are substituted in Eq. (1), a linear equation is found that may be expressed in the form $y = A_{yx}x + B_{yx}$, where

$$A_{yx} = \sqrt{(s_y^2 + s_{y_\beta}^2)/(s_{x_\alpha}^2 + s_{x_\beta}^2)}, \tag{14}$$

and

$$B_{yx} = \tfrac{1}{2}(M_{y_\alpha} + M_{y_\beta}) - \tfrac{1}{2}A_{yx}(M_{x_\alpha} + M_{x_\beta}). \tag{15}$$

A variant of this procedure is to combine the data on Form x for the two half-groups, combine the data on Form y for the two half-groups, and equate by the method of Design I.

3.2. Equipercentile Analog—Equally Reliable Tests

Combine data, as just previously described, for Form x, also for Form y, and equate by applying the equipercentile procedure described in Design I.

3.3. Linear Procedure—Unequally Reliable Tests

When Forms x and y are unequally reliable, the average within-group variances are calculated, not on the basis of observed scores as shown in Eqs. (12) and (13), but on the basis of true scores, as in Eqs. (16) and (17):

$$s_x^2 = \tfrac{1}{2}(s_{x_\alpha}^2 r_{xx_\alpha} + s_{x_\beta}^2 r_{xx_\beta}), \tag{16}$$

$$s_y^2 = \tfrac{1}{2}(s_{y_\alpha}^2 r_{yy_\alpha} + s_{y_\beta}^2 r_{yy_\beta}). \tag{17}$$

4. Design III—Random Groups, One Test Administered to Each Group; Common Equating Test Administered to Both Groups*

4.1. Linear Procedure

In order to introduce greater control over the equating than random administration will permit, a test score v is used, based on a set of items administered to both groups. As before, examinees are assigned to groups α and β at random. Group α is given Form x, group β is given Form y, and the additional (or common) Form v is given to both groups. Maximum likelihood estimates are then made of the population means and variances on Forms x and y:

$$\hat{\mu}_x = M_{x_\alpha} + b_{xv_\alpha}(\mu_v - M_{v_\alpha}), \tag{18}$$

$$\hat{\mu}_y = M_{x_\beta} + b_{yv_\beta}(\mu_v - M_{v_\beta}), \tag{19}$$

$$\hat{\sigma}_x^2 = s_{x_\alpha}^2 + b_{xv_\alpha}^2(\sigma_v^2 - s_{v_\alpha}^2), \tag{20}$$

$$\hat{\sigma}_y^2 = s_{y_\beta}^2 + b_{yv_\beta}^2(\sigma_v^2 - s_{v_\beta}^2), \tag{21}$$

where $\mu_v = M_{v_t}$ and $\sigma_v^2 = s_{v_t}^2$. As before, these values are substituted in Eq. (1) to yield a linear equation of the form $y = A_{yx}x + B_{yx}$, where here $A_{yx} = \hat{\sigma}_y/\hat{\sigma}_x$ and $B_{yx} = \hat{\mu}_y - A_{yx}\hat{\mu}_x$.

4.2. Linear Procedure—Unequally Reliable Tests— Major Axis Equations†

When Forms x and y are unequally reliable, the slope A_{yx} of the conversion equation $y = A_{yx}x + B_{yx}$ is taken as the ratio of the population values of the true-score standard deviations, $\sigma_y\sqrt{\rho_{yy'}}/\sigma_x\sqrt{\rho_{xx'}}$. Now when Form v is separate from and exclusive of, and parallel in function to, Form x, $\sqrt{\rho_{xx'}} = \rho_{xv}/\sqrt{\rho_{vv'}}$ [Eq. (54), Appendix 6.1]. Under corresponding conditions for Form v and Form y, $\sqrt{\rho_{yy'}} = \rho_{yv}/\sqrt{\rho_{vv'}}$, and $A_{yx} = \sigma_y\rho_{yv}/\sigma_x\rho_{xv} = \sigma_{yv}/\sigma_{xv}$. It has been shown (Lord, 1955) that the maximum likelihood estimates of σ_{xv} and σ_{yv} may be written as $b_{xv}\sigma_v^2$ and $b_{yv}\sigma_v^2$, respectively. Therefor, the slope A_{yx} of the conversion equation $y = A_{yx}x + B_{yx}$ for equating unequally reliable tests, when Form v is parallel to and exclusive of Forms x and y, is written as follows:

$$A_{yx} = b_{yv_\beta}/b_{xv_\alpha}. \tag{22}$$

* Lord (1955).
† Levine (1955).

When Form v is included in Forms x and y, $\sqrt{\rho_{xx'}} = \sqrt{\rho_{vv'}}/\rho_{xv}$ and $\sqrt{\rho_{yy'}} = \sqrt{\rho_{vv'}}/\rho_{yv}$ [Eq. (58), Appendix 6.1], and the ratio $\sigma_y\sqrt{\rho_{yy'}}/\sigma_x\sqrt{\rho_{xx'}}$ becomes $\sigma_y^2\sigma_{xv}/\sigma_x^2\sigma_{yv}$. As indicated in the preceding paragraph, maximum likelihood estimates of σ_{xv} and σ_{yv} may be written as $b_{xv_\alpha}\sigma_v^2$ and $b_{yv_\beta}\sigma_v^2$, respectively. Maximum-likelihood estimates of σ_x^2 and σ_y^2 are given in Eqs. (20) and (21). The result is that the slope A_{yx} of the conversion equation $y = A_{yx}x + B_{yx}$, when Form v is parallel to and included in Forms x and y, is written as

$$A_{yx} = \frac{b_{xv_\alpha}[s_{y_\beta}^2 + b_{yv_\beta}^2(\sigma_v^2 - s_{v_\beta}^2)]}{b_{yv_\beta}[s_{x_\alpha}^2 + b_{xv_\alpha}^2(\sigma_v^2 - s_{v_\alpha}^2)]}. \tag{23}$$

As before, $B_{yx} = \hat{\mu}_y - A_{yx}\hat{\mu}_x$, in which $\hat{\mu}_x$ and $\hat{\mu}_y$ take their values from Eqs. (18) and (19). In this instance, A_{yx}, as indicated, takes its value from Eq. (23).

5. Design IV—Nonrandom Groups, One Test to Each Group; Common Equating Test Administered to Both Groups

5.1. Linear Procedure for Groups Not Widely Different in Ability

Form x is given to group α; Form y is given to group β; and a third test, Form v, is given to both groups. The derivation is based on three assumptions:

(1) The intercept of x on v is the same for group t ($t = \alpha + \beta$) as it is for group α:

$$M_{x_t} - b_{xv_t}M_{v_t} = M_{x_\alpha} - b_{xv_\alpha}M_{v_\alpha}; \tag{24}$$

(2) The regression coefficient of x on v is the same for group t as it is for group α:

$$b_{xv_t} = b_{xv_\alpha}; \tag{25}$$

(3) The variance error of estimate of x from v is the same for group t as it is for group α:

$$s_{x_t}^2(1 - r_{xv_t}^2) = s_{x_\alpha}^2(1 - r_{xv_\alpha}^2). \tag{26}$$

Substituting (25) in (24) and solving for $\hat{M}_{x_t}$,

$$\hat{M}_{x_t} = M_{x_\alpha} + b_{xv_\alpha}(M_{v_t} - M_{v_\alpha}). \tag{27}$$

Since $r_{xv}s_x$ may be written as $b_{xv}s_v$, Eq. (26) may be written

$$s_{x_t}^2 - b_{xv_t}^2 s_{v_t}^2 = s_{x_\alpha}^2 - b_{xv_\alpha}^2 s_{v_\alpha}^2. \tag{28}$$

Substituting (25) in (28) and solving for $\hat{s}^2_{x_t}$,

$$\hat{s}^2_{x_t} = s^2_{x_\alpha} + b^2_{xv_\alpha}(s^2_{v_t} - s^2_{v_\alpha}). \tag{29}$$

Parallel assumptions and development may be made for the relationship between Forms y and v and groups β and t, resulting in Eqs. (30) and (31), which parallel, respectively, Eqs. (27) and (29):

$$\hat{M}_{y_t} = M_{y_\beta} + b_{yv_\beta}(M_{v_t} - M_{v_\beta}), \tag{30}$$

$$\hat{s}^2_{y_t} = s^2_{y_\beta} + b^2_{yv_\beta}(s^2_{v_t} - s^2_{v_\beta}). \tag{31}$$

The values $\hat{M}_{x_t}$, $\hat{M}_{y_t}$, $\hat{s}_{x_t}$, and $\hat{s}_{y_t}$ from (27), (29), (30), and (31) are then substituted into Eq. (1) to yield the conversion equation $y = A_{yx}x + B_{yx}$, where $A_{yx} = \hat{s}_{y_t}/\hat{s}_{x_t}$ and $B_{yx} = \hat{M}_{y_t} - A_{yx}\hat{M}_{x_t}$.

5.2. Equipercentile Analog—Frequency Estimation Method

The intent here is to estimate the frequencies of the distributions of Forms x and y for group t:

(1)   Combine the distributions of groups α and β on Form v to form a distribution for group t. If any of these frequencies f_{i_t} is less than, say, 10, combine with the next interval(s) toward the center of the distribution. Combine similarly the frequencies in the corresponding intervals in the distribution for groups α and β.

(2)   Working with the scatterplot of v versus x for group α, and considering the interval v_i, multiply the ratio of the frequencies f_{i_t}/f_{i_α} by each of the frequencies in the array for the score interval, v_i. (Note: If any ratio f_{i_t}/f_{i_α} exceeds, say, 5, combine f_{i_t} with the next interval toward the center of the distribution. Combine similarly the frequencies in the corresponding intervals in the distribution for group α.) Repeat for all intervals v_i. This operation yields a new scatterplot of v versus x with $N = N_{t(=\alpha+\beta)}$.

(3)   Sum all cell frequencies for each value of x. This yields a new distribution of Form x scores for group t.

(4)   Repeat steps 2 and 3 for the scatterplot of v versus y for groups t and β.

Carry out an equipercentile equating of Form x versus Form y, as described above for Design I, using the two sets of frequencies estimated for group t.

5.3. Equipercentile Equating—Forms x and y Equated to Common Test v

Equate Form x and Form v by equipercentile method, with data taken from group α. Equate Form y and Form v by equipercentile method, with data taken from group β. For each score on Form v, find the equivalent scores on x and y. Plot and smooth to yield the conversion of scores on Form x to equivalent scores on Form y.

5.4. Linear Procedures for Samples of Different Ability— Major Axis Equations*

5.4.1. EQUALLY RELIABLE TESTS

When groups α and β are of widely different ability and fallible measures are used, the assumptions basic to classical selection theory are not appropriate. But when Form v is parallel to Form x, the following assumptions may be made:

(1) The intercept of the line relating true scores on Form x to true scores on Form v ($r_{\tilde{x}\tilde{v}} = 1.00$) is the same for group t as for group α:

$$M_{x_t} - (s_{\tilde{x}_t}/s_{\tilde{v}_t})M_{v_t} = M_{x_\alpha} - (s_{\tilde{x}_\alpha}/s_{\tilde{v}_\alpha})M_{v_\alpha}, \tag{32}$$

where $s_{\tilde{x}} = s_x\sqrt{r_{xx'}}$ and $s_{\tilde{v}} = s_v\sqrt{r_{vv'}}$;

(2) the slope of the line of relationship is the same for group t as for group α:

$$s_{\tilde{x}_t}/s_{\tilde{v}_t} = s_{\tilde{x}_\alpha}/s_{\tilde{v}_\alpha}; \tag{33}$$

and the variance error of measurement for Form x is the same for group t as for group α:

$$s_{x_t}^2(1 - r_{xx'_t}) = s_{x_\alpha}^2(1 - r_{xx'_\alpha}). \tag{34}$$

Substituting (33) in (32) and solving for $\hat{M}_{x_t}$,

$$\hat{M}_{x_t} = M_{x_\alpha} + (s_{\tilde{x}_\alpha}/s_{\tilde{v}_\alpha})(M_{v_t} - M_{v_\alpha}). \tag{35}$$

Substituting (33) in (34) we find

$$\hat{s}_{x_t}^2 = s_{x_\alpha}^2 + (s_{\tilde{x}_\alpha}^2/s_{\tilde{v}_\alpha}^2)(s_{\tilde{v}_t}^2 - s_{\tilde{v}_\alpha}^2). \tag{36}$$

However, since

$$s_{v_t}^2(1 - r_{vv'_t}) = s_{v_\alpha}^2(1 - r_{vv'_\alpha}), \tag{37}$$

we can see that

$$s_{\tilde{v}_t}^2 - s_{\tilde{v}_\alpha}^2 = s_{v_t}^2 - s_{v_\alpha}^2, \tag{38}$$

and therefore that

$$\hat{s}_{x_t}^2 = s_{x_\alpha}^2 + (s_{\tilde{x}_\alpha}^2/s_{\tilde{v}_\alpha}^2)(s_{v_t}^2 - s_{v_\alpha}^2). \tag{39}$$

* Levine (1955).

Parallel assumptions and development are appropriate for the relationship between Forms y and v and groups β and t. Their use will result in Eqs. (40) and (41), which parallel, respectively, Eqs. (35) and (39):

$$\hat{M}_{y_t} = M_{y_\beta} + (s_{\tilde{y}_\beta}/s_{\tilde{v}_\beta})(M_{v_t} - M_{v_\beta}), \tag{40}$$

$$\hat{S}_{y_t}^2 = S_{y_\beta}^2 + (s_{\tilde{y}_\beta}^2/s_{\tilde{v}_\beta}^2)(s_{v_t}^2 - s_{v_\beta}^2). \tag{41}$$

Equations (35), (39), (40), and (41) are the appropriate ones to use in connection with Eq. (1). However, they are not always convenient for use in operational work. It is noted [Eq. (48), Appendix 6.1] that $s_{\tilde{x}}/s_{\tilde{v}} = n_{xv}$, the ratio of effective test lengths of Form x to Form v. When Form v is separate from and exclusive of Form x, n_{xv} may be written as $(s_x^2 + s_{xv})/(s_v^2 + s_{xv})$ [see Eq. (53), Appendix 6.1]. When Form v is included in Form x, n_{xv} may be written as $1/b_{vx}$ [see Eq. (57), Appendix 6.1]. Similar applications of formulas may be made for data involving Forms y and v. Note that in using these alternative expressions for n_{xv} and n_{yv} it is important that scores on Forms x and v (or, correspondingly, on Forms y and v) be derived in the same way—e.g., both scored rights or both scored by formula (the *same* formula).

As before, scores on Form x are equated to scores on Form y by substituting in Eq. (1), $y = A_{yx}x + B_{yx}$, where $A_{yx} = \hat{S}_{y_t}/\hat{S}_{x_t}$ and $B_{yx} = \hat{M}_{y_t} - A_{yx}\hat{M}_{x_t}$. The estimated statistics are given in Eqs. (35), (39), (40), and (41).

5.4.2. UNEQUALLY RELIABLE TESTS

The slope A_{yx} of the conversion equation $y = A_{yx}x + B_{yx}$, where Forms x and y are unequally reliable, is

$$A_{yx} = s_{\tilde{y}_t}/s_{\tilde{x}_t}. \tag{42}$$

Assuming, as in Eq. (33), that

$$s_{\tilde{x}_t}/s_{\tilde{v}_t} = s_{\tilde{x}_\alpha}/s_{\tilde{v}_\alpha}, \tag{33}$$

and, correspondingly, that

$$s_{\tilde{y}_t}/s_{\tilde{v}_t} = s_{\tilde{y}_\beta}/s_{\tilde{v}_\beta}, \tag{43}$$

we find that

$$A_{yx} = s_{\tilde{y}_t}/s_{\tilde{x}_t} = (s_{\tilde{y}_\beta}/s_{\tilde{v}_\beta})(s_{\tilde{v}_\alpha}/s_{\tilde{x}_\alpha}) = n_{yv_\beta}/n_{xv_\alpha}, \tag{44}$$

$$B_{yx} = \hat{M}_{y_t} - A_{yx}\hat{M}_{x_t}, \tag{45}$$

where A_{yx} is as shown in Eq. (44),

$$\hat{M}_{x_t} = M_{x_\alpha} + n_{xv_\alpha}(M_{v_t} - M_{v_\alpha}), \quad \text{as in Eq. (35),}$$

and

$$\hat{M}_{y_t} = M_{y_\beta} + n_{yv_\beta}(M_{v_t} - M_{v_\beta}), \quad \text{as in Eq. (40).}$$

When Form v is exclusive of Forms x and y,

$$n_{xv_\alpha} = (s_{x_\alpha}^2 + s_{xv_\alpha})/(s_{v_\alpha}^2 + s_{xv_\alpha}) \quad \text{and} \quad n_{yv_\beta} = (s_{y_\beta}^2 + s_{yv_\beta})/s_{v_\beta}^2 + s_{yv_\beta}),$$

as given in Eq. (53), Appendix 6.1.

When Form v is included in Forms x and y,

$$n_{xv_\alpha} = b_{vx_\alpha}^{-1} \quad \text{and} \quad n_{yv_\beta} = b_{vy_\beta}^{-1},$$

as given in Eq. (57), Appendix A.

6. Appendixes

Appendix 6.1 contains derivations of expressions for the determination of the reliability coefficients of the full-length tests (Forms x and y—the tests to be equated) and of the short equating test (Form v), which is parallel to Forms x and y. These expressions are useful in connection with the major axis equations in making adjustments for observed differences in the groups taking Forms x and y. They are taken and further developed from Angoff (1953).

Appendix B contains some suggestions for smoothing the data used in connection with equipercentile equating.

6.1. Formulas Derived from Considerations of Effective Test Length

The reliability of a test x may be expressed as

$$r_{xx'} = r_{(v_1 + v_2 + \cdots + v_i + \cdots + v_n)(v_1' + v_2' + \cdots + v_i' + \cdots + v_n')}, \tag{46}$$

where there are n parallel subtests v of equal length contained in test x and each parallel in function to test x. Therefore,

$$r_{xx'} = n^2 r_{vv'} s_v^2 / s_x^2, \tag{47}$$

and

$$n_{xv} = s_x \sqrt{r_{xx'}} / s_v \sqrt{r_{vv'}}, \tag{48}$$

the ratio of effective test length, x to v.

Excluded Case.   The correlation of a test v with a parallel test x, of which test v is *not* a part, may be given as

$$r_{vx} = r_{v(v_1' + v_2' + \cdots + v_i' + \cdots + v_n')}$$

$$= nr_{vv'}s_v/s_x \tag{49}$$

and

$$n_{xv} = r_{xv}s_x/r_{vv'}s_v = b_{xv}/r_{vv'}. \tag{50}$$

The variance of test x may be given as

$$s_x^2 = ns_v^2 + n(n-1)r_{vv'}s_v^2. \tag{51}$$

Substituting (49) in (51),

$$s_x^2 = ns_v^2 + (n-1)r_{xv}s_x s_v, \tag{52}$$

from which we find that

$$n_{xv} = (s_x^2 + s_{xv})/(s_v^2 + s_{xv}). \tag{53}$$

Substituting (48) in (50),

$$r_{xv} = \sqrt{r_{xx'}}\sqrt{r_{vv'}}. \tag{54}$$

Included Case.   The correlation of a test v with a parallel test x, of which v *is* a part may be given as

$$r_{vx} = r_{v_i(v_1' + v_2' + \cdots + v_i + \cdots + v_n')}, \tag{55}$$

where, as before, there are n parallel subtests of length v contained in the total test x.

$$r_{vx} = [s_v^2 + (n-1)r_{vv'}s_v^2]/s_v s_x. \tag{56}$$

Substituting (56) in (51), we observe that

$$s_x^2 = ns_v^2 + n(r_{xv}s_v s_x - s_v^2),$$

and, finally, that

$$n_{xv} = \frac{s_x}{r_{xv}s_v} = \frac{1}{b_{vx}}. \tag{57}$$

Finally, setting (57) and (48) equal,

$$s_x/r_{xv}s_v = s_x\sqrt{r_{xx'}}/s_v\sqrt{r_{vv'}},$$

we find

$$r_{xv} = \sqrt{r_{vv'}}/\sqrt{r_{xx'}}, \tag{58}$$

which, it is noted, applies to the "included" case, and corresponds to Eq. (54), which applies to the "excluded" case.

Mixed Case. Suppose the common items v contain two subtests of unequal length: Form u, exclusive of Form x, and Form w, included in Form x, Form u and Form w parallel in function to each other and both parallel to Form x. We can write, as a restatement of Eq. (48),

$$s_x^2 r_{xx'} = n_{xu}^2 s_u^2 r_{uu'},$$ (59)

$$s_x^2 r_{xx'} = n_{xw}^2 s_w^2 r_{ww'}.$$ (60)

We can also write

$$s_x^2 r_{xx'} = s_x^2 r_{(u_1 + u_2 + \cdots + u_i + \cdots + u_{n_{xu}})(w_1 + w_2 + \cdots + w_i + \cdots + w_{n_{xw}})},$$ (61)

where n_{xu} and n_{xw} are the numbers of subtests of effective length u and w, respectively, contained in Form x.

From (61) we find

$$s_x^2 r_{xx'} = n_{xu} n_{xw} s_{uw}.$$ (62)

We can also write

$$s_v^2 r_{vv'} = s_u^2 r_{uu'} + s_w^2 r_{ww'} + 2s_{uw}.$$ (63)

Substituting (59), (60), and (62) in (63),

$$s_v^2 r_{vv'} = (s_x^2 r_{xx'}/n_{xu}^2) + (s_x^2 r_{xx'}/n_{xw}^2) + (2s_x^2 r_{xx'}/n_{xu}n_{xw}).$$ (64)

From Eqs. (48) and (64),

$$\frac{s_v^2 r_{vv'}}{s_x^2 r_{xx'}} = \frac{1}{n_{xv}^2} = \frac{1}{n_{xu}^2} + \frac{1}{n_{xw}^2} + \frac{2}{n_{xu}n_{xw}},$$ (65)

from which

$$\frac{1}{n_{xv}^2} = \frac{n_{xu}^2 + n_{xw}^2 + 2n_{xu}n_{xw}}{n_{xu}^2 n_{xw}^2}.$$ (66)

Finally,

$$n_{xv} = \frac{n_{xu}n_{xw}}{(n_{xu} + n_{xw})},$$ (67)

in which

$$n_{xu} = (s_x^2 + s_{xu})/(s_u^2 + s_{xu}) \quad \text{and} \quad n_{xw} = s_x/r_{xw}s_w,$$

as in Eqs. (53) and (57), respectively.

6.2. Smoothing

When the percentiles of a distribution are plotted on arithmetic graph paper, they will typically describe an S-shaped curve, the shape of the curve reflecting the characteristics of the frequency polygon. Since S-shaped curves are usually more difficult to smooth than straight lines, it is recommended that ogives be plotted on arithmetic probability paper, which tends to rectify bell-shaped distributions; it is designed to yield a straight line for distributions that are strictly normal. There are very few guidelines available to achieve the desired results of smoothing except to say that the smoothed curve should sweep through the succession of points in such a way as to equalize the discrepancies of the points on either side of the curve, ideally preserving all the moments of the original, unsmoothed distribution. Beyond this general rule it is left to the experience and judgment of the person working with the data to determine the degree to which irregularities in the data are to be so considered and smoothed out.

Analytical methods of smoothing may be preferred to hand smoothing because they *are* analytical and do not depend on the subjective judgment of the statistician. One such method, developed by Cureton and Tukey, which preserves the parabolic and cubic trends within successive sets of points, involves a rolling weighted average of frequencies. In order to determine a new smoothed frequency, f'_i, in interval i, the method calls for multiplying the weights $-3/35$, $12/35$, $17/35$, $12/35$, and $-3/35$, respectively, by the frequencies f_{i-2}, f_{i-1}, f_i, f_{i+1}, and f_{i+2}, and summing the products. An alternative set of weights, to be applied over a seven-point span of frequencies, f_{i-3}, f_{i-2}, f_{i-1}, f_i, f_{i+1}, f_{i+2}, and f_{i+3}, in order to find a smoothed frequency f'_i for the raw frequency f_i is the following: $-2/21$, $3/21$, $6/21$, $7/21$, $6/21$, $3/21$, and $-2/21$. In the application of either set of weights, negative rounded frequencies necessarily result in intervals beyond the limits of the observed frequencies. Because of their small size it has been found expedient to ignore them and to readjust all rounded frequencies by the factor $\Sigma f_i / \Sigma f'_i$.

Another analytical method of smoothing, which at present is appropriate only to rights-scored tests, is derived from the negative hypergeometric distribution, described by Keats and Lord (1962).

REFERENCES

Angoff, W. H. (1953). Test reliability and effective test length. *Psychometrika, 18*, 1–14.
Angoff, W. H. (1971). Scales, norms, and equivalent scores. In R. L. Thorndike (Ed.), *Educational Measurement* (2nd ed.). Washington, D. C.: American Council on Education, 508–600.

Cureton, E. E., and Tukey, J. W. (1951). Smoothing frequency distributions, equating tests, and preparing norms. *American Psychologist, 6,* 404. (Abstract).

Keats, J. A., and Lord, F. M. (1962). A theoretical distribution for mental test scores. *Psychometrika, 27,* 59–72.

Levine, R. S. (1955). *Equating the Score Scales of Alternate Forms Administered to Samples of Different Ability* (RB-55-23). Princeton, N. J.: Educational Testing Service.

Lord, F. M. (1950). *Notes on Comparable Scales for Test Scores* (RB-50-48). Princeton, N. J.: Educational Testing Service.

Lord, F. M. (1955). Equating test scores—a maximum likelihood solution. *Psychometrika, 20,* 193–200.

A Test of the Adequacy
of Linear Score Equating Models*

Nancy S. Petersen

Gary L. Marco

E. Elizabeth Stewart

1. Introduction

In many common testing situations it is necessary to compare the test scores of examinees who have taken different forms of a test (i.e., tests made to the same content specifications but containing different questions). In practice, two forms of a test cannot be expected to be of exactly equal difficulty for examinees at all ability levels. Therefore, a comparison of raw scores on the two forms of the test will be unfair to the examinees who have taken the more difficult form. Statistical procedures that have been developed to deal with this problem are referred to as *equating methods*.

In an ideal psychometric world, tests on which scores need to be equated would be parallel in all important respects; an anchor test, if used, would be

* The study was funded jointly by the College Entrance Examination Board and Educational Testing Service. The authors gratefully acknowledge the advice and assistance provided by William H. Angoff, Rebecca A. Bullock, Samuel A. Livingston, Frederic M. Lord, Ruth C. Mroczka, Joanne S. Peng, William B. Schrader, Ledyard R Tucker, Marilyn S. Wingersky, and R L Wood.

71

parallel to the total tests; and random samples on which to base the equating would always be available. In actual testing practice, however, these conditions are seldom satisfied. In some nationally administered testing programs, program policy and legislative requirements preclude administering different test forms to random samples of examinees on the same test date. The composition of successive groups of examinees who elect to take a particular test tends to shift over time. There usually are differences even among the groups of examinees who are tested on different national administration dates within a single year, differences that occur for generally unknown but undoubtedly complex reasons. Furthermore, the specifications that guide test construction evolve over time, so that test forms used during different periods of a testing program's history may differ in important respects. Even when two test forms are themselves parallel, various practical constraints may require that their scores be equated through scores on an anchor test that does not faithfully reflect the characteristics of the total test.

Because test scores sometimes must be equated under less-than-optimum conditions, an extensive study was designed to examine the adequacy of various score-equating models when certain sample and test characteristics were systematically varied. This study is the second part of a large-scale investigation. The emphasis in this part of the study is on linear models. The first part focused on curvilinear models (Marco *et al.*, 1979). The study is more comprehensive than previous studies of equating models (e.g., Levine, 1955; Tucker, 1974; Rentz and Bashaw, 1975; Slinde and Linn, 1977, 1978) in that it includes a greater variety of equipercentile, linear, and ICC models and investigates equatings based on dissimilar samples as well as on random samples.

2. Equating Models

An equating method is an *empirical* procedure for determining a transformation to be applied to the scores on one of two forms of a test. Its purpose is (ideally) to transform the scores in such a way that it makes no difference to the examinee which form of the test he or she takes. This ideal can be reached only if the two forms of the test measure exactly the same latent trait (ability or skill) and yield scores that are equally reliable.

The purpose of equating—to establish, as nearly as possible, an effective equivalence between test scores—requires that the equating transformation be invertible. That is, if score x_0 on test X equates to score y_0 on test Y, then y_0 must equate to x_0. Therefore, the equating problem is not simply a regression problem, since, in general, the regression of x on y does not coincide with the regression of y on x.

Because an equating method is an empirical procedure, it involves a design for data collection and a rule for determining the transformation. There are three basic designs for data collection and three general rules for determining the transformation. Any of the three designs for data collection can be used with any of the three transformation rules.

The three designs for data collection are the single-group method, the equivalent-group method, and the anchor-test method (Lord, 1975). All of the equating procedures used in this study use anchor-test data. An anchor-test design requires administering one form of a total test to one group of examinees, a second form to a second group of examinees, and a common anchor test to both groups. The anchor test can be internal or external to the tests to be equated. An internal anchor test is a subset of items contained in both of the tests to be equated. An external anchor test is a separately timed test that each examinee takes in addition to taking one or the other of the tests to be equated. The anchor test is used to reduce equating bias resulting from differences in ability between the two groups. Its effectiveness will depend on the correlation between it and the two tests to be equated.

The three general rules for determining the transformation can be described as follows:

A. *Equipercentile equating.* Choose a transformation such that scores from the two tests will be equated if they correspond to the same percentile rank in some group of examinees.

B. *Linear equating.* Choose a linear transformation such that scores from the two tests will be equated if they correspond to the same number of standard deviations from the mean in some group of examinees.

C. *Item characteristic curve (ICC) equating.* Choose a transformation such that true scores from the two tests will be equated if they correspond to the same estimated level of the latent trait underlying both tests.

In this study, equipercentile and linear equating were used for the experimental equatings and ICC equating was used to establish the criterion against which some of the experimental equatings were compared.

When an equipercentile model is used with anchor-test data, scores on each test and on the anchor test are first equated separately within each group. Then scores on the two tests to be equated are said to be equivalent if they correspond to the same score on the anchor test.

Linear equating methods all produce an equating transformation of the form $T(x) = Ax + B$, where T is the equating transformation, x is the test score to which it is applied, and A and B are parameters estimated from the data. Some of the authors who have developed equating methods (e.g., Potthoff, 1966) have estimated the parameters of the equating transformation directly from the observed data. Others (e.g., Lord, 1955) have estimated them by means of an equation that expresses the idea of equating in standard-score terms:

$$(x - m_x)/s_x = (y - m_y)/s_y,$$

where x and y refer to the tests to be equated and m and s refer to their means and standard deviations in some group of examinees. Methods using the above equation differ in their identification of the means and standard deviations to be estimated. Some of the methods are based on the estimated means and standard deviations of true scores; others are based on observed scores.

Equating methods also differ in the array of explicit assumptions that they make about the data. The weakest type of assumption used is constancy of regression; e.g., the regression of the test form on the anchor test is invariant across samples. A stronger assumption is that scores on the tests are congeneric; i.e., the nonrandom components of the scores are functionally related. Slightly stronger yet is the assumption that the tests are parallel except for length; i.e., the tests are congeneric and the ratio of error variances of the two tests is the square root of the ratio of their true-score variances. The strongest assumptions used as a basis for deriving equating methods are the multivariate-normal distributional assumptions used to derive maximum-likelihood methods.

Many of the equating methods used require error variance estimates, and the different methods of estimating error variances that may be used involve still additional assumptions. Three methods of estimating error variances were used in the full study: Angoff's method (1953), which assumes that the test to be equated and the anchor test are parallel except for length; Feldt's method (1975), which assumes that the tests to be equated and the anchor test can each be divided into two parts that are parallel except for length; and coefficient alpha, which assumes that all items within a test are congeneric with equal true-score variances. Angoff's method is of particular interest in the context of linear equating because several of the different linear equating methods become mutually equivalent when Angoff error variance estimates are used. Thus in this study only the Angoff method of estimating error variances was considered in order to collapse the volume of data to a manageable size. The effect of using different methods of estimating error variances will be the focus of further study.

The various equating methods used in the full study are described briefly in Table 1. The models are categorized by whether the procedure results in a linear or curvilinear transformation between observed or true scores. The table provides references and information on the major assumption underlying each model, the kind of data required, and whether specific error variance estimates are needed. If the codes for the models differ for an external and an internal anchor test, then the formulas for computing the transformation parameters differ in the two cases.

Except for the frequency estimation, one-parameter logistic, and three-parameter logistic models, all the models described in Table 1 were used in this study. In this study, even though a number of the models can only be applied strictly when the anchor test and the total test are congeneric, i.e., when the nonrandom components of the scores are functionally related, the methods were also applied when noncongeneric anchor tests were used in order to study the robustness of the procedures.

A variety of linear equating models were used in this study. The Tucker 1, the Tucker 2, the Tucker 3, and the Levine models are based on univariate selection sampling theory. Scores on the relevant selection attribute (the attribute on which the equating samples vary) are assumed to be collinear with scores on the anchor test in the case of the Tucker 1 model; with true scores on the anchor test in the case of the Tucker 2 model; with true scores on the test form in the case of the Tucker 3 model; and with true scores on both the anchor test and the test form in the case of the Levine model. The Tucker 2, Tucker 3, and Levine models were formulated in terms of both observed-score and true-score relationships.

The Tucker–Modified Levine model is a modification of the Levine model developed by Tucker. This procedure assumes that the anchor test and the test form are congeneric. However, it allows for departures from unity in the slope of the linear function relating scores on the test form to scores on an ideal test that is congeneric to but longer than the anchor test.

The Lord XY, the Lord V, and the Lord Maximum Likelihood models assume that true scores on the anchor test and the test forms are linearly related. This assumption leads to a set of 10 equations in 12 unknowns. Hence further assumptions are needed in order to solve the equations. In the case of the Lord XY model, the error variances for scores on the test forms are assumed to be known; in Lord V, the error variances for scores on the anchor test are assumed to be known; and in Lord Maximum Likelihood, the error variances on the test forms and on the anchor test are assumed to be known.

The Lord Congeneric Subtests model assumes that the test forms can each be split into two half-tests. The true scores on the anchor test and the half-tests are assumed to be linearly related. This procedure can only be used with an external anchor test.

The Potthoff models are based on various assumptions about the distribution of scores on the tests to be equated and the anchor test. Potthoff used a maximum-likelihood approach that estimates the equating parameters directly by expressing the assumed distribution function in terms of the

TABLE 1

Description of Equating Models

Major assumption	Data	Error variances known		Code		Model	Reference
		External anchor	Internal anchor	External anchor	Internal anchor		
I. Linear observed-score models							
A. Constancy of regression							
1. X on v and Y on v							
a. Curvilinear	Bivariate distributions	None	None	FE-O	FE-O	Frequency Estimation	Angoff (1979), 581–582
b. Linear	Cross-products matrices	None	None	T1	T1	Tucker 1	Angoff (1971), 576–577
2. X on v' and Y on v' are linear	Cross-products matrices	v_a, v_b	v_a, v_b	T2O-E[a]	T2O-I[b]	Tucker 2 Observed	Tucker (1974)
3. V on x' and V on y' are linear	Cross-products matrices	x_a, y_b	x_a, v_a, y_b, v_b	T3O-E[a]	T3O-I[b]	Tucker 3 Observed	Tucker (1974)
B. X and V and Y and V congeneric and constancy of regression of X on v' and of Y on v'	Cross-products matrices	x_a, v_a, y_b, v_b	x_a, v_a, y_b, v_b	LER[a]	LER[b]	Levine Equally Reliable	Levine (1955)
C. Distributional assumptions							
1. X|v and Y|v are normal	Cross-products matrices	None	None	PAC[h]	PAC[i]	Potthoff A	Potthoff (1966)
2. V|x and V|y are normal	Cross-products matrices	None	None	PB	PB[f,g]	Potthoff B	Potthoff (1966)
3. X and V and Y and V are bivariate normal	Cross-products matrices	None	None	PAC[h]	PAC[i]	Potthoff C	Potthoff (1966)
4. X and V and Y and V are more general than bivariate normal to allow for group differences	Cross-products matrices	None	None	PD	PD	Potthoff D	Potthoff (1966)

II. Curvilinear observed-score models							
A. Constancy of regression- X on v and Y on v are curvilinear	Bivariate distributions	None	None	FE-EQ	FE-EQ	Frequency Estimation	Angoff (1971), 581–582
B. No explicit assumptions	Marginal distributions	None	None	EQ	EQ	Equipercentile	
III. Linear true-score models							
A. Constancy of regression							
1. X on v' and Y on v'							
a. Curvilinear	Bivariate distributions	x_a, y_b	x_a, y_b	FE-T	FE-T	Frequency Estimation	Angoff (1971), 581–582
b. Linear	Cross-products matrices	x_a, v_a, y_b, v_b	x_a, v_a, y_b, v_b	T2T-E[c]	T2T-I[d]	Tucker 2 True	Tucker (1974)
2. V on x' and V on y' are linear	Cross-products matrices	x_a, y_b	x_a, y_b	T3T-E[c]	T3T-I[d]	Tucker 3 True	Tucker (1974)
B. X and V and Y and V congeneric							
1. Constancy of regression of X' on v' and of Y' on v'	Cross-products matrices	x_a, v_a, y_b, v_b	x_a, v_a, y_b, v_b	LUR[e]	LUR[f]	Levine Unequally Reliable	Levine (1955)
2. Observed scores on test form are linear function of observed scores on ideal test. True scores on ideal test are multiplicative function of true scores on anchor test.	Cross-products matrices	x_a, v_a, y_b, v_b	None	TML-E[e]	TML-E[f,g]	Tucker–Modified Levine	Tucker (1974)
3. True scores on test form are linear function of true scores on anchor test.							
a. As stated above	Cross-products matrices	x_a, y_b	x_a, y_b	LXY-E[e]	LXY-I[f]	Lord XY	Lord (1976)
b. As stated above	Cross-products matrices	v_a, v_b	v_a, v_b	LV-E[e]	LV-I[f]	Lord V	Lord (1976)

TABLE I (Continued)

Major assumption	Data	Error variances known External anchor	Error variances known Internal anchor	Code External anchor	Code Internal anchor	Model	Reference
c. As stated above	Cross-products matrices	$x_a, v_a,$ y_b, v_b	x_a, v_a y_b, v_b	LML-E[e]	LML-I[f]	Lord Maximum Likelihood	Lord (1976), Birch, 1964
4. True scores on part test are linear function of true scores on anchor test.	Cross-products matrices	None	—	LCS	—	Lord Congeneric Subtests	Lord (1976)
IV. Curvilinear true-score models A. X and V and Y and V congeneric							
1. Item response function represented by a one-parameter logistic function	Item responses	None	None	ICC-1	ICC-1	One-Parameter Logistic ICC	Lord (1975)
2. Item response function represented by a three-parameter logistic function	Item responses	None	None	ICC-3	ICC-3	Three-Parameter Logistic ICC	Lord (1975)

[a–f] These models yield identical conversion parameters when the Angoff method of estimating error variances is used.

[g–i] These models always yield identical conversion parameters.

Notation: x, y, v = observed score on new form, old form, and anchor test, respectively. x', y', v' = true scores. a, b, c = group taking tests X and V; group taking tests Y and V; combined group.

equating parameters. The estimates are those values of the equating parameters under which the set of scores actually observed would be most likely to occur.

3. Study Design

ETS computer files for two national administrations, April 1975 and November 1975, of the Scholastic Aptitude Test (SAT) and its companion the Test of Standard Written English (TSWE) served as the source of the data for the study. Twenty-two pairs of total test scores were equated on the basis of the data from each of the two administrations. Records of test scores, responses to test items, and responses to the Student Descriptive Questionnaire (SDQ), filled out as part of the registration process for the SAT and TSWE, were used. Samples having specified characteristics were constructed, selected scores for a number of special-purpose total tests and anchor tests were extracted or calculated, and item statistics and other data needed for the various equating models were computed.

3.1. Samples

Six base samples were used for the study, three each from the April 1975 and November 1975 Saturday administrations of the SAT. The April base samples were selected from those candidates who took verbal equating tests ee, fe, and fm; the November base samples were selected from those who took verbal equating tests eg, fg, and fo.

Each base sample consisted of 4731 cases from which three nonoverlapping subsamples of 1577 cases each were created in three different ways (making a total of nine subsamples). Three of the samples ("random" samples) were selected by use of an IBM recursive random number generator; three ("similar" samples), by use of an algorithm designed to yield samples similar in mean verbal ability; and three ("dissimilar" samples), by an algorithm designed to yield samples dissimilar in mean verbal ability. Two variables from the Student Descriptive Questionnaire, level of educational aspiration and amount of high school foreign language training, were used to select the similar and dissimilar samples. These variables were known from prior information to have a high relationship with SAT-Verbal scores.

Thus a total of 54 subsamples of 1577 cases each, nine for each of the six base samples, was used in the study. Tables 2 and 3 give the means and standard deviations of the verbal equating test raw scores and the SAT-Verbal raw and scaled scores for the 54 subsamples. Although the sub-

TABLE 2

Descriptive Statistics on April 1975 Samples[a]

	Verbal Equating Test raw score[b]		SAT-Verbal			
			Raw score[c]		Scaled score	
Sample	M	SD	M	SD	M	SD
Base sample that took Verbal Equating Test fe (sample No. 12)						
1. Random	13.61	7.91	35.11	15.26	436	106
2. Random	14.02	7.75	35.72	15.09	441	105
3. Random	13.56	7.97	34.95	15.33	435	106
4. Similar	13.63	8.31	34.76	16.42	434	114
5. Similar	13.03	7.55	33.87	14.67	428	102
6. Similar	14.53	7.68	37.15	14.32	450	99
7. Dissimilar	11.09	7.07	30.14	14.41	402	100
8. Dissimilar	13.09	7.42	33.95	14.41	428	100
9. Dissimilar	17.01	7.94	41.69	14.53	482	101
Base sample that took Verbal Equating Test ee (sample No. 22)						
1. Random	14.35	7.59	35.47	15.54	439	108
2. Random	13.96	7.24	34.84	15.02	434	104
3. Random	14.20	7.45	34.92	15.15	435	105
4. Similar	14.16	7.87	34.49	16.40	432	114
5. Similar	13.37	7.24	33.41	14.72	425	102
6. Similar	14.97	7.08	37.32	14.26	452	99
7. Dissimilar	12.32	7.00	30.38	14.35	404	99
8. Dissimilar	13.14	7.02	33.08	14.67	422	102
9. Dissimilar	17.05	7.40	41.77	14.31	482	99
Base sample that took Verbal Equating Test fm (sample No. 32)						
1. Random	17.00	8.03	34.47	15.48	432	107
2. Random	16.64	8.03	34.67	15.57	433	108
3. Random	16.34	7.99	33.60	14.85	426	103
4. Similar	16.63	8.49	33.87	16.57	428	115
5. Similar	16.23	7.73	33.13	14.63	423	101
6. Similar	17.13	7.80	35.73	14.52	441	101
7. Dissimilar	14.44	7.61	29.51	14.61	398	101
8. Dissimilar	16.02	7.75	32.95	14.48	421	100
9. Dissimilar	19.52	7.85	40.28	14.82	472	103

[a] The sample size was 1577 in each case.

[b] The Verbal Equating Test score identification codes were as follows: fe, FE2FE; ee, FE2EE; fm, FE2FM.

[c] The SAT-Verbal score identification code was FT2XX.

TABLE 3

Descriptive Statistics on November 1975 Samples[a]

| | Verbal Equating Test raw score[b] | | SAT-Verbal | | | |
| | | | Raw score[c] | | Scaled score | |
Sample	M	SD	M	SD	M	SD
Base sample that took Verbal Equating Test fg (sample No. 44)						
1. Random	14.38	8.32	36.34	15.77	445	107
2. Random	14.15	8.02	35.97	14.92	442	101
3. Random	14.54	8.25	37.18	15.60	450	106
4. Similar	14.04	8.69	35.69	16.72	440	113
5. Similar	14.90	8.13	37.59	15.47	453	105
6. Similar	14.14	7.72	36.21	13.95	444	95
7. Dissimilar	11.29	7.44	30.60	14.25	406	97
8. Dissimilar	14.57	7.80	36.70	14.74	447	100
9. Dissimilar	17.22	8.24	42.18	15.10	484	102
Base sample that took Verbal Equating Test eg (sample No. 54)						
1. Random	14.62	8.05	36.36	15.78	445	107
2. Random	14.26	7.90	35.80	15.64	441	106
3. Random	14.39	7.85	35.97	15.40	442	104
4. Similar	14.12	8.37	35.11	16.72	436	113
5. Similar	15.08	8.00	37.42	15.64	452	106
6. Similar	14.08	7.36	35.60	14.27	440	97
7. Dissimilar	11.45	7.22	30.06	14.17	402	96
8. Dissimilar	14.16	7.50	35.64	14.59	440	99
9. Dissimilar	17.66	7.81	42.43	15.51	486	105
Base sample that took Verbal Equating Test fo (sample No. 64)						
1. Random	16.96	8.63	36.14	15.55	443	106
2. Random	16.35	8.69	35.33	15.49	438	105
3. Random	16.61	8.81	35.90	15.88	442	108
4. Similar	15.96	9.25	34.40	16.73	431	113
5. Similar	17.76	8.76	37.93	15.87	455	108
6. Similar	16.21	7.97	35.05	13.99	436	95
7. Dissimilar	13.19	8.19	29.57	14.19	399	96
8. Dissimilar	16.40	8.33	35.37	14.89	438	101
9. Dissimilar	20.33	8.11	42.44	15.11	486	103

[a] The sample size was 1577 in each case.

[b] The Verbal Equating Test score identification codes were as follows: fg, FE4FG; eg, FE4EG; fo, FE4FO.

[c] The SAT-Verbal score identification code was FT4XX.

sampling procedures produced random samples that were very similar in mean ability, they resulted in a few similar samples that were more dissimilar than intended and a few dissimilar samples that were more similar than intended, particularly for the April samples. The similar samples for both months had discrepant standard deviations.

3.2. Equating Design

The total-test–anchor-test combinations used in the various equatings for each of the two administrations can be classified into eleven categories, referred to here as test variations. The test variations studied are outlined in Table 4.

TABLE 4

Test Variations Incorporated in Equating Design

Test variation	Comparison of total tests[a]			Anchor test in relation to total tests		
	Content	Difficulty	Length	Location	Content	Difficulty
1	Identical	Equal	Equal	External	Similar	Similar
2	Identical	Equal	Equal	External	Dissimilar	Similar
3	Identical	Equal	Equal	External	Dissimilar	Dissimilar
4	Identical	Equal	Equal	Internal	Similar	Similar
5	Identical	Equal	Equal	Internal	Similar	Dissimilar
6	Identical	Equal	Equal	Internal	Dissimilar	Similar
7	Similar	Similar	Unequal	Internal	Similar	Similar
8	Similar	Dissimilar	Equal	Internal	Similar	Similar to one total test
9	Similar	Dissimilar	Equal	Internal	Similar	Between total tests
10	Dissimilar	Similar	Equal	Internal	Dissimilar	Similar
11	Similar	Similar	Equal	Internal	Similar	Similar

[a] Test variations 1–6 involved equating a test to itself; hence, the content, difficulty, and length were exactly the same. Test variations 7–11 involved equating a test to a different test.

3.2.1. EQUATING A TEST TO ITSELF

In this part of the study a single form of the operational verbal portion of the SAT was treated as if it represented two different forms and it was equated to itself. (This type of design was first used by Levine, 1955.) Six test variations for equating a test to itself were defined in terms of the characteristics of the anchor test in relation to the total test with regard to location (external or internal), content (similar or dissimilar), and difficulty (similar or dissimilar).

This part of the study was designed to investigate the effects of varying (a) the degree of similarity between the total test and the anchor test and (b) the degree of similarity between the two samples on which the equating operations were based. The three test-related variables (location, difficulty, and content of the anchor test in relation to the total test) were systematically manipulated one at a time. The equatings of SAT-Verbal to itself through a specified anchor test were carried out for random samples, similar samples, and dissimilar samples selected from each of the six base samples used in the study.

All specific test combinations that displayed the same pattern of relations among the tests were considered to represent the same test variation. Thus some test variations included distinguishable test combinations.

3.2.2. EQUATING A TEST TO A DIFFERENT TEST

In this part of the study, a number of total tests were constructed from an item pool comprising the operational SAT-Verbal items, the items in a nonoperational section of verbal material, and the items in the TSWE. Five test variations for equating a test to a different test through an internal anchor test were defined by the relations between the total tests and the relations of the anchor test to the total tests.

This part of the study was designed to investigate the effects of varying (a) the degree of similarity between the two total tests on which scores were to be equated and (b) the degree of similarity between the two equating samples. The dimensions of test-related variation were level of difficulty, length (number of items), and content composition. One internal anchor test was used with all pairs of tests that were similar in content; another was used with pairs of tests that were dissimilar in content. When the total tests in a pair were similar in difficulty, an anchor test of similar difficulty was used. When the total tests differed in difficulty, the anchor test could be similar in difficulty to one of them, or it could be of intermediate difficulty. The equatings of a test to a different test were carried out for random, similar, and dissimilar samples selected from one of the three base samples for each administration

(No. 32 and No. 44). Again, all specific test combinations that displayed the same pattern of relations among the tests were considered to represent the same test variation.

3.3. Tests

Several scores calculated as part of the normal processing for SAT administrations were included in the study. Included also were scores computed on a number of tests constructed retrospectively solely for use in the study.

The general approach followed in constructing the special-purpose tests entailed (a) developing content and statistical specifications for all special-purpose tests and (b) identifying sets of items in accordance with these specifications.

Test content was specified only in terms of distributions of item types, although more detailed specifications are followed in developing operational forms of SAT-Verbal. SAT-Verbal is composed of three types of discrete five-choice items—antonyms, analogies, and sentence completions—and of reading comprehension passages, each of which is followed by five five-choice items based on the passage.

Statistical specifications were stated in terms of the item statistics conventionally used in the development of ETS tests (described by Angoff and Dyer, 1971, pp. 9–10). The equated delta (Δ_e) served as the index of item difficulty, and the biserial correlation (r_b) of the item score with the total score on the operational test in which the item appeared served as the index of item discrimination. The statistic Δ_e is an estimate of the difficulty of the item for a standard reference group. It ranges in value from about 6 (very easy) to about 18 (very hard). If a test composed of five-choice items were of middle difficulty for the reference group, its mean Δ_e would be 12.0.

The item statistics used to construct the tests were taken from the results of item analyses routinely conducted for each new form of the SAT. The analyses were based on systematic samples of approximately 1700–2000 examinees each.

3.3.1. Tests Used in Equating a Test to Itself

The content and statistical characteristics of SAT-Verbal and of the anchor tests used in equating SAT-Verbal to itself are described in Tables 5 and 6. Tables 5 and 6 also give the five-digit codes used to identify the tests.

For any one of the three base samples for a given administration, scores were available on only one of three external anchor tests (a nonoperational section of verbal material given with the SAT and TSWE) similar in content to SAT-Verbal. The external anchor tests each contained equal numbers of

TABLE 5

Codes, Content, and Statistical Characteristics for Tests Used in Equating a Test to Itself, April 1975 Administration

Test description	ID code	Distribution of item types[a]						Item summary statistics			
		ANT	ANA	RDG	SCP	USG	SCR	n	$\bar{\Delta}_e$	$\sigma_{\Delta e}$	$\bar{r}_b$
Total test											
SAT-Verbal	FT2XX	25	20	25	15	0	0	85	11.40	3.38	0.47
Anchor tests											
Similar in content to SAT-V											
External											
ee	FE2EE	10	10	10	10	0	0	40	11.94	2.99	0.40
fe	FE2FE	10	10	10	10	0	0	40	12.05	2.85	0.42
fm	FE2FM	10	10	10	10	0	0	40	11.14	3.06	0.43
Internal											
Easy	FE2DE	10	6	10	8	0	0	34	9.39	2.90	0.49
Medium	FE2DM	10	6	10	8	0	0	34	11.40	3.26	0.49
Hard	FE2DH	10	6	10	8	0	0	34	12.96	2.49	0.47
Dissimilar in content to SAT-V											
Vocabulary (internal)	FE2CV	18	12	0	0	0	0	30	11.38	3.24	0.48
Test of Standard Written English											
E4	FE2CW	0	0	0	0	35	15	50	9.31	1.93	0.41
E6	FE2CE	0	0	0	0	35	15	50	9.64	2.48	0.46
SAT-Mathematical[b]	FE2CM	0	0	0	0	0	0	60	12.16	3.27	0.53

[a] ANT—antonym, ANA—analogy, RDG—reading comprehension, SCP—sentence completion, USG—usage, SCR—sentence correction.
[b] SAT-Mathematical contains 40 five-choice regular mathematics items and 20 four-choice quantitative comparison items.

TABLE 6

Codes, Content, and Statistical Characteristics for Tests Used in Equating a Test to Itself, November 1975 Administration

Test description	ID code	Distribution of item types[a]						Item summary statistics			
		ANT	ANA	RDG	SCP	USG	SCR	n	$\bar{\Delta}_e$	σ_{Δ_e}	$\bar{r}_b$
Total test											
SAT-Verbal	FT4XX	25	20	25	15	0	0	85	11.36	3.40	0.48
Anchor tests											
Similar in content to SAT-V											
External											
eg	FE4EG	10	10	10	10	0	0	40	11.99	2.64	0.43
fg	FE4FG	10	10	10	10	0	0	40	12.01	2.95	0.43
fo	FE4FO	10	10	10	10	0	0	40	11.46	2.80	0.48
Internal											
Easy	FE4DE	10	8	10	8	0	0	36	9.40	2.68	0.51
Medium	FE4DM	10	8	10	8	0	0	36	11.44	3.43	0.49
Hard	FE4DH	10	8	10	8	0	0	36	13.29	2.18	0.48
Dissimilar in content to SAT-V											
Vocabulary (internal)	FE4CV	18	14	0	0	0	0	32	11.44	3.42	0.48
Test of Standard Written English											
E5	FE4CE	0	0	0	0	35	15	50	9.38	2.14	0.47
E7	FE4CW	0	0	0	0	35	15	50	9.07	2.01	0.51
SAT-Mathematical[b]	FE4CM							60	11.77	3.29	0.58

[a] ANT—antonym, ANA—analogy, RDG—reading comprehension, SCP—sentence completion, USG—usage, SCR—sentence correction.

[b] SAT-Mathematical contains 40 five-choice regular mathematics items and 20 four-choice quantitative comparison items.

items of the four types included in the SAT-Verbal. The difficulties of the external anchor tests were not subject to experimentation. The mean difficulties of the external anchor tests were within about half a Δ_e point of the mean for SAT-Verbal, but both the standard deviations of the Δ_e's (σ_{Δ_e}) and the mean r_b's tended to be somewhat lower for the anchor tests.

Two operational tests administered with SAT-Verbal were also used as external anchor tests, namely, SAT-Mathematical and the TSWE. The content of each, of course, is dissimilar to that of SAT-Verbal. SAT-Mathematical and TSWE raw scores were used in the equating analyses.

For each administration three internal anchor tests for equating SAT-Verbal to itself were specially constructed for the study from the pool of 85 operational items in each form. The internal anchor tests were constructed to be similar in content to SAT-Verbal but to vary with regard to each other in mean difficulty. A target value of two Δ_e points was established for the mean difference between easy and medium-difficulty tests and between medium-difficulty and hard tests. For the medium-difficulty anchor tests, $\bar{\Delta}_e$, σ_{Δ_e}, and $\bar{r}_b$ were made to match the corresponding values for SAT-Verbal as closely as possible, given the prescribed content distribution. The σ_{Δ_e}'s for the easy and hard anchor tests were, of necessity, smaller than those for SAT-Verbal and the medium-difficulty anchor test.

A fourth internal anchor test was assembled solely from antonym and analogy items, both of which are designed to measure vocabulary. This test was considered dissimilar to SAT-Verbal because it did not include any items primarily intended to measure reading skills. The vocabulary anchor test was made as statistically similar as possible to the medium-difficulty internal anchor test that reflected the full content of SAT-Verbal. With regard to content, the vocabulary anchor test, the TSWE, and SAT-Mathematical were considered to be progressively distant from SAT-Verbal.

3.3.2. TESTS USED IN EQUATING DIFFERENT TESTS

The content and statistical characteristics of the total tests and anchor tests used in equating a test to a different test are described in Tables 7 and 8. Tables 7 and 8 also give the five-digit codes used to identify the tests. The specific combinations of total tests and anchor tests included in each test variation for each administration are listed in Table 9.

For each administration an expanded item pool consisting of the 85 SAT-Verbal items, the 40 items in one of the verbal external anchor tests, and the 50 items in one form of TSWE formed the basis for defining four sets of three total tests each. The tests in one set, Set P, all contained the same number of items, identically distributed with regard to item type, and were very similar in difficulty. The tests within each of the other three sets were systematically different in one respect, either in average difficulty (Set D), in

TABLE 7

Codes, Content, and Statistical Characteristics for Tests Used in Equating a Test to a Different Test, April 1975 Administration

Test description	ID code	Distribution of item types[a]						Item summary statistics			
		ANT	ANA	RDG	SCP	USG	SCR	n	$\bar{\Delta}_e$	σ_{Δ_e}	$\bar{r}_b$
Total tests											
Set P: Similar content, similar difficulty, equal in length											
Version 1	FT2P1	15	13	15	11	0	0	54	11.32	3.32	0.46
Version 2	FT2P2	15	13	15	11	0	0	54	11.31	3.26	0.46
Version 3	FT2P3	15	13	15	11	0	0	54	11.31	3.31	0.46
Set D: Similar content, dissimilar difficulty, equal in length											
Easy	FT2DE	15	13	15	11	0	0	54	9.34	2.89	0.46
Medium	FT2DM	15	13	15	11	0	0	54	11.34	2.80	0.46
Hard	FT2DH	15	13	15	11	0	0	54	13.26	2.95	0.44
Set L: Similar content, similar difficulty, unequal in length											
Short	FT2LS	11	9	10	9	0	0	39	11.34	3.26	0.46
Intermediate	FT2P3	15	13	15	11	0	0	54	11.31	3.31	0.46
Long	FT2LL	19	17	20	13	0	0	69	11.30	3.31	0.46
Set C: Dissimilar content, similar difficulty, equal in length											
Reading	FT2CR	4	3	24	3	5	2	41	10.92	2.59	0.45
Vocabulary	FT2CV	15	12	4	3	5	2	41	10.83	2.55	0.46
English	FT2CE	4	3	4	3	19	8	41	10.81	2.50	0.46
Anchor tests											
Internal—used to equate tests in sets P, D, and L	FE2PA	6	5	5	4	0	0	20	11.31	3.26	0.46
Internal—used to equate tests in set C	FE2NP	4	3	4	3	5	2	21	11.83	2.63	0.46

[a] ANT—antonym, ANA—analogy, RDG—reading comprehension, SCP—sentence completion, USG—usage, SCR—sentence correction.

TABLE 8

Codes, Content, and Statistical Characteristics for Tests Used in Equating a Test to a Different Test, November 1975 Administration

Test description	ID code	Distribution of item types[a]						Item summary statistics			
		ANT	ANA	RDG	SCP	USG	SCR	n	$\bar{\Delta}_e$	σ_{Δ_e}	$\bar{r}_b$
Total tests											
Set P: Similar content, similar difficulty, equal in length											
Version 1	FT4P1	15	13	15	11	0	0	54	11.60	3.27	0.47
Version 2	FT4P2	15	13	15	11	0	0	54	11.59	3.12	0.47
Version 3	FT4P3	15	13	15	11	0	0	54	11.59	3.12	0.47
Set D: Similar content, dissimilar difficulty, equal in length											
Easy	FT4DE	15	13	15	11	0	0	54	9.59	2.94	0.50
Medium	FT4DM	15	13	15	11	0	0	54	11.69	2.49	0.46
Hard	FT4DH	15	13	15	11	0	0	54	13.51	2.77	0.45
Set L: Similar content, similar difficulty, unequal in length											
Short	FT4LS	11	9	10	9	0	0	39	11.41	3.07	0.47
Intermediate	FT4P3	15	13	15	11	0	0	54	11.59	3.12	0.47
Long	FT4LL	19	17	20	13	0	0	69	11.58	3.27	0.47
Set C: Dissimilar content, similar difficulty, equal in length											
Reading	FT4CR	4	3	24	3	5	2	41	11.00	2.49	0.48
Vocabulary	FT4CV	15	12	4	3	5	2	41	10.91	2.82	0.49
English	FT4CE	4	3	4	3	19	8	41	10.86	2.17	0.48
Anchor tests											
Internal—used to equate tests in Sets P, D, and L	FE4PA	6	5	5	4	0	0	20	11.58	3.05	0.47
Internal—used to equate tests in Set C	FE4NP	4	3	4	3	5	2	21	10.92	2.68	0.49

[a] ANT—antonym, ANA—analogy, RDG—reading comprehension, SCP—sentence completion, USG—usage, SCR—sentence correction.

TABLE 9

Combination of Total Tests and Anchor Tests within Test Variations

| Test variation | Equating a test to itself | | | | | |
| | April 1975 administration | | | November 1975 administration | | |
	New form	Old form	Anchor	New form	Old form	Anchor
1	FT2XX	FT2XX	FE2EE	FT4XX	FT4XX	FE4EG
	FT2XX	FT2XX	FE2FE	FT4XX	FT4XX	FE4FG
	FT2XX	FT2XX	FE2FM	FT4XX	FT4XX	FE4FO
2	FT2XX	FT2XX	FE2CM	FT4XX	FT4XX	FE4CM
3	FT2XX	FT2XX	FE2CW	FT4XX	FT4XX	FE4CW
	FT2XX	FT2XX	FE2CE	FT4XX	FT4XX	FE4CE
4	FT2XX	FT2XX	FE2DM	FT4XX	FT4XX	FE4DM
5	FT2XX	FT2XX	FE2DE	FT4XX	FT4XX	FE4DE
	FT2XX	FT2XX	FE2DH	FT4XX	FT4XX	FE4DH
6	FT2XX	FT2XX	FE2CV	FT4XX	FT4XX	FE4CV

| Test variation | Equating a test to a different test | | | | | |
| | April 1975 administration | | | November 1975 administration | | |
	New form	Old form	Anchor	New form	Old form	Anchor
7	FT2LS	FT2P3	FE2PA	FT4LS	FT4P3	FE4PA
	FT2LS	FT2LL	FE2PA	FT4LS	FT4LL	FE4PA
	FT2P3	FT2LL	FE2PA	FT4P3	FT4LL	FE4PA
8	FT2DE	FT2DM	FE2PA	FT4DE	FT4DM	FE4PA
	FT2DM	FT2DH	FE2PA	FT4DM	FT4DH	FE4PA
9	FT2DE	FT2DH	FE2PA	FT4DE	FT4DH	FE4PA
10	FT2CR	FT2CV	FE2NP	FT4CR	FT4CV	FE4NP
	FT2CR	FT2CE	FE2NP	FT4CR	FT4CE	FE4NP
	FT2CV	FT2CE	FE2NP	FT4CV	FT4CE	FE4NP
11	FT2P1	FT2P2	FE2PA	FT4P1	FT4P2	FE4PA
	FT2P1	FT2P3	FE2PA	FT4P1	FT4P3	FE4PA
	FT2P2	FT2P3	FE2PA	FT4P2	FT4P3	FE4PA

length (Set L), or in content (Set C), but were similar in the other two respects. When equating tests in Set D using dissimilar samples, the low-ability sample was assigned to the easy test; the middle-ability sample, the medium-difficulty test; and the high-ability sample, the hard test.

For each administration a 20-item internal anchor test of medium difficulty, similar in content to SAT-Verbal, was constructed for use in equating the tests in Sets P, D, and L. A 21-item internal anchor test, composed of equal numbers of vocabulary, reading, and TSWE items, was built to equate the tests in Set C. Exclusive of the equating items, each test in Set C contained items from only one of the three content categories. Like the total tests in Set C, the anchor test was relatively easy.

3.4. *Evaluative Procedures*

3.4.1. DISCREPANCY INDICES

For each raw score x there is a corresponding criterion score t and an estimated criterion score t' derived from a specific equating model. The smaller the difference d between t' and t, the smaller the equating error and the more appropriate the equating model.

The standardized weighted mean-square difference and squared bias were selected as the most useful summary indices for evaluating the effectiveness of the various models. The weighted mean-square difference gives the greatest weight to those values of x that are most likely to occur and is consistent with what is used to represent total error in the statistical literature. The indices were standardized (expressed as a proportion of the criterion standard deviation) so that results could be compared across equating situations as well as across equating models.

The standardized weighted mean-square difference or total error is equal to the variance of the difference plus the squared bias, that is,

$$\sum_j f_j d_j^2 / n s_t^2 = \sum_j f_j (d_j - \bar{d})^2 / n s_t^2 + \bar{d}^2 / s_t^2$$

(total error) = (variance of difference) + (squared bias),

where $d_j = (t_j' - t_j')$, t_j' is the estimated criterion score for raw score x_j, t_j is the criterion score for x_j, $\bar{d} = \sum_j f_j d_j / n$, s_t is the standard deviation of the criterion scores t_j, f_j is the frequency of x_j, $n = \sum_j f_j$ and the summation was over that range of x for which extrapolation was unnecessary for any of the models studied. If the ratio of the squared bias to the total error is 1, then the criterion line and the conversion line are parallel. If the difference is less than 1, then there is an interaction between the model and the criterion.

3.4.2. CRITERION EQUATINGS

In the case where a test was equated to itself, the criterion for the various equatings was the test score itself. The new and old forms were treated as different tests, when in reality they were the same test. The ideal equating would reproduce the score on the old form exactly; that is, the conversions from raw to scaled scores would be the same for the new form as for the old form. The criterion was not so simply established in the case where a test was equated to a different test. In these instances it was necessary to calculate "true" conversions by equating the tests in as ideal a manner as possible.

The criterion equatings were accomplished using data from all the cases in the two base samples (No. 32 and No. 44) from which subsamples were selected. Since all 4731 cases in each base sample had scores on the tests being equated, it was possible to equate the scores using a single sample, an ideal equating situation. The scores could be linked directly without involving an anchor test.

Two equating methods were used to establish the two criteria against which to compare the results of the experimental equatings; equipercentile equating of estimated true scores derived from the three-parameter logistic test model, the "ICC equipercentile criterion," and equipercentile equating of observed scores, the "direct equipercentile criterion." The latter equating was done as a check, and results were used as a means of checking on criterion bias in the first study where all equating models (linear, equipercentile, frequency estimation, one-parameter logistic, and three-parameter logistic) were compared (Marco et al., 1979). In this study of linear and equipercentile models, only the ICC equipercentile criterion was used.

To determine the ICC equipercentile criterion, the three-parameter logistic test model was applied separately to reading (reading comprehension and sentence completion), vocabulary (antonym and analogy), and TSWE (usage and sentence correction) items, so that unidimensionality did not have to be assumed across all item types. The 12 pairs of 11 scores that were equated for each administration are listed in Table 9.

The following steps were required to accomplish the three-parameter logistic item characteristic curve criterion equatings:

1. Use LOGIST (Wood and Lord, 1976; Wood et al., 1976) to calculate item parameter estimates and examinee ability estimates separately for each of the three sets of relatively homogeneous items (reading, vocabulary, and TSWE).

2. For each of the 11 scores, calculate for each examinee an estimated true raw $(R - W/4)$ score across the item types represented in each score. The true raw score (R_a) was estimated as follows:

$$R_a = \sum_i P_i(\hat{\theta}_{ia}) - [\sum_i Q_i(\hat{\theta}_{ia})]/4,$$

where $\hat{\theta}_{ia}$ is the estimated ability of examinee a on the item type represented by item i (if item i is a reading item, then $\hat{\theta}_{ia}$ is the estimate of the examinee's reading ability, for example), P_i is the probability of examinee a answering item i correctly (as calculated from the three-parameter logistic test model), $Q_i = 1 - P_i$, and 1/4 is the correction factor for guessing for five-choice multiple-choice items. Each summation was over only those items to which examinee a actually responded.

3. For each of the 12 pairs of scores, equate the scores directly by the equipercentile method. Raw and scaled score equivalents were generated for each integral score on the new form for which it was possible to establish a conversion. (The true raw scores did not usually extend over the possible score range, thus making it impossible to establish a conversion for some scores.)

4. Results and Discussion

Results for this part of the overall study are presented in Table 10 and in Figs. 1–8. As noted in the discussion on equating models, several of the different linear models become equivalent when Angoff error-variance estimates are used. These were the only error variance estimates considered in this part of the study. Thus when the anchor test is external (Test Variations 1–3), results are given only for the Tucker 1, Levine Equally Reliable (same as Tucker 2 Observed and Tucker 3 Observed), Tucker 2 True (same as Tucker 3 True), Levine Unequally Reliable (same as Tucker–Modified Levine, Lord XY, Lord V, and Lord Maximum Likelihood), Lord Congeneric Subtests, Potthoff AC, Potthoff B, Potthoff D, and equipercentile models. When the anchor test is internal (Test Variations 4 through 11), results are given for all the above models except for the Lord Congeneric Subtests model (not applicable) and Potthoff B model (same as Levine Unequally Reliable).

Table 10 summarizes the weighted mean-squared error (total error) between the experimental and criterion equatings for each test variation by model and type of samples used in the experimental equating. To make the results comparable for different tests, the total error is expressed on a scale on which the standard deviation of the criterion scores would be 100 for a standard reference group. For each equating, the total error is classified as to whether it is less than or equal to 25 (zone A), greater than 25 but less than or equal to 100 (zone B), greater than 100 but less than or equal to 225 (zone C), greater than 225 but less than or equal to 400 (zone D), or greater than 400 (zone E). The zones were established arbitrarily. However, suppose the total error was due entirely to bias (i.e., total error equals bias squared). Then, in practice, most psychometricians would probably feel that the equating results were acceptable if the mean differed from "truth" by no more than five points on a scale with a standard deviation of 100; and all psychometricians would probably feel that the equating results were totally unacceptable if the mean differed from "truth" by at least 20 points. Also it would seem reasonable to use random sample results for equating a test to itself through an anchor test of similar content and difficulty as a baseline for

TABLE 10

Total Weighted Mean-Squared Error in Various Zones (A–E)[a] for Experimental Equatings

Test Variation	Model	Random Samples — No. of Equatings	A	B	C	D	E	Similar Samples — No. of Equatings	A	B	C	D	E	Dissimilar Samples — No. of Equatings	A	B	C	D	E
1 — Equating SAT-V to an External Anchor Test Similar in Content and Difficulty	T1	18	16	2	-	-	-	18	4	13	1	-	-	18	1	8	6	2	1
	LER	18	16	2	-	-	-	18	13	5	-	-	-	18	9	7	2	1	-
	T2T	18	15	3	-	-	-	18	12	6	-	-	-	18	9	7	2	-	-
	LUR	18	16	2	-	-	-	18	13	5	-	-	-	18	9	7	2	-	-
	LCS	18	16	2	-	-	-	18	13	5	-	-	-	18	7	9	2	-	-
	PAC	18	7	10	1	-	-	18	-	-	-	3	15	18	-	-	-	1	17
	PB	18	17	1	-	-	-	18	15	3	-	-	-	18	5	9	3	1	-
	PD	18	16	2	-	-	-	18	13	5	-	-	-	18	9	8	1	1	-
	EQ	18	9	9	-	-	-	18	6	12	-	-	-	18	8	8	1	1	-
2 — Equating SAT-V to Itself through SAT-M	T1	18	17	1	-	-	-	18	1	8	7	2	-	18	-	-	5	2	11
	LER	18	17	1	-	-	-	18	11	7	-	-	-	18	4	9	4	1	-
	T2T	18	13	5	-	-	-	18	6	12	-	-	-	18	4	9	4	1	-
	LUR	18	17	1	-	-	-	18	8	10	-	-	-	18	4	9	4	1	-
	LCS	18	13	5	-	-	-	18	3	5	6	3	1	18	1	1	8	4	4
	PAC	18	6	10	2	-	-	18	-	-	-	1	17	18	-	-	-	1	17
	PB	18	13	4	1	-	-	18	-	6	8	2	2	18	1	2	6	3	6
	PD	18	17	1	-	-	-	18	11	6	1	-	-	18	2	7	4	3	2
	EQ	18	5	13	-	-	-	18	3	12	3	-	-	18	2	15	1	-	-
3 — Equating SAT-V to Itself through TSWE	T1	18	14	4	-	-	-	18	6	6	3	3	-	18	-	2	6	4	6
	LER	18	12	6	-	-	-	18	1	5	9	2	1	18	3	6	4	3	2
	T2T	18	11	7	-	-	-	18	1	4	10	2	1	18	3	3	5	3	4
	LUR	18	13	5	-	-	-	18	-	6	9	2	1	18	3	5	3	4	3
	LCS	18	14	4	-	-	-	18	3	1	9	2	3	18	1	5	3	4	5
	PAC	18	8	9	1	-	-	18	-	-	-	3	15	18	-	-	-	1	17
	PB	18	14	4	-	-	-	18	2	2	3	7	4	18	1	2	5	2	8
	PD	18	14	4	-	-	-	18	1	8	6	3	-	18	2	7	4	3	2
	EQ	18	5	13	-	-	-	18	1	9	6	2	-	18	2	6	6	2	2
4 — Equating SAT-V to Itself through an Internal Anchor	T1	18	18	-	-	-	-	18	16	2	-	-	-	18	8	8	2	-	-
	LER	18	18	-	-	-	-	18	18	-	-	-	-	18	17	1	-	-	-
	T2T	18	18	-	-	-	-	18	18	-	-	-	-	18	17	1	-	-	-
	LUR	18	18	-	-	1	-	18	18	-	-	-	-	18	17	1	-	-	-
	PAC	18	7	9	2	-	-	18	-	-	-	4	14	18	-	-	-	1	17

Condition	Stat	N					N						N		N						
Test Similar in Content and Difficulty	PD	18	18	—	—		18	—	—	—	—		18	—	18	18	15	3	3	—	—
	EQ	18	16	2	—		18	13	5	—	—		18	—	18	18	9	9	9	—	—
5 Equating SAT-V to Itself through an Internal Anchor Test Similar in Content but Dissimilar in Difficulty	T1	36	34	2	—	—	36	20	16	—	—		36	3	36	36	3	11	14	5	3
	LER	36	34	2	—	—	36	20	15	1	—		36	3	36	36	3	11	9	8	5
	T2T	36	33	3	—	—	36	17	17	2	—		36	1	36	36	1	12	10	4	9
	LUR	36	33	3	—	—	36	20	16	—	4		36	3	36	36	3	10	11	9	3
	PAC	36	14	18	2	—	36	—	—	1	31		36	4	36	36	4	1	—	2	33
	PD	36	34	2	—	—	36	23	13	—	—		36	12	36	36	12	21	12	5	3
	EQ	36	32	4	—	—	36	20	16	—	—		36	—	36	36	—	3	3	—	—
6 Equating SAT-V to Itself through an Internal Vocabulary Test Similar in Difficulty	T1	18	18	—			18	15	3				18	—	18	18	3	11	5	—	—
	LER	18	18	—			18	18	—				18	—	18	18	11	11	—	—	—
	T2T	18	18	—			18	18	—				18	—	18	18	10	12	—	—	—
	LUR	18	18	—			18	18	—				18	—	18	18	12	6	—	—	—
	PAC	18	8	10			18	—	3				18	3	18	18	—	7	—	1	17
	PD	18	18	—			18	18	—				18	—	18	18	11	7	—	—	—
	EQ	18	15	3			18	15	3				18	—	18	18	5	13	—	—	—
7 Equating Different Tests Similar in Content and Difficulty but Unequal in Length through an Internal Anchor Test Similar in Content and Difficulty	T1	10	8	2	—		6	3	3				6	6	6	6	—	3	3	—	—
	LER	10	9	1	—		6	6	—				6	6	6	6	3	3	—	—	—
	T2T	10	8	2	—		6	5	1				6	6	6	6	4	2	—	—	—
	LUR	10	8	2	—		6	5	1	2			6	6	6	6	5	1	—	—	—
	PAC	10	—	1	1		6	—	2	1			6	6	6	6	—	—	6	—	—
	PD	10	9	1	—		6	4	2				6	—	6	6	—	6	—	6	—
	EQ	10	6	4	—		6	4	2				6	—	6	6	—	5	1	—	—
8 Equating Different Tests Similar in Content but Somewhat Dissimilar in Difficulty through an Internal Anchor Test Similar in Content	T1	6	—				4	—				4	4	—	4	4	—	—	—	—	4
	LER	6	—				4	—				4	4	—	4	4	—	—	—	—	4
	T2T	6	—				4	—				4	4	—	4	4	—	—	—	—	4
	LUR	6	—				4	—				4	4	—	4	4	—	—	—	—	4
	PAC	6	—				4	—				4	4	—	4	4	—	—	—	6	4
	PD	6	—				4	—				4	4	—	4	4	—	—	—	—	4
	EQ	6	6				4	4				—	4	—	4	4	—	—	—	—	—

TABLE 10 (Continued)

Test Variation	Model	Random Samples						Similar Samples						Dissimilar Samples					
		No. of Equatings	A	B	C	D	E	No. of Equatings	A	B	C	D	E	No. of Equatings	A	B	C	D	E
9 — Equating Different Tests Similar in Content but Very Dissimilar in Difficulty through an Internal Anchor Test Similar in Content	T1	4	—	—	—	—	4	2	—	—	—	—	2	2	—	—	—	—	2
	LER	4	—	—	—	—	4	2	—	—	—	—	2	2	—	—	—	—	2
	T2T	4	—	—	—	—	4	2	—	—	—	—	2	2	—	—	—	—	2
	LUR	4	—	—	—	—	4	2	—	—	—	—	2	2	—	—	—	—	2
	PAC	4	—	—	—	—	4	2	—	—	—	—	2	2	—	—	—	—	2
	PD	4	—	2	2	—	4	2	—	1	1	—	2	2	—	1	1	—	2
	EQ	4	—	—	—	—	—	2	—	1	1	—	—	2	—	1	1	—	—
10 — Equating Different Tests Dissimilar in Content but Similar in Difficulty through an Internal Anchor Test Dissimilar in Content but Similar in Difficulty	T1	10	6	4	—	—	—	6	1	5	—	—	—	6	2	3	1	—	—
	LER	10	5	5	—	—	—	6	4	2	—	—	—	6	2	4	—	—	—
	T2T	10	4	6	—	—	—	6	4	2	—	—	—	6	2	4	—	—	—
	LUR	10	6	4	—	—	—	6	3	3	—	—	—	6	2	4	—	—	—
	PAC	10	1	8	—	1	—	6	—	1	—	3	2	6	—	—	—	—	6
	PD	10	6	4	—	—	—	6	3	3	—	—	—	6	3	3	—	—	—
	EQ	10	3	7	—	—	—	6	3	3	—	—	—	6	2	4	—	—	—
11 — Equating Different Tests Similar in Content and Difficulty through an Internal Anchor Test Similar in Content and Difficulty	T1	10	10	—	—	—	—	6	4	2	—	—	—	6	—	3	3	—	—
	LER	10	9	1	—	—	—	6	5	1	—	—	—	6	4	2	—	—	—
	T2T	10	9	1	—	—	—	6	5	1	—	—	—	6	4	2	—	—	—
	LUR	10	10	—	—	—	—	6	5	1	—	—	—	6	4	2	—	—	—
	PAC	10	4	3	3	—	—	6	—	—	—	4	2	6	—	—	—	—	6
	PD	10	10	—	—	—	—	6	4	2	—	—	—	6	1	4	1	—	—
	EQ	10	8	2	—	—	—	6	2	4	—	—	—	6	1	4	1	—	—

[a] Zone A: total error ≤ 25, B: 25 < total error ≤ 100, C: 100 < total error ≤ 225, D: 225 < total error ≤ 400, E: 400 < total error.

what can be achieved, and zone A was the range within which most of the results for Test Variations 1 and 4 fell.

Figures 1 through 8 depict the two components of total error, bias and standard deviation of the difference, between the experimental and criterion equatings for each test variation by model and type of samples used in the experimental equating. Bias is denoted on the horizontal axis and the standard deviation of the difference is denoted on the vertical axis. The semicircles denote total error equal to 25, 100, 225, and 400 points. For Test Variations 1 through 6, results are plotted for only one-third of the experimental equatings actually performed, selected to represent the possible sample combinations. For Test Variations 7 through 11, results are plotted for all the experimental equatings.

Both exploratory and confirmatory data analyses were conducted in the study. It was anticipated that the following hypotheses would be confirmed:

1. All models, but particularly the T1 model, yield similar and satisfactory results when random samples are used and nonlinear relationships are not expected.

2. When dissimilar samples are used, the total error is smaller for models using error variance estimates.

3. Models using Angoff error variance estimates do not give satisfactory results when the anchor test is not measuring the same content as the total tests.

4. Equipercentile equating yields larger total errors than linear models when both are appropriate.

5. Equipercentile equating is superior to linear equating when a nonlinear relationship is expected between the total tests or between the anchor test and the total tests.

4.1. Comparisons of Equating Models

4.1.1. Equating a Test to Itself

Test Variation 1. In Test Variation 1, SAT-Verbal was equated to itself through an external anchor test of similar content and difficulty (differences in difficulty ranged from 0.1 to 0.6 Δ_e units). This is an ideal anchor-test design, in which one might expect all models to yield satisfactory results when random samples are used. The equipercentile model is known to require larger samples than the linear models and thus would be expected to yield more variable results for a given sample size. Results are given in Table 10 and depicted in Fig. 1. Except for the PAC and equipercentile models, the results for most equatings using random samples did fall in zone A as hypothesized. The results for the PAC model were unacceptable for similar and dissimilar samples. The results for the equipercentile model tended to fall in zones A and B regardless of type of samples used. The results for the T1 model became increasingly less satisfactory as the samples became more dissimilar in ability. The results for the LER, T2T, LUR, and PD models were similar and satisfactory for most equatings.

Test Variation 2.   In Test Variation 2, SAT-Verbal was equated to itself through SAT-Mathematical. SAT-Verbal and SAT-Mathematical differed in difficulty by 0.8 Δ_e units for the April form and by 0.4 Δ_e units for the November form. Here use of Angoff error variance estimates was not expected to yield satisfactory results. Results are given in Table 10 and depicted in Fig. 1. Except for the PAC and equipercentile models, the results for most equatings, including those using Angoff estimates, fell in zone A when random samples were used. The results for the PAC model were

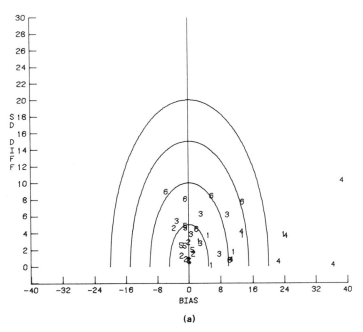

(a)

Fig. 1   Total weighted mean-squared error for representative experimental equatings for test variations (TV) 1 and 2. Shown are the following models: (a) Tucker 1 (T1), (b) Levine equally reliable (LER), (c) Tucker 2 true (T2T), (d) Levine unequally reliable (LUR), (e) Lord congeneric subtests (LCS), (f) Potthoff AC (PAC), (g) Potthoff B (PB), (h) Potthoff D (PD), and (i) Equipercentile (EQ). 1: TV1, dissimilar samples; 2: TV1, random samples; 3: TV1, similar samples; 4: TV2, dissimilar samples; 5: TV2, random samples; 6: TV2, similar samples. Concentric circles denote total error (squared SD DIFF plus squared BIAS) of 25, 100, 225, and 400 points.

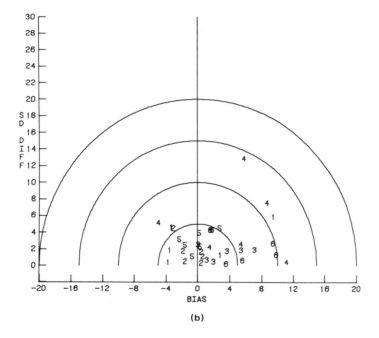

(b)

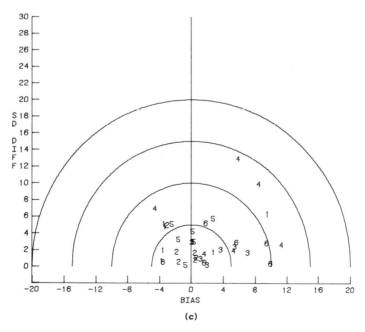

(c)

Fig. 1   (*Continued*)

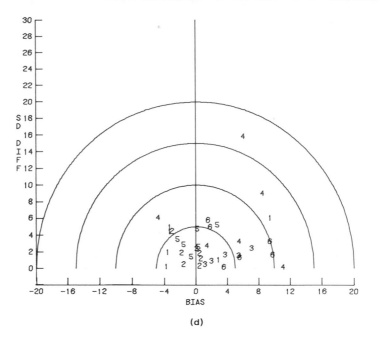

(d)

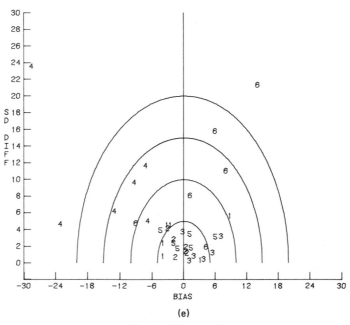

(e)

Fig. 1 *(Continued)*

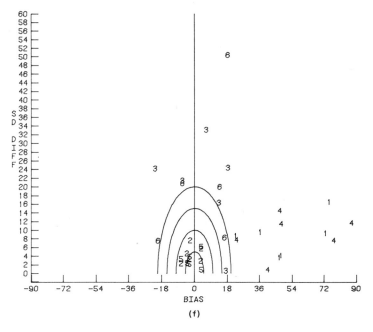

(f)

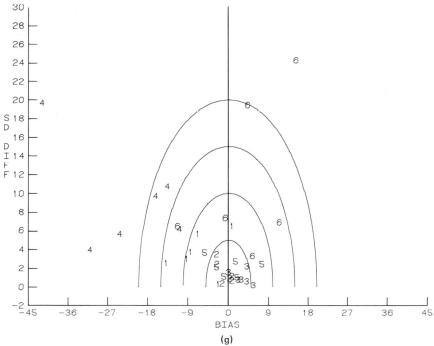

(g)

Fig. 1 (*Continued*)

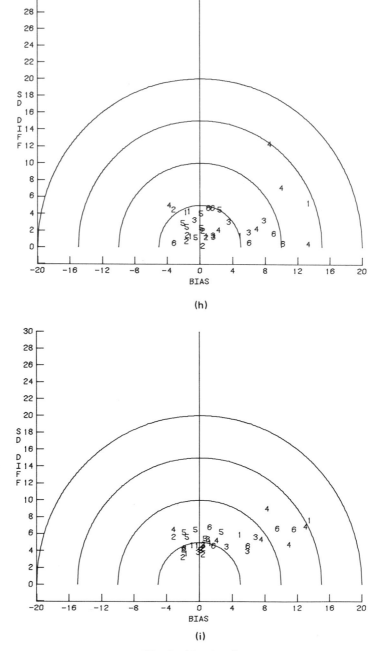

(h)

(i)

Fig. 1 (*Continued*)

unacceptable when nonrandom samples were used. The results for the T1, LCS, and PB models became increasingly unsatisfactory as the samples became more dissimilar in ability. Except for the LCS and PB models, as can be seen in Fig. 1, the bias tended to be positive and accounted for a large proportion of the total error when nonrandom samples were used. This effect is confounded with the way in which the nonrandom samples were assigned to the "old" and "new" forms in that the less able sample was assigned to the new form and the more able sample to the old form. Interestingly, despite the difference in content between the anchor test and the total test, the models using Angoff error variance estimates (LER, T2T, and LUR) performed at least as well as any of the other models regardless of type of samples used.

Figure 1 compares the results obtained for Test Variations 1 and 2. Test Variations 1 and 2 differ in whether the anchor test is similar or dissimilar in content to SAT-Verbal. The results for the T1, LER, LUR, and PD models were satisfactory for both test variations when random samples were used. When nonrandom samples were used, the total errors were smaller for Test Variation 1 than for Test Variation 2, as expected. Also, when nonrandom samples were used, the bias accounted for a larger proportion of the total error for Test Variation 2 than for Test Variation 1. The results confirm the hypothesis that an anchor test that is constructed to be a miniature of the total tests is required when nonrandom samples, particularly dissimilar samples, are used.

Test Variation 3. In Test Variation 3, SAT-Verbal was equated to itself through TSWE. SAT-Verbal and TSWE differed in difficulty by 1.8– 2.3 Δ_e units. Results are given in Table 10 and depicted in Fig. 2. Except for the PAC and equipercentile models, the results for the majority of the equatings using random samples fell in zone A. The total error for all the linear models increased substantially when nonrandom samples were used. In general, the T1 model yielded the smallest total error with similar samples and the equipercentile model yielded the smallest total error with dissimilar samples.

Figure 2 compares the results obtained for Test Variations 2 and 3. In Test Variation 2, SAT-Mathematical was used as the anchor test. In Test Variation 3, TSWE was used as the anchor test. TSWE is more dissimilar in difficulty to SAT-Verbal than is SAT-Mathematical, but TSWE is more similar in content to SAT-Verbal than is SAT-Mathematical. The total errors tended to be greater for all models when TSWE was used as the anchor test than when SAT-Mathematical was used as the anchor test, suggesting that equating results may be affected more by differences in difficulty between the anchor test and the total test than by differences in content. Since these results are confounded by the type of content differences between the anchor test and SAT-Verbal, it would be of interest to conduct a further study using anchor tests of similar content (different from SAT-Verbal) but of varying difficulty levels.

Test Variation 4. In Test Variation 4, SAT-Verbal was equated to itself through an internal anchor test of similar content and difficulty. Since a parallel internal-anchor test would correlate more highly with the total test than a parallel external-anchor test, this is the ideal anchor-test design, from a correlation standpoint, and one would expect all models to yield satisfactory results for random samples. Results are given in Table 10 and

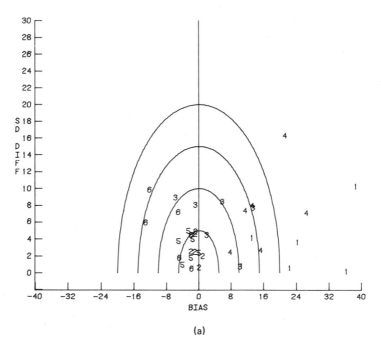

(a)

Fig. 2 Total weighted mean-squared error for representative experimental equatings for test variations (TV) 2 and 3. Shown are the following models: (a) Tucker 1 (T1), (b) Levine equally reliable (LER), (c) Tucker 2 true (T2T), (d) Levine unequally reliable (LUR), (e) Lord congeneric subtests (LCS), (f) Potthoff AC (PAC), (g) Potthoff B (PB), (h) Potthoff D (PD), and (i) equipercentile (EQ). 1: TV2, dissimilar samples; 2: TV2, random samples; 3: TV2, similar samples; 4: TV3, dissimilar samples; 5: TV3, random samples; 6: TV3, similar samples. Concentric circles denote total error (squared SD DIFF plus squared BIAS) of 25, 100, 225, and 400 points.

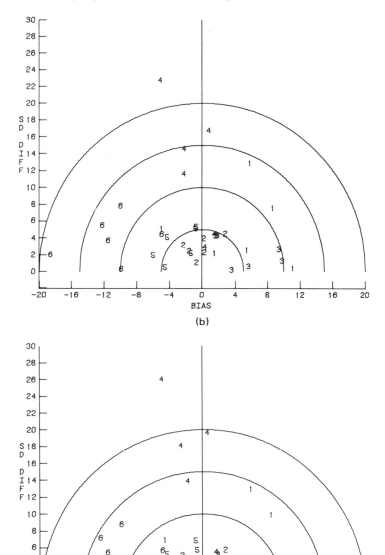

Fig. 2 (*Continued*)

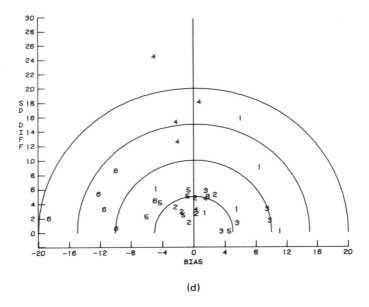

(d)

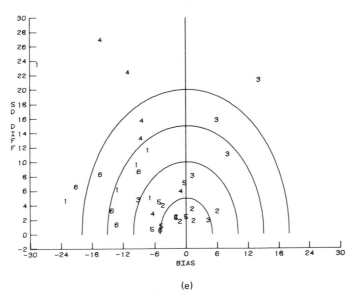

(e)

Fig. 2 (*Continued*)

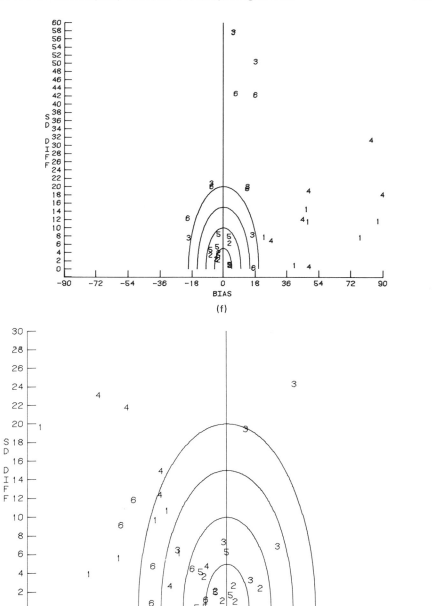

Fig. 2 (*Continued*)

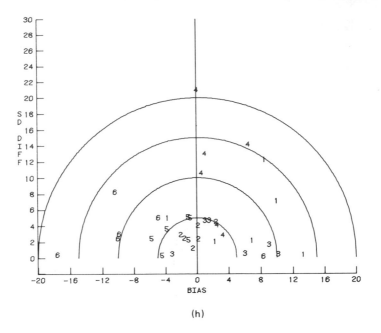

(h)

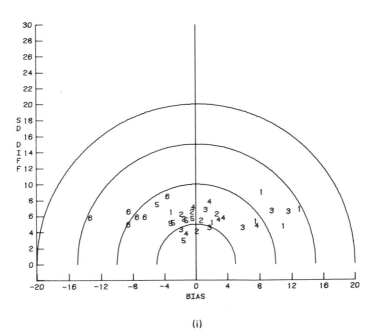

(i)

Fig. 2　(*Continued*)

depicted in Fig. 3. Except for the PAC and equipercentile models, the results for all equatings using random samples did fall in zone A as hypothesized. The results for the PAC model were unacceptable for nonrandom samples. The results for the T1 and the equipercentile models became less satisfactory as the samples became more dissimilar in ability. Models using Angoff error variance estimates (LER, T2T, and LUR) seemed to provide appropriate adjustments for the sample differences when nonrandom samples were used (results for all but one equating fell in zone A). These results suggest that true score models are to be preferred when nonrandom samples are used, at least when Angoff error-variance estimates are appropriate.

Figure 3 compares the results obtained for Test Variations 1 and 4. Test Variations 1 and 4 differ in whether the anchor test is internal or external. The total error was less for the equatings involving an internal anchor. This may simply be due to the fact that the external anchor tests were not quite as

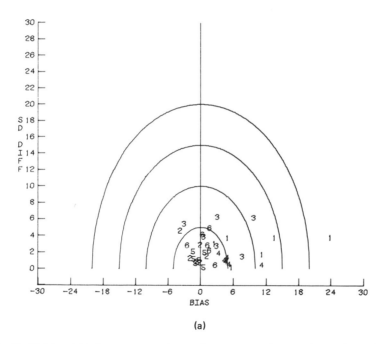

(a)

Fig. 3 Total weighted mean-squared error for representative experimental equatings for test variations (TV) 1 and 4. Shown are the following models: (a) Tucker 1 (T1), (b) Levine equally reliable (LER), (c) Tucker 2 true (T2T), (d) Levine unequally reliable (LUR), (e) Potthoff AC (PAC), (f) Potthoff B (PB), (g) Potthoff D (PD), and (h) equipercentile. 1: TV1, dissimilar samples; 2: TV1, random samples; 3: TV1, similar samples; 4: TV4, dissimilar samples; 5: TV4, random samples; 6: TV4, similar samples. Concentric circles denote total error (squared SD DIFF plus squared BIAS) of 25, 100, 225, and 400 points.

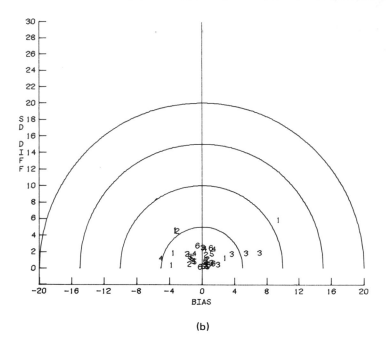

(b)

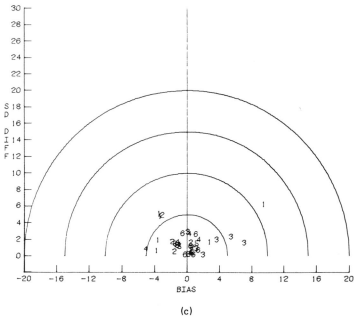

(c)

Fig. 3 (*Continued*)

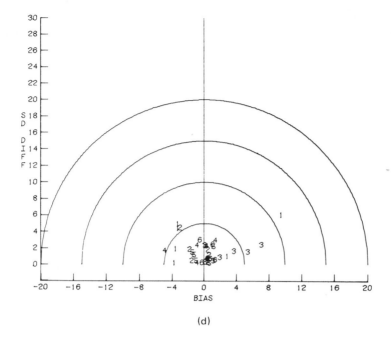

(d)

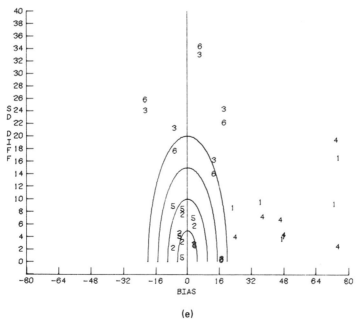

(e)

Fig. 3 (*Continued*)

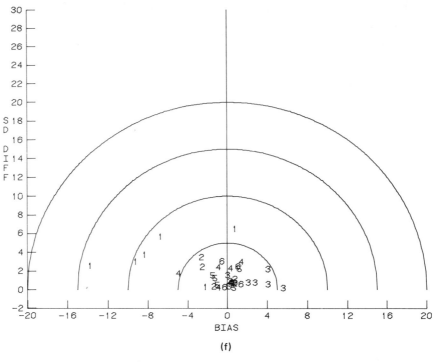

(f)

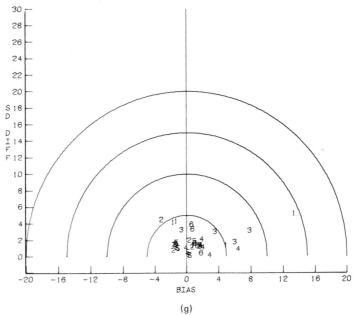

(g)

Fig. 3 (*Continued*)

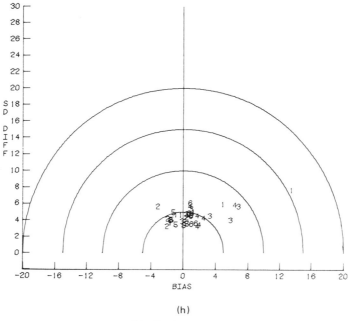

(h)

Fig. 3 (*Continued*)

similar to SAT-Verbal in content and difficulty as were the internal anchor tests.

Test Variation 5. In Test Variation 5, SAT-Verbal was equated to itself through an internal anchor test of similar content but dissimilar difficulty. Half of the equatings used anchor tests that were approximately 2 Δ_e units easier than SAT-Verbal and half of the equatings used anchor tests that were approximately 1.7 Δ_e units harder than SAT-Verbal. The equipercentile model might be expected to perform better than the linear models in these instances. Results are given in Table 10 and depicted in Fig. 4. Except for the PAC model, the results for most equatings using random samples fell in zone A and the results for most equatings using similar samples fell in zones A and B. Except for the equipercentile model, all the models tended to yield a high proportion of unsatisfactory results when dissimilar samples were used.

Figure 4 compares the results obtained for Test Variations 4 and 5. Test Variations 4 and 5 differ in whether the difficulty level of the anchor test is similar or dissimilar to that of the total test. As expected, the total error was less for those equatings that involved an anchor test of similar difficulty to the total test. However, except for the PAC model, all the models tended to yield satisfactory results for both test variations when random and similar

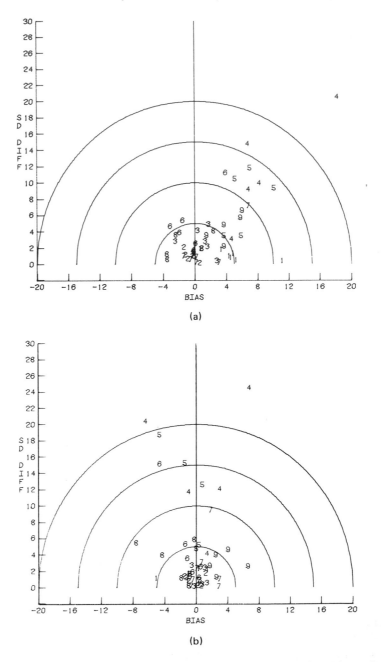

Fig. 4 Total weighted mean-squared error for representative experimental equatings for test variations (TV) 4 and 5. Shown are the following models: (a) Tucker 1 (T1), (b) Levine equally reliable (LER), (c) Tucker 2 true (T2T), (d) Levine unequally reliable (LUR), (e) Potthoff AC (PAC), (f) Potthoff D (PD), and (g) equipercentile (EQ). 1: TV4, dissimilar samples, middle anchor; 2: TV4, random samples, middle anchor; 3: TV4, similar samples, middle anchor;

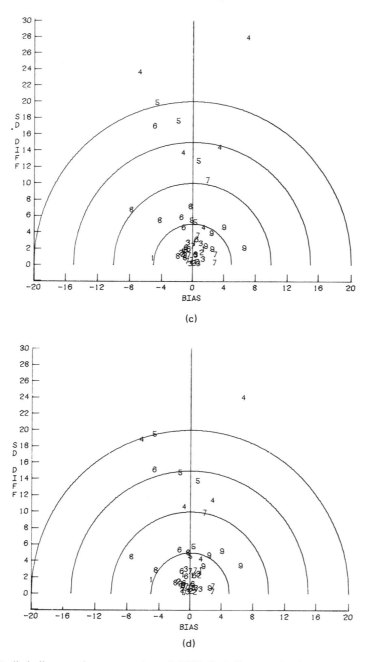

(c)

(d)

4: TV5, dissimilar samples, easy anchor; 5: TV5, dissimilar samples, hard anchor; 6: TV5, random samples, easy anchor; 7: TV5, random samples, hard anchor; 8: TV5, similar samples, easy anchor; 9: TV5, similar samples, hard anchor. Concentric circles denote total error (squared SD DIFF plus squared BIAS) of 25, 100, 225, and 400 points.

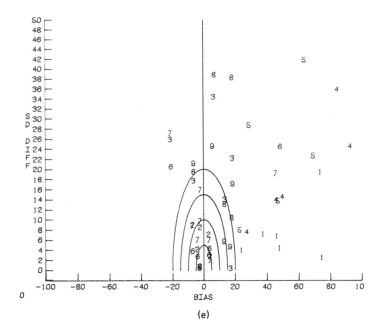

(e)

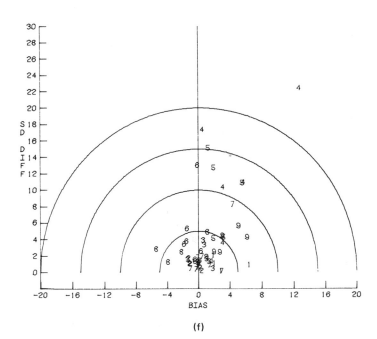

(f)

Fig. 4 (*Continued*)

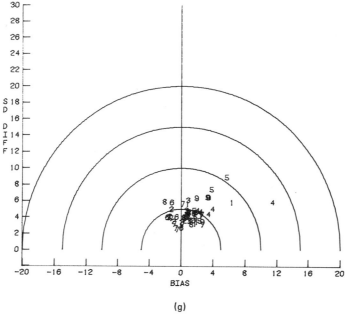

(g)

Fig. 4 (*Continued*)

samples were used. When dissimilar samples were used, all the linear models, except for PAC, gave satisfactory results when the anchor test was similar in difficulty to the total test; but none of the linear models performed well when the anchor test was different in difficulty from the total test.

Test Variation 6. In Test Variation 6, SAT-Verbal was equated to itself through an internal vocabulary test of similar difficulty. Results are given in Table 10 and depicted in Fig. 5. Except for the PAC and equipercentile models, the results for all equatings using random samples fell in zone A. The results for the PAC model were unacceptable for nonrandom samples. The results for the T1 and equipercentile models became less satisfactory as the samples became more dissimilar in ability. The LER, T2T, LUR, and PD models performed well for all types of samples.

Figure 5 compares the results obtained for Test Variations 4 and 6. Test Variations 4 and 6 differ in whether the content of the anchor test is similar or dissimilar to that of the total test. The total error was similar for both test variations when random and similar samples were used. When dissimilar samples were used, the total error tended to be smaller when the anchor test was similar in both content and difficulty to the total test. A comparison of Figs. 4 and 5 suggests that equating results may be affected more by differences in difficulty level between the anchor test and the total test than by

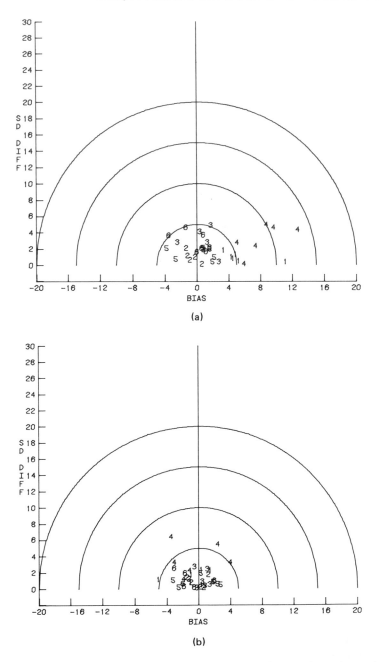

Fig. 5 Total weighted mean-squared error for representative experimental equatings for test variations (TV) 4 and 6. Shown are the following models: (a) Tucker 1 (T1), (b) Levine equally reliable (LER), (c) Tucker 2 true (T2T), (d) Levine unequally reliable (LUR), (e) Potthoff AC (PAC), (f) Potthoff D (PD), and (g) Equipercentile (EQ). 1: TV4, dissimilar samples; 2:

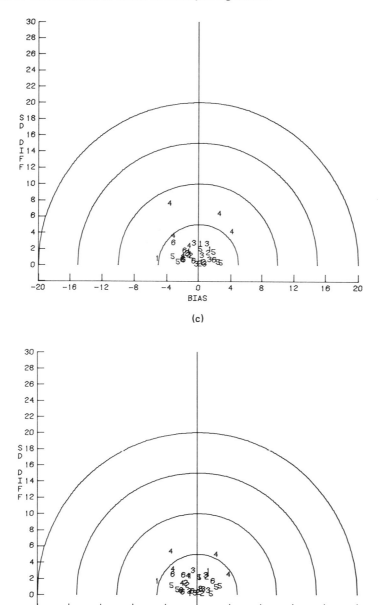

(c)

(d)

TV4, random samples; 3: TV4, similar samples; 4: TV6, dissimilar samples; 5: TV6, random samples; 6: TV6, similar samples. Concentric circles denote total error (squared SD DIFF plus squared BIAS) of 25, 100, 225, and 400 points.

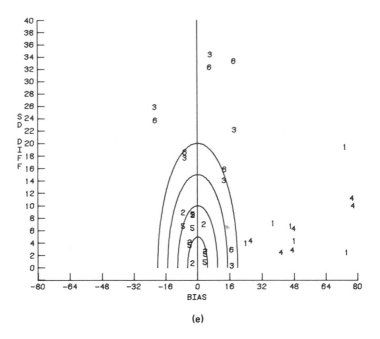

(e)

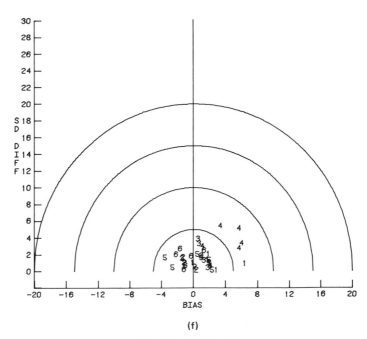

(f)

Fig. 5 (*Continued*)

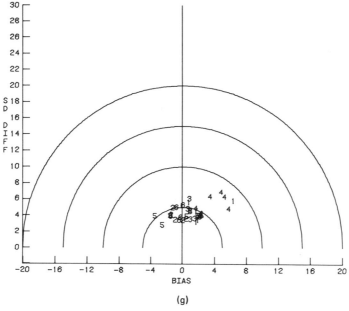

(g)

Fig. 5 (*Continued*)

differences in content. As noted earlier, these results are confounded by the type of content differences between the anchor test and SAT-Verbal. It would be of interest to conduct a further study using anchor tests similar in content (different from SAT-Verbal) but of varying difficulty levels.

4.1.2. Equating a Test to a Different Test

Test Variation 7. In Test Variation 7, tests of similar content and difficulty but of unequal length were equated through an internal anchor test of similar content and difficulty. Total tests of 39 and 54 items, of 54 and 69 items, and of 39 and 69 items were equated. Since test length is related to reliability, one would expect true score models to yield more satisfactory results than observed score models. Results are given in Table 10 and depicted in Fig. 6. The PAC model gave unacceptable results for all types of samples. Except for PAC, all the models tended to yield satisfactory results for random and similar samples. However, the true score models did yield slightly smaller total errors, on the average, than the observed score models when similar samples were used. When dissimilar samples were used, the total error was greater for observed score models than for true score models. Also, from Fig. 6, it can be seen that the total error for equatings using dissimilar samples with observed score models (except for LER) contained a large positive bias component.

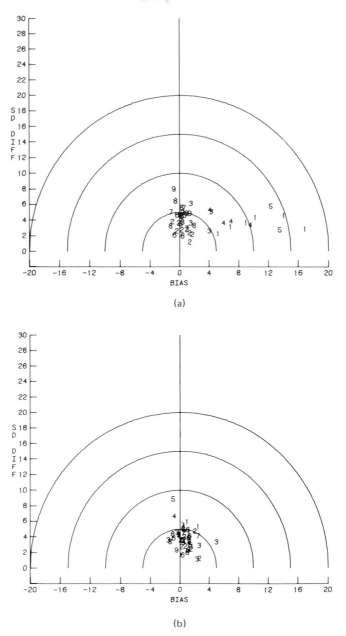

Fig. 6 Total weighted mean-squared error for experimental equatings for test variations (TV) 7 and 11. Shown are the following models: (a) Tucker 1 (T1), (b) Levine equally reliable (LER), (c) Tucker 2 true (T2T), (d) Levine unequally reliable (LUR), (e) Potthoff AC (PAC), (f) Potthoff D (PD), and (g) equipercentile (EQ). 1: TV11, dissimilar samples, middle tests; 2: TV11, random samples, middle tests: 3: TV11, similar samples, middle tests; 4: TV7, dis-

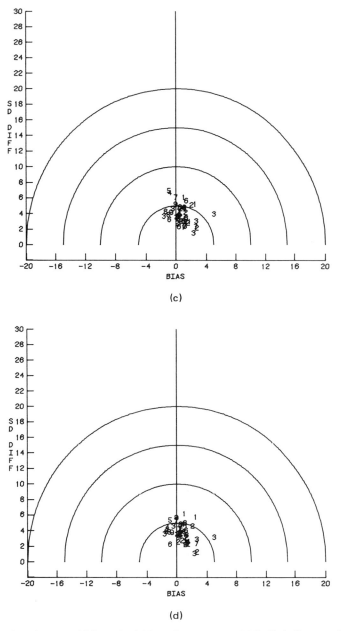

(c)

(d)

similar samples, short-to-middle and middle-to-long tests; 5: TV7, dissimilar samples, short-to-long tests; 6: TV7, random samples, short-to-middle and middle-to-long tests; 7: TV7, random samples, short-to-long tests: 8: TV7, similar samples, short-to-middle and middle-to-long tests; 9: TV7, similar samples, short-to-long tests. Concentric circles denote total error (squared SD DIFF plus squared BIAS) of 25, 100, 225, and 400 points.

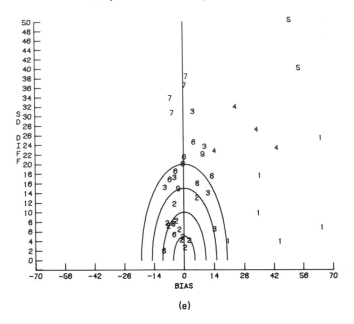

(e)

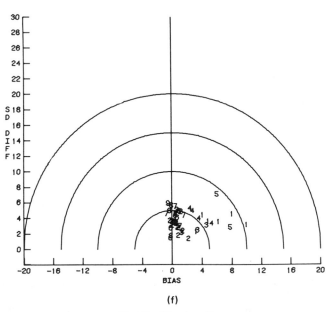

(f)

Fig. 6 (*Continued*)

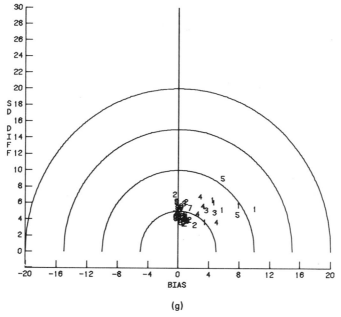

(g)

Fig. 6 (*Continued*)

Test Variations 8 and 9. In Test Variations 8 and 9, tests of similar content and dissimilar difficulty were equated through an internal anchor test of similar content. In Test Variation 8, the two tests to be equated differ in difficulty by approximately 2 Δ_e units. In Test Variation 9, the two tests to be equated differ in difficulty by approximately 4 Δ_e units. When the two tests to be equated differ substantially in difficulty, one would expect the true relationship between the scores on the two tests to be curvilinear. Linear models would presumably yield unsatisfactory results. Results are given in Table 10 and depicted in Fig. 7. The linear models gave unacceptable results for all types of samples. The equipercentile model yielded the only acceptable results, but none of the results fell in zone A. From Fig. 7 it can be seen that the greater the difference in difficulty between the two tests to be equated, the greater the total error. Also, for the linear models, the bias component was small in comparison to the standard deviation of the difference between the experimental and criterion equatings.

Test Variation 10. In Test Variation 10, a reading test, a vocabulary test, and an English test of equal length and similar difficulty were equated to each other through an internal anchor test of dissimilar content but similar difficulty. This variation represents a situation in which equating is strictly impossible but that occurs sometimes in practice. Results are given in Table

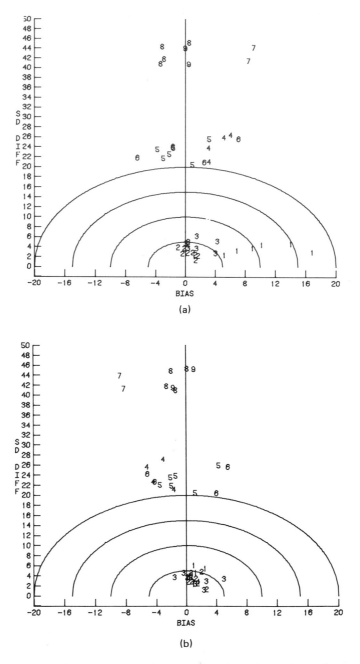

Fig. 7 Total weighted mean-squared error for experimental equatings for test variations (TV) 8, 9, and 11. Shown are the following models: (a) Tucker 1 (T1), (b) Levine equally reliable (LER), (c) Tucker 2 true (T2T), (d) Levine unequally reliable (LUR), (e) Potthoff AC (PAC), (f) Potthoff D (PD), and (g) equipercentile (EQ). 1: TV11, dissimilar samples, middle tests; 2: TV11, random samples, middle tests; 3: TV11, similar samples, middle tests; 4: TV8, Dis-

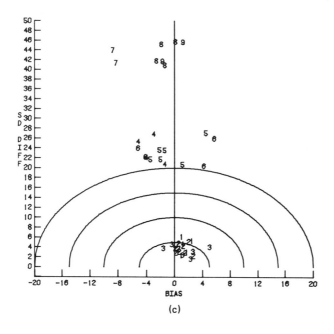

(c)

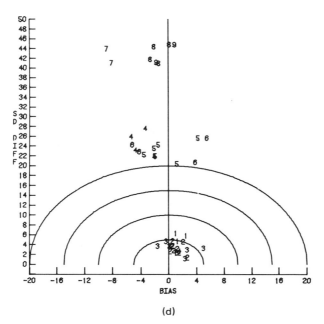

(d)

similar samples, easy-to-middle and middle-to-hard tests; 5: TV8, random samples, easy-to-middle and middle-to-hard tests; 6: TV8, similar samples, easy-to-middle and middle-to-hard tests; 7: TV9, dissimilar samples, easy-to-hard tests; 8: TV9, random samples, easy-to-hard tests; 9: TV9, similar samples, easy-to-hard tests. Concentric circles denote total error (squared SD DIFF plus squared BIAS) of 25, 100, 225, and 400 points.

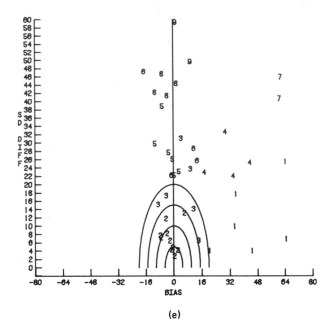

(e)

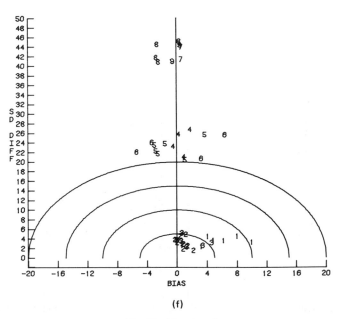

(f)

Fig. 7 (*Continued*)

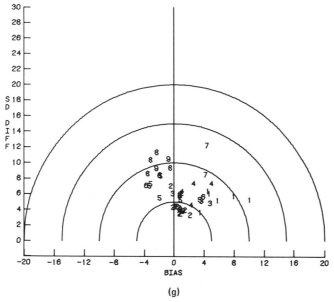

(g)

Fig. 7 (*Continued*)

10 and depicted in Fig. 8. Except for the PAC model, all the models tended to yield satisfactory and similar results for all types of samples. The T1 model performed less satisfactorily than these other models when similar samples were used.

A comparison of Figs. 7 and 8 suggests that differences in difficulty level between the two tests to be equated lead to greater equating error than moderate differences in content. It would be of interest to conduct a further study in which tests of dissimilar content and difficulty were equated in order to separate the effects of content and difficulty differences on the amount of total error.

Test Variation 11.   In Test Variation 11, 54-item tests of similar content and difficulty were equated to each other through an internal anchor test of similar content and difficulty. This test variation is the closest to the ideal anchor-test design that one can achieve in practice, and, as in the case of Test Variation 4, all models would be expected to yield satisfactory results for random samples. Results are given in Table 10 and depicted in Figs. 6 and 7. Except for the PAC model, the results for most equatings using random samples did fall in zone A. The results for the PAC model were unacceptable for nonrandom samples. The results for the equipercentile, PD, and T1 models became increasingly less satisfactory as the samples became more dissimilar in ability. The total error in these instances contained a large

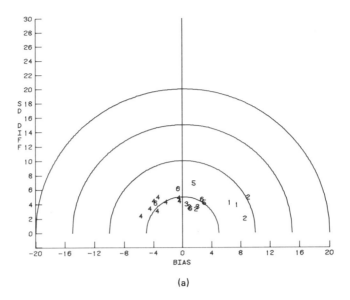

(a)

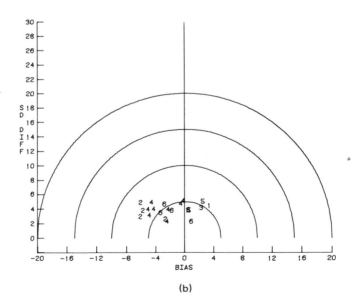

(b)

Fig. 8 Total weighted mean-squared error for experimental equatings for test variation (TV) 10. Shown are the following models: (a) Tucker 1 (T1), (b) Levine equally reliable (LER), (c) Tucker 2 true (T2T), (d) Levine unequally reliable (LUR), (e) Potthoff AC (PAC), (f) Potthoff D (PD), and (g) equipercentile (EQ). 1: dissimilar samples, reading to vocab.; 2: dissimilar

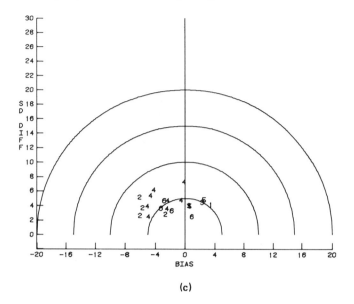

(c)

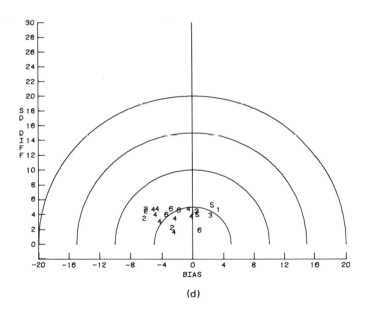

(d)

samples, vocab. to English and reading to English; 3: random samples, reading to vocab.; 4: random samples, vocab. to English and reading to English; 5: similar samples, reading to vocab.; 6: similar samples, vocab. to English and reading to English. Concentric circles denote total error (squared SD DIFF plus squared BIAS) of 25, 100, 225, and 400 points.

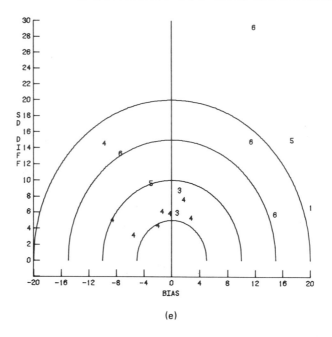

(e)

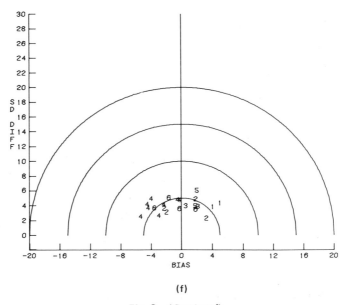

(f)

Fig. 8 *(Continued)*

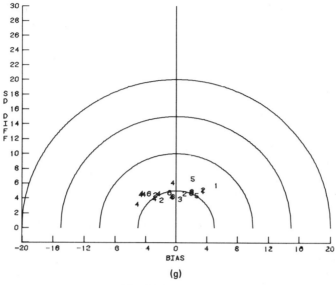

(g)

Fig. 8 (*Continued*)

positive bias component. As in the case of Test Variation 4, the models using Angoff error variance estimates (LER, T2T, and LUR) again seemed to provide satisfactory adjustments for differences in samples.

Figure 6 compares the results obtained for Test Variations 7 and 11. Test Variations 7 and 11 differ in whether or not the two tests to be equated are of the same length. The total error increased for all models as the two tests became more unequal in length. However, the LER, T2T, and LUR models gave satisfactory results for all types of samples for both test variations.

Figure 7 compares the results obtained for Test Variations 8, 9, and 11. Test Variations 8, 9, and 11 differ in the size of the discrepancy in difficulty level between the two tests to be equated. The total error increases for all models as the difference in difficulty level increases between the two tests to be equated. The equipercentile model was the only model that gave fairly satisfactory results for Test Variations 8 and 9 regardless of the type of samples used in the equating.

5. Conclusions

One should be careful in generalizing from the results of the study because (1) only verbal tests were equated in the study, (2) further work is needed to separate fully the effects of content and difficulty differences between the anchor test and total tests, and (3) the criterion equatings may be biased in Test Variations 7–11. However, since the results for Test Variation 11 were almost identical to those of Test Variation 4, criterion bias is unlikely to be serious for those situations in which the tests to be equated were similar in content.

The following conclusions constitute a reasonable summary of the results of this study, which considered the appropriateness of various linear models and one equipercentile model for a variety of equating situations:

1.   The Potthoff AC model gives unsatisfactory results in most of the equating situations studied.

2.   For a given sample size, equipercentile equating yields larger total errors than linear models when both are appropriate.

3.   Except for the Potthoff AC model, use of the Tucker 1 model results in larger total errors than use of the other linear models when dissimilar samples are used.

4.   Linear models perform poorly when a nonlinear relationship due to differences in difficulty is expected between the total tests to be equated.

5.   Equipercentile equating is superior to linear equating when a nonlinear relationship, resulting from differences in difficulty, is expected between the anchor test and the total tests and dissimilar samples are used.

6.   When dissimilar samples are used, the total error is smaller for those models that require error variance estimates.

7.   Differences in difficulty level between the anchor test and the total test lead to greater equating error than moderate differences in content between the anchor test and total test.

8.   Differences in difficulty between the two tests to be equated lead to greater equating error than moderate differences in content.

9.   An anchor test constructed to be a miniature of the total tests gives the best equating results.

Table 10 can be used as a decision table in practical applications. It can help practioners choose the equating models that are appropriate for a given situation. The test variation and the sample characteristics must, of course, be known to enter the table. The practioner may wish to consider using models requiring error variance estimates for similar as well as for dissimilar samples. Further research is needed to develop detailed decision rules that indicate which models should be selected for particular test and sample variations.

REFERENCES

Angoff, W. H. (1953). Test reliability and effective test length. *Psychometrika, 18*, 1–14.
Angoff, W. H. (1971). Scales, norms, and equivalent scores. In R. L. Thorndike (Ed.), *Educational Measurement* (2nd ed.). Washington, D.C.: American Council on Education, 508–600.

Angoff, W. H., and Dyer, H. S. (1971). The Admissions Testing Program. In W. H. Angoff (Ed.), *The College Board Admissions Testing Program: A Technical Report on Research and Development Activities Relating to the Scholastic Aptitude Test and Achievement Tests*. New York: College Entrance Examination Board, 1–13.

Birch, M. W. (1964). A note on the maximum likelihood estimation of a linear structural relationship. *Journal of the American Statistical Association, 54*, 1175–1178.

Feldt, L. S. (1975). Estimation of the reliability of a test divided into two parts of unequal length. *Psychometrika, 40*, 557–561.

Levine, R. S. (1955). *Equating the Score Scales of Alternate Forms Administered to Samples of Different Ability* (RB-55-23). Princeton, N.J.: Educational Testing Service.

Lord, F. M. (1955). Equating test scores—A maximum likelihood solution. *Psychometrika, 20*, 193–200.

Lord, F. M. (1975). *A Survey of Equating Methods Based on Item Characteristic Curve Theory* (RB-75-13). Princeton, N.J.: Educational Testing Service.

Lord, F. M. (1976). Personal communication, February 24.

Marco, G. L., Petersen, N. S., and Stewart, E. E. (1979). *A Test of the Adequacy of Curvilinear Score Equating Models*. Paper presented at the Computerized Adaptive Testing Conference, Minneapolis.

Potthoff, R. F. (1966). Equating of grades or scores on the basis of a common battery of measurements. In P. R. Krishnaiah (Ed.), *Multivariate Analysis*. New York: Academic Press, 541–559.

Rentz, R. R., and Bashaw, W. L. (1975). *Equating Reading Tests with the Rasch Model* (Vols. 1 and 2). Athens: Educational Research Laboratory, College of Education, University of Georgia.

Slinde, J. A., and Linn, R. L. (1977). Vertically equated tests: Fact or phantom? *Journal of Educational Measurement, 14*, 23–32.

Slinde, J. A., and Linn, R. L. (1978). An exploration of the adequacy of the Rasch model for the problem of vertical equating. *Journal of Educational Measurement, 15*, 23–35.

Tucker, L. R (1974). Personal communication, August 7.

Wood, R L, and Lord, F. M. (1976). *A User's Guide to LOGIST* (RM-76-4). Princeton, N.J.: Educational Testing Service.

Wood, R L, Wingersky, M. S., and Lord, F. M. (1976). *LOGIST: A Computer Program for Estimating Examinee Ability and Item Characteristic Curve Parameters* (RM-76-6). Princeton, N.J.: Educational Testing Service.

Discussion of "A Test of the Adequacy of Linear Score Equating Models"

T. W. F. Stroud

In the midst of all the discussion about linear, equipercentile, and IRT methods of equating, it is valuable to have an empirical study that compares how well each of these methods work. Let me summarize what I consider to be the most important conclusions of this study.

1.   For equating two different tests of similar content and difficulty (Variation 11), which (we hope) is the usual test-equating situation, the linear methods LER, LUR, and T2T all work well, showing a slight edge over the direct equipercentile method.

2.   All linear methods, as well as the one-parameter IRT method, work badly when equating different tests of different difficulties. In this situation the three-parameter IRT method works best according to the ICC equipercentile criterion, whereas the direct equipercentile method works best according to the direct equipercentile criterion.

3.   The frequency estimation equipercentile method does not work well for dissimilar samples.

4.   The one-parameter IRT method works well in equating a test to itself; it seems to work better than any other method when the anchor is

either easier or harder than the test being equated. It does not do well at equating different tests of different difficulties.

Whether the best linear methods work well seems to depend strongly on whether or not the two tests are of similar difficulty. If a discrepancy in difficulty should exist, it is clear that the desired equating transformation is nonlinear. Of the methods tried, the choice seems to be between the direct equipercentile method and the three-parameter IRT method. Both of these methods should be given further study. One drawback sometimes cited as pertaining to the equipercentile method is the discreteness problem. This could be partially alleviated if, as Dr. Potthoff has suggested, formula scores were not rounded before equating. The equipercentile method does not really solve the problem of equating tests with different difficulties and unequal reliabilities. But then no other method does either.

The study has presented a good deal of information. As in any study of this kind, the need for certain additional information not provided by the study has been revealed. The following suggestions are given:

1. Because of the large discrepancies in the results for equating tests of similar difficulties versus dissimilar difficulties, it would be desirable to compare the various methods when the difficulties of the two tests being equated are different from each other, but closer together than in Variation 8.

2. In this study the results using the direct equipercentile criterion are biased in favor of the direct equipercentile method. This is because some of the same examinees were involved both in establishing the criterion and in the equating. In a future study, perhaps the criterion could be established using a completely separate population.

3. The one-parameter IRT method should be tried out on different tests of similar difficulties.

One final comment. Professor Olkin has mentioned the possibility of considering criteria for equating that use the joint distribution of scores. For equating test Y to test X, the criterion score for equated test X' could be the score X itself, in situations like the one used here, where all items are available for all examinees. For this kind of criterion the definition of the discrepancy function would have to be modified, since the criterion is now no longer a function of the raw score. However, it would be useful to consider this criterion along with the others.

Part II

ITEM RESPONSE THEORY EQUATING METHODS

Item Response Theory and Equating—
A Technical Summary

Frederic M. Lord

1. Introduction

Item response theory models the response of an examinee to a test item (a test question is called an *item*). The theory is useful for designing tests (selecting the items), for describing and evaluating items and tests, for optimal scoring of the examinee's responses, for predicting the test scores of examinees and groups of examinees on a variety of tests, and for manipulating and interpreting examinees' test scores.

Examinee performance on a pool of test items is assumed to depend on a limited number (k) of psychological traits. These traits are called *abilities* (a general term without restrictive implications). Each examinee is characterized by a vector $\{\theta_1, \ldots, \theta_k\}$ of k real numbers representing his standing on each of the k abilities. In item response theory (IRT), the probability of a given response u_i to a given test item i is assumed to be some specified function $f_i(u_i|\theta_1, \ldots, \theta_k; \boldsymbol{\beta}_i)$, where $\boldsymbol{\beta}_i$ is a vector of parameters describing test item i.

2. Unidimensionality

Tests and test scores are most meaningful when all the test items depend only on a single trait or ability θ ($-\infty < \theta < \infty$). This is the case to be con-

141

sidered here ($k = 1$). In practice, this unidimensionality condition is usually approximately met by various tests of spatial ability and by such tests as the following: vocabulary, spelling, arithmetic reasoning, number series, matrices, verbal analogies, reading comprehension (only one item per reading passage).

A necessary condition for unidimensionality is that for fixed θ, the distribution f_i is independent of the distribution f_j for any i and $j (i \neq j)$ in the pool of items. (If f_i and f_j were not independent for fixed θ, this would mean that responses to items i and j depend on some trait other than θ shared by the two items, a contradiction of unidimensionality.)

3. The Item Response Function

The present treatment will be limited to the case in which each test item is scored $u_i = 0$ ("wrong") or $u_i = 1$ ("right"). In this case the probability of a correct answer, $P_i \equiv P_i(\theta) \equiv \mathrm{Prob}(u_i = 1 | \theta, \boldsymbol{\beta}_i)$, considered as a function of θ, is called the *item response function*.

It is usual and natural to assume that $P_i(\theta)$ is a monotonic increasing function of θ. Now θ varies from $-\infty$ to ∞, whereas P_i varies from a lower limit $c_i \geq 0$ to an upper limit of 1; thus $P_i(\theta)$ must in a general way have an ogive shape. It is common to assume that

$$P_i(\theta) = c_i + (1 - c_i)F_i(\theta), \tag{1}$$

where $F_i(\theta)$ is some cumulative distribution function, for example, a normal or a logistic distribution function. Figure 1 shows estimated logistic item response functions for 39 representative vocabulary and reading comprehension items from the Test of English as a Foreign Language.

4. Frequency Distribution of Item Scores

The frequency distribution of the item score u_i for fixed θ is necessarily Bernouilli. It may be written as

$$f(u_i | \theta, \boldsymbol{\beta}_i) = P_i^{u_i} Q_i^{1 - u_i}, \tag{2}$$

where $u_i = 0$ or 1 and $Q_i \equiv 1 - P_i$. Since u_i and u_j are conditionally independent, we have for any entire test composed of n items

$$f(\mathbf{u} | \theta, \mathbf{B}) = \prod_{i=1}^{n} P_i^{u_i} Q_i^{1 - u_i}, \tag{3}$$

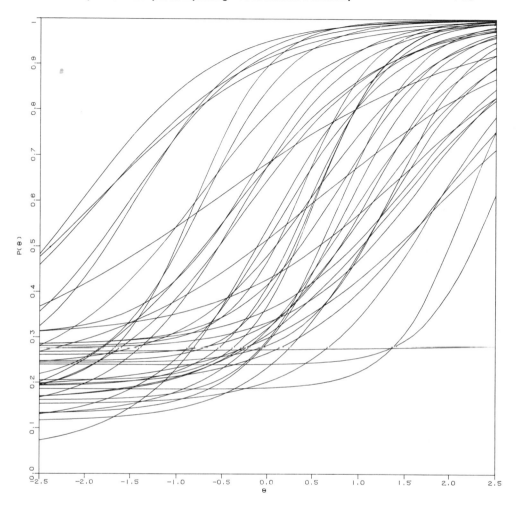

Fig.1 Item characteristic curves.

where $\mathbf{u} \equiv \{u_i\}$ is the vector of responses given by the examinee and $\mathbf{B}$ is the matrix of all item parameters. If N examinees are drawn at random from a population of examinees, the N vectors $\mathbf{u}$ will be independently distributed; thus the $n \times N$ matrix $\mathbf{U}$ of all item responses has the distribution

$$L(\mathbf{U}|\boldsymbol{\theta}, \mathbf{B}) = \prod_{i=1}^{n} \prod_{a=1}^{N} P_{ia}^{u_{ia}} Q_{ia}^{1-u_{ia}}, \tag{4}$$

where $\boldsymbol{\theta}$ is the vector $\{\theta_a\}$ and the subscript a is used to distinguish symbols relating to examinee a ($a = 1, 2, \ldots, N$).

5. Estimation of Parameters

The joint frequency distribution of the observations ($\mathbf{U}$) depends on N examinee parameters and on rn item parameters, where r is the number of parameters required to describe a single test item. In current practice usually $r = 1, 2,$ or 3. Long experience with test items suggests that items measuring the same trait typically differ statistically from each other in at least two respects: in difficulty and in degree of correlation with the trait.

For sufficiently large n and N, experience has shown that all or most of the $N + rn$ parameters can be satisfactorily estimated from the responses by maximum likelihood. If a prior distribution for θ and for $\boldsymbol{\beta}$ is available, Bayesian estimation methods can be used.

If an examinee answers all items correctly or answers all items incorrectly, his θ cannot be accurately estimated from the data. Problems may also arise in estimating certain item parameters. The item response function $P_i(\theta)$ can be accurately estimated for the range of θ that is well represented in the data. This need not mean that all item parameters can be accurately estimated. The lower asymptote c_i, for example, cannot be accurately estimated if item i is administered only to people who usually answer correctly.

6. Arbitrary Origin and Scale

There is no accepted origin and unit of measurement for measuring "ability." As a reflection of this, the function $F_i(\theta)$ in (1) is ordinarily chosen to have the form $F[a_i(\theta - b_i)]$. Thus any change in the origin and unit of measurement used to measure θ can be absorbed by a compensatory change in the scales used to measure a_i and b_i; the item response function remains essentially unchanged whether we decide to measure θ in feet or in inches, in Fahrenheit or in Celsius.

In practice, the scale for measuring θ is often chosen so that the mean and variance of θ are 0 and 1 (approximately) for some convenient group of examinees.

7. Three-Parameter Item Response Functions

We now see from (1) that a common form for the item response function is

$$P_i(\theta) = c_i + (1 - c_i)F(a_i(\theta - b_i)), \tag{5}$$

where F is a cumulative distribution function, such as logistic or normal. The parameter b_i is a location parameter and represents the *difficulty* of item i. Parameter a_i is a slope parameter, termed the *discriminating power* of item i. Parameter c_i is the height of the lower asymptote, sometimes called the *guessing parameter*, since $c_i > 0$ whenever item i can be answered correctly by guessing. (There is no assumption here that guessing is random.)

8. Parameter Estimation for Two Groups

If one group of examinees takes test x and a different group takes test y, there is in general no way to use the results to compare examinees in group 1 with examinees in group 2, nor to compare items in test x with items in test y. Both kinds of comparisons can be accomplished if either

1. Tests x and y contain some common items.
2. Groups 1 and 2 overlap.
3. Groups 1 and 2 are representative samples from the same population.

Without overlapping tests or overlapping groups or matching samples, estimated item parameters β_i and person parameters θ_a will not be on the same scale for different tests or for different groups. For equating work some overlap of tests or groups or some matching of groups must be arranged in order to provide meaningful comparisons.

9. Invariance

Once the scale for measuring θ is (arbitrarily) fixed, the response function for each item is also fixed; consequently, the item parameters for all items are fixed, regardless of the group to which the test is administered. Given that all items are chosen from the same unidimensional pool, it is readily seen that the ability parameter θ for each individual is similarly invariant, regardless of the set of items administered to him.

This important feature of IRT deserves emphasis: *Item parameters are invariant across groups. Ability parameters are invariant across items or tests from a given unidimensional pool.*

10. Test Scores

Although optimal scoring methods can be devised on the basis of item response theory, in practical testing the examinee's test score is most often taken to be the number of test items that he answers correctly. If all n items had the same response function $P(\theta)$, the examinee's *number-right score x* would have a binomial distribution:

$$f(x|\theta) = \binom{n}{x}(P(\theta))^x(Q(\theta))^{n-x}. \tag{6}$$

When each item has its own $P_i(\theta)$, the distribution of x is a generalized binomial defined by the probability generating function

$$\sum_{x=0}^{n} f(x|\theta)t^x \equiv \prod_{i=1}^{n} [Q_i(\theta) + tP_i(\theta)]. \tag{7}$$

The mean and variance of this distribution are

$$\mu_{x|\theta} = \sum_{i=1}^{n} P_i(\theta), \tag{8}$$

$$\sigma_{x|\theta}^2 = \sum_{i=1}^{n} P_i(\theta)Q_i(\theta), \tag{9}$$

11. True Scores

The expected value of an examinee's score is called his *true score*. In the case of the number-right score

$$x \equiv \sum_{i=1}^{n} u_i; \tag{10}$$

we have by this definition

$$\xi \equiv \mathscr{E}(x) \equiv \sum_{i=1}^{n} \mathscr{E}(u_i) = \sum_{i=1}^{n} P_i(\theta), \tag{11}$$

where ξ denotes the examinee's (number-right) true score.

Note that true score ξ is a monotonic increasing transformation of θ. This means that *true score and θ are equivalent measures of ability, differing only in the measurement scale used.*

12. True-Score Equating

Suppose $x \equiv \sum^n u_i$ and $y \equiv \sum^m u_j$ are number-right scores on two tests measuring the same ability θ. Denoting true score on test y by η, we have

$$\xi = \sum_{i=1}^{n} P_i(\theta),$$

$$\eta = \sum_{j=1}^{m} P_j(\theta).$$

(12)

These two equations define the functional relationship between ξ and η. Clearly, corresponding values of ξ and η have identical meaning. They are said to be *equated*.

In practice, true-score equating is carried out by substituting estimated parameters $\hat{\beta}_i$ or $\hat{a}_i$, $\hat{b}_i$, and $\hat{c}_i$ into (12). Paired values of ξ and η are then computed for a series of arbitrary θ values. Note that the actual θ values in the group of examinees tested are not needed for this purpose: The equating is necessarily invariant across groups of examinees.

13. Practical Use of True-Score Equating

If the model holds, the foregoing procedure equates ξ and η within the limits imposed by sampling fluctuations in the item parameter estimates. The practical problem, however, is not to equate true scores ξ and η but to equate *observed scores x and y.*

In actual practice, we only know the examinee's observed score x or y; we *cannot* know his true score ξ or η. To use (12) we act as if relationship (12) applied between x and y as well as between ξ and η. This is not really satisfactory. Just one symptom of the basic problem is seen in the fact that ξ is never less than $\Sigma_i c_i$, whereas x may go down to zero; hence (12) supplies no equating for examinees with observed scores x below $\Sigma_i c_i$.

14. Observed-Score "Equating"

The following method of observed-score "equating" has been tried but not extensively used. It will be carefully studied in the near future.

1.   For each examinee in some convenient group, estimate $f(x|\theta)$ by substituting estimated θ and estimated item parameters into (7).

2.   Sum the resulting $\hat{f}(x|\theta)$ across examinees to obtain an estimate of the distribution of the N observed scores x.

3.   Repeat 1 and 2 for test y, for the same group of examinees.

4.   Assert that an x score and a y score having the same rank (each in its appropriate distribution) are to be treated as equivalent (equated).

Although this approach seems very plausible, it also is flawed, as is evident from the following:

1.   Since the $\hat{\theta}$ contain errors of estimation, probably $\sigma_{\hat{\theta}} \neq \sigma_\theta$. If the $\hat{\theta}$ estimated from test x contain larger errors of estimation than the $\hat{\theta}$ estimated from test y, $\hat{f}(x|\theta)$ will not be properly comparable to the corresponding estimated distribution for y.

2.   It can be shown that changing the group used in the foregoing procedure will usually change the resulting "equating."

A proper equating needs to be invariant against arbitrary changes in the group. Equatings of observed scores in general fail to meet this requirement. This is not a failure of IRT equating; with trivial exceptions it is characteristic of all equatings of observed scores by any method.

Item-Response-Theory Pre-Equating
in the TOEFL Testing Program

William R. Cowell

1. Introduction

Item characteristic curve (ICC) methods of item precalibration and test pre-equating have recently been applied operationally in the TOEFL (Test of English as a Foreign Language) testing program. New forms (i.e., editions) of TOEFL are given each month at testing centers around the world. Scores for each new form are equated to scores on previous forms of the test. Beginning with the form introduced in September 1978, all new forms of TOEFL have been equated using the three-parameter logistic curve model. The ICC equating methods have been found to be appropriate and practical for TOEFL. The principal practical advantages are improved security of test items and a reduction in the time between test administration and score reporting. In addition, new forms can be properly linked to the TOEFL score scale even if administered to small or unusual groups of examinees.

2. The TOEFL Program

The TOEFL (Test of English as a Foreign Language) testing program is sponsored jointly by ETS, the College Board, and the GRE Board. The

149

testing program is designed to provide a measure of the English language proficiency of foreign students applying for admission to colleges and universities in the United States. A number of other institutions and agencies also use the TOEFL testing program. New forms of the test are given each month. Tests are administered at testing centers in 135 countries around the world. The tests are also used in separate institutional testing programs in the United States and overseas. Approximately 190,000 applicants were tested in the TOEFL program in 1977–1978.

3. The TOEFL Examination

The TOEFL examination consists of three separately timed sections:

Section I.   Listening Comprehension
Section II. Structure and Written Expression
Section III. Reading Comprehension and Vocabulary

A score is reported for each section and a total score is also reported. To provide a means of adjusting scores for differences in difficulty among the various forms and to provide a convenient scale for reporting, the number-right scores for each section are converted to a scale ranging from 20 to 80. The total score is $3\frac{1}{3}$ times the sum of the converted scores for the three sections. Therefore, the total score can range from 200 to 800.

4. Equating

In order to determine equivalent scores for persons of equal ability, regardless of the level of difficulty of the particular form of the test they happened to have been given or the level and distribution of English proficiency of those with whom they were tested, each new form of TOEFL is equated to the previous forms of the test. Prior to September 1978, equating had been done by including in each new form of the test a set of items from a single previously administered (and equated) form. Different "old forms" were selected for each new form being equated, but it was necessary to take all the equating items for a particular new form from the same old form. To reduce the security problems involved in the reuse of test items, at least two versions of the test were prepared for each test administration. One version included the equating items from the old form. This version was given at selected centers where security was considered best. Another version was prepared for use at other test centers at the same administration. In that version the equating items were replaced with new items. The remaining items were common to the two versions and provided an equating link between the two new forms. The first new version was equated to an old form; then the second new version was equated to the first. Since the equating items had appeared in a previous test, there was a risk that they might have been taken by some of the same examinees at the earlier administration or that some of the items might have appeared in coaching materials. In discussing ways to further tighten security of equating items, and thus provide better controls for insuring the accuracy of the equating process, the advantages of using item characteristic curve (ICC) equating methods became apparent.

5. ICC Equating

The Summer 1977 issue of the *Journal of Educational Measurement* included articles by Lord and by Marco describing how item characteristic curve theory could be used to pre-equate tests. ICC equating methods permit the use of a few items from many different sources instead of a fairly large set of items from a single source. And perhaps more important, the ICC equating data are obtained from various pretests, which can be given to relatively small groups at the most secure testing centers, rather than from previously used final test forms. The improvement in test security may be the most important practical advantage of using ICC equating methods in the TOEFL program.

Another advantage has been the shortening of the time required to prepare score reports following a test administration. For tests consisting entirely of precalibrated items, the equating can be done before the test is given, removing the test-equating procedures from the schedule of operations required during the critical period between test administration and score reporting.

ICC equating methods also provide a convenient means of dealing with some exceptional situations. It is possible quickly and easily to modify the score conversion data to adjust for certain types of irregularities in test administration (e.g., a few items deleted from the Listening Comprehension section because of a faulty tape or faulty equipment at a test center). Also, a new form can be equated to the TOEFL scale with the same precision as other TOEFL tests even if it is administered to a small or unusual group. An almost unlimited number of different test forms for special purposes could be assembled (and equated) from the pool of precalibrated items with very little likelihood that any given examinee would have seen more than one or two items at a previous administration.

Both item analysis data and score-equating data can be obtained simultaneously from pretest samples of 1000–2000 examinees. In addition to measures of item difficulty and discrimination, the ICC item parameters can provide estimates of the mean, variance, reliability, and true-score and observed-score frequency distributions for any test constructed from the precalibrated item pool.

6. An Alternative Method

A procedure has been devised for equating new forms of the test consisting of a mixture of precalibrated and noncalibrated items. This was done to make use of the existing pool of items that had been pretested but not precalibrated. In this procedure the time reduction in the reporting cycle is sacrificed in order to make use of the existing item pool without pretesting the items again to obtain the ICC item parameters. After the new form has been administered, the ICC item parameters are computed for the previously noncalibrated items; then the whole test is equated to the TOEFL scale. Computer programs are now available that can produce these results within the period usually required for conventional equating procedures.

7. A Feasibility Study

In July 1977, an extensive investigation of the feasibility of using ICC methods of item precalibration and test pre-equating in the TOEFL program was initiated. In the first phase of the study all the TOEFL test forms that had been administered between September 1976 and May 1977, in addition to three experimental test forms, were equated using ICC methods. All of these forms had been previously equated by conventional methods. The items in these forms, plus items from several pretests that had been administered in this same period, were all calibrated to a common scale. The three-parameter logistic curve model was used for item calibration. The calibrations were performed separately for each of the three sections of the test. Score conversion tables were also obtained for each section. Conventional item analysis and score equating had also been done separately by section.

In the second phase, new forms administered between September 1977 and August 1978 were equated by both ICC and conventional methods. Scores were reported based on the conventional equating procedures. Data were collected and analyzed comparing the equating methods. Operational and pretest items were calibrated and the item data added to the item data file.

The TOEFL score reporting system was modified to accommodate nonlinear score conversion and to provide the flow of data for item calibration and ICC test equating. Modifications were made in the item calibration and the equating programs to increase speed and reduce costs. The procedures for equating test forms containing a mixture of precalibrated and noncalibrated items were implemented.

Extensive and careful examination of the results of the feasibility study (and the recent operational use) indicate that ICC methods of equating are indeed appropriate and practical for the TOEFL testing program.

8. Results of the Feasibility Study

For each of the three sections of all the test forms that have been equated by both conventional and ICC methods, score conversion tables obtained by ICC methods have been compared with those obtained by conventional equating methods. In general, over most of the score range, the scaled score assigned by ICC equating matches closely that obtained by conventional methods. Graphs comparing the conversion data for a representative sample of these forms are presented in Figs. 1–9 of the Appendix to this report. The straight lines represent the conventional (linear) equating results and the curves represent the corresponding ICC results. The results are nearly identical over most of the score range. The largest differences are typically in the neighborhood of two points, except for the "chance-score" range, where differences may be as great as four points. Since we have no criteria to judge the superiority of either method, the differences cannot be considered error for either method. However, the fact that the differences are relatively small, particularly in the more critical mid-scale range, indicates that ICC equating methods can be used for TOEFL with little effect on the score scale. The ICC equating method has one clear advantage in that error due to equating is not cumulative over forms, as is the case with conventional methods.

9. Stability of Conversion Data

If the characteristics of the test items change from form to form, one would expect differences to appear in the conversion tables reflecting these changes. However, over a range of several successive forms, the TOEFL forms are assembled to very similar specifications. That is, the forms could be expected to be nearly parallel. Consequently, the conversion tables would be expected to remain fairly stable. Graphs showing the distribution of converted scores corresponding to number-right scores in five-point increments are shown in Figs. 10–12 of the Appendix to this report. Data are plotted for the first three forms equated by ICC methods and for three recent forms equated by conventional methods. In general, the graphs show a consistency across equating methods that compares favorably with consistency across forms within equating method.

10. Related Research and Development

The availability of a large pool of item data has facilitated a large-scale study of item bias across several selected native language groups in TOEFL. Data at the item level and at the test-form level have been analyzed and a TOEFL Research Report is expected to be released soon.

At the present time, both conventional item analysis data (i.e., percent correct, deltas, and biserial correlations) and ICC item data are obtained. The TOEFL test forms are being assembled using the conventional item data. A cooperative effort involving the test development, statistical analysis, and research areas is planned for the next testing year leading to the use of ICC item data in the assembly of future forms of TOEFL. At that time it is expected that the collection of conventional item analysis data will be discontinued.

The availability of ICC data for a large pool of items also facilitates the potential operational use of interactive tailored testing for the TOEFL program.

11. Appendix

Figures 1–9 present representative samples of graphs comparing ICC and linear score conversion data. Straight lines represent conventional linear conversion data from the number-right score to the TOEFL score scale. Curves represent the corresponding ICC conversion data. Graphs for each of the three sections of TOEFL are shown for three test forms (December 1977, February 1978, and April 1978).

Figures 10–12 give histograms showing frequency of occurrence of scaled scores corresponding to number-right scores in five-point increments

0, 5, 10, etc. The letters A, B, and C denote forms equated by conventional linear methods. R, S, and T denote forms equated by item characteristic curve methods. These letters are used to identify the form(s) contributing to each bar of the histogram. Heights (or areas) correspond to the number of forms for which a given rights score (0, 5, 10, etc.) converts to a given scaled score (20, 21, ..., 70). For example, the histogram at the lower left corner of Fig. 10 shows that a right score of 0 converts to a scaled score of 25 for four forms (A, R, S, and T) and scaled scores of 27 and 28 for one form each (B and C). That is, for the three forms equated by ICC methods, a raw score of zero converts to a scaled score of 25, whereas the scaled scores corresponding to a raw score of zero varied from 25 to 28 for forms equated by conventional linear methods.

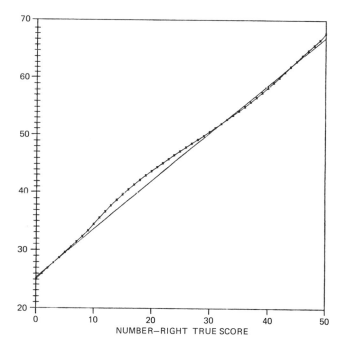

Fig. 1 Graph comparing ICC and linear equating results. December 1977, Section I: listening comprehension. ⇢⇢⇢, true-score ICC equating scale score; ——, conventional linear equating scale score.

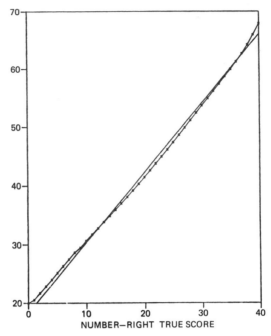

Fig. 2 Graph comparing ICC and linear equating results. December 1977, Section II: structure and written expression. •—•—•, true-score ICC equating scale score; ——, conventional linear equating scale score.

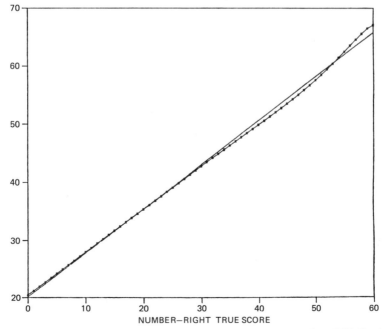

Fig. 3 Graph comparing ICC and linear equating results. December 1977, Section III: reading comprehension and vocabulary. •—•—•, true-score ICC equating scale score; ——, conventional linear equating scale score.

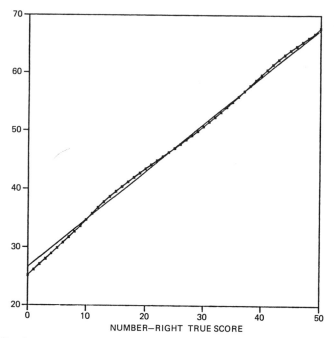

Fig. 4 Graph comparing ICC and linear equating results. February 1978, Section I: listening comprehension. ⊶, true-score ICC equating scale score; ────, conventional linear equating scale score.

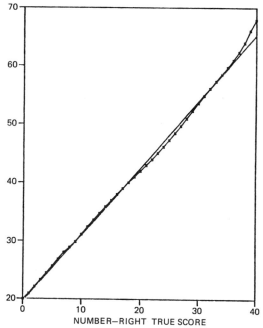

Fig. 5 Graph comparing ICC and linear equating results. February 1978, Section II: structure and written expression. ⊶ true-score ICC equating scale score; ────, conventional linear equating scale score.

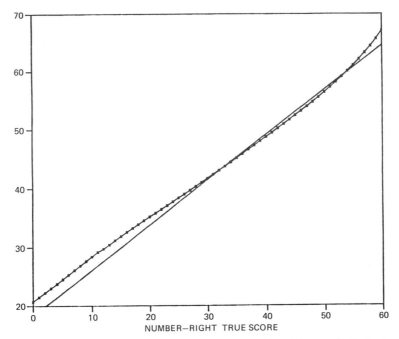

Fig. 6 Graph comparing ICC and linear equating results. February 1978, Section III: reading comprehension and vocabulary. ●—●—●, true-score ICC equating; ——, conventional linear equating scale score.

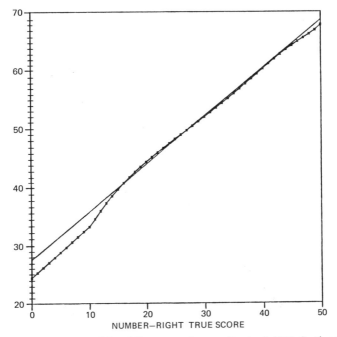

Fig. 7 Graph comparing ICC and linear equating results. April 1978, Section I: listening comprehension. ●—●—●, true-score ICC equating scale score; ——, conventional linear equating scale score.

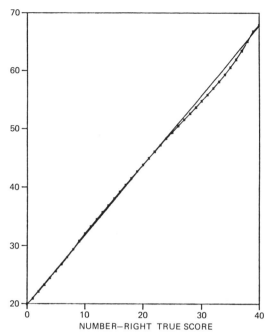

Fig. 8 Graph comparing ICC and linear equating results. April 1978, Section II: structure and written expression. ➤➤➤, true score ICC equating scale score; ——, conventional linear equating scale score.

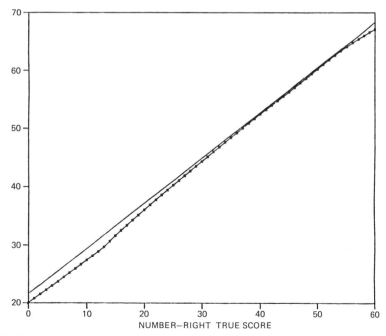

Fig. 9 Graph comparing ICC and linear equating results. April 1978, Section III: reading comprehension and vocabulary. ➤➤➤, true-score ICC equating scale score; ——, conventional linear equating scale score.

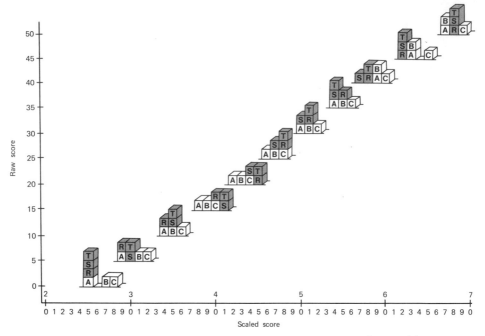

Fig. 10 Histograms showing frequency of scaled scores corresponding to rights scores of 0, 5, 10, . . . , 50. Section I, listening comprehension. □: A, B, C denote forms equated by conventional linear methods. ■: R, S, T denote forms equated by item characteristic curve methods.

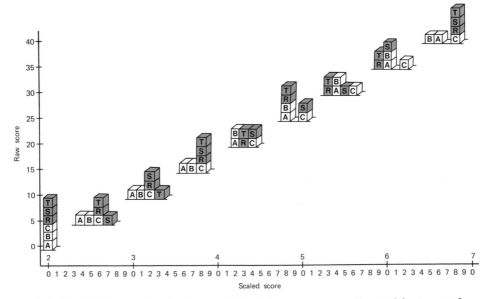

Fig. 11 Histograms showing frequency of scaled scores corresponding to rights scores of 0, 5, 10, . . . , 40. Section II, structure and written expression. □: A, B, C denote forms equated by conventional linear methods. ■: R, S, T denote forms equated by item characteristic curve methods.

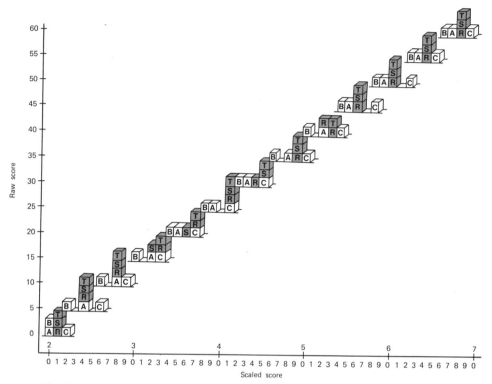

Fig. 12 Histograms showing frequency of scaled scores corresponding to rights scores of 0, 5, 10, ..., 60. Section III, reading comprehension and vocabulary. □: A, B, C denote forms equated by conventional linear methods. ■: R, S, T denote forms equated by item characteristic curve methods.

REFERENCES

Lord, F. M. (1977). Practical applications of item characteristic curve theory. *Journal of Educational Measurement, 14,* 117–138.

Lord, F. M. (1980). *Applications of Item Response Theory to Practical Testing Problems.* Hillsdale, N. J.: Lawrence Erlbaum Associates.

Marco, G. L. (1977). Item characteristic curve solutions to three intractable testing problems. *Journal of Educational Measurement, 14,* 139–159.

Discussion of Item Response Theory

Robert H. Berk

IRT equating appears to have been carried out, in practice, by using a three- (or fewer) parameter logistic ICC. It is perhaps surprising that the results thus obtained seem to agree so well with linear equating. This may be because the method of test construction yields tests that are already quite similar (parallel).

Nevertheless, one cannot expect to find close agreement in every instance. One question that then arises concerns the robustness of IRT equating: Is it sensitive to the functional form (e.g., logistic) of the ICC used? Since a logistic ICC is an arbitrary choice, without any real theoretical justification, it would certainly be disturbing if other choices of ICC give substantially different answers.

It would thus be of interest to study the results given by other kinds of ICCs. One possibility would be to use a different location/scale family (with, e.g., a guessing parameter) that is really different from the logistic. The Cauchy distribution is one such. Location/scale families have the attractive feature that the (location and scale) parameters have nice interpretations: The location parameter b is the difficulty. Changing b changes by an equal amount the median of the ICC (the level of ability θ needed for a 50% chance of correct response). This median is analogous to the LD-50 in bioassay, and we might refer to it as the LA-50 (LA = level of ability). Of course, all other

163

"percentiles" of the ICC are similarly affected by changing b. The scale parameter a is a measure of item discrimination and is (apart from a factor not depending on the item) proportional to the slope of the ICC at the LA-50 and also to the standard deviation of the ICC.

A location/scale family has one further property: It leaves the origin and scale of the ability θ undetermined. For items not previously calibrated (i.e., for which parameters have not been estimated), this indeterminancy can be removed by an arbitrary convention [such as scaling so that the (estimated) levels of ability across a group of calibrating examinees have mean zero and variance one]. Whether or not this last property of location/scale families is desirable is not entirely clear.

Another possibility for studying robustness is to use something other than a location/scale family. A possible alternative is suggested by the fact that b controls the "stochastic size" of the ICC. That is, increasing b results in a stochastically larger distribution function (ICC). A similar role is played by the exponent in a Lehmann alternative. Thus one might consider a family of ICCs of the form

$$P(\theta; a, b) = [F(a\theta)]^b \tag{1}$$

or of the dual form

$$1 - P(\theta; a, b) = [1 - F(a\theta)]^{1/b}. \tag{2}$$

Here F is a specified distribution function and the parameter b again controls the stochastic size of the ICC. Here too b may be interpreted as difficulty, for the LA-50 increases with b. In fact, for (1),

$$\text{LA-50} = F^{-1}(2^{-1/b})/a.$$

The scale parameter a adds more flexibility to the family. Thus if the discrimination is thought of as the slope of the ICC at the LA-50, we obtain for (1)

$$\text{discrimination} = ab2^{1-b}f[F^{-1}(2^{-1/b})],$$

where f is the density of F. Unlike the location/scale case, the slope at the LA-50 depends on b as well as a. For a given value of b, a may be chosen to give the desired slope.

Some important properties of this Lehmann alternative/scale family should be noted. One is that even if F is symmetric, the ICC in general will not be so. Also, since there is no location parameter, the interval on which ability is measured can be taken to be $(0, \infty)$ rather than $(-\infty, \infty)$, thereby fixing an absolute origin for ability. This also allows for convenient choices of the generator F, such as the negative exponential distribution in (1). [This is not a good choice for the dual form (2).]

Another possible family where b controls stochastic size is the exponential family generated by F:

$$p(\theta; b) = e^{\theta b - c(b)} f(\theta). \tag{3}$$

Here p is the density of the ICC $P(\theta; b)$ and $c(b)$ gives the normalizing factor needed to have p integrate to 1. A scale parameter may also be introduced to give

$$p(\theta; a, b) = a e^{\theta ab - c(b)} f(a\theta). \tag{4}$$

A possible drawback of these last families is that closed-form expressions for $P(\theta; a, b)$ may not be available. This might make them analytically intractable (since, for example, the likelihood for a set of responses then cannot be written in closed form). However, the same objection can apply to a location/ scale family, for example, the one generated by the normal distribution. Numerical methods may prove feasible for these families.

One might also contemplate using a two-parameter family that would completely fix the scale for ability. One such is the family of beta distributions on $[0, 1]$. There are then no entirely obvious choices for the difficulty or discrimination parameters, but the mean and standard deviation might serve the purpose. It would then be possible to examine the stability of the estimated item parameters across different groups of examinees.

To simplify the foregoing discussion, guessing parameters have been omitted, but there is no reason why they cannot be used with any of the above families.

1. Measuring How Well Two Tests Can Be Equated

Suppose Y is equated to X via the conversion $X^* = J(Y)$. One criterion for saying that J equates is that the joint distribution of X and X^* across a (common) group of examinees be exchangeable. Exchangeability provides a notion of fairness for the group as a whole, although it must be invoked with some common sense. For example, if X is uniformly distributed on $[0, 1]$, X and $1 - X$ are exchangeable, but can hardly be said to be fairly equated. Thus one wants, in addition, a positive association between X and X^*. In other words, one wants their joint distribution to be concentrated as close to the main diagonal as possible.

There are two separate but related notions involved here. One is the idea of fairness: How close to exchangeable are X and X^*? In this connection, note that if X and X^* are exchangeable, they have the same marginal distribution, so that J is necessarily an equipercentile equating. The other idea involves concentration about the main diagonal and is clearly related to the quality of the equating. Historically, (rank) correlations have been used to measure this trait, but that is usually not directly possible in the situations contemplated here, since paired observations are required (i.e., there would have to be examinees that took both the X and the Y test). An equating between X and Y can be entirely fair (if X and X^* are exchangeable) but not very good (if the joint distribution scatters greatly about the main diagonal). A bivariate normal distribution with low correlation provides an example of a fair but poor equating. Clearly, nothing can be done to improve matters in such a case (except perhaps to rescore the tests).

There are various measures of the two traits mentioned above that can be put forth. Consider first fairness (exchangeability). Let $F(x)$ and $G(y)$ denote the distribution functions of X and Y and let $H(x, y)$ be their joint distribution (if both were measured on a hypothetical population). Below we discuss ways to estimate H. Then $X^* = F^{-1}G(Y)$ and the joint distribution of (X, X^*) is

$$H^*(x, x^*) = H(x, G^{-1}F(x^*)).$$

If X and X^* are exchangeable,

$$H^*(x, x^*) = H^*(x^*, x),$$

and thus a measure of departure from exchangeability for X and X^* is the Kolmogorov-like quantity

$$\psi = \sup_{x, x^*} |H^*(x, x^*) - H^*(x^*, x)|$$

$$= \sup_{0 < u, v < 1} |H(F^{-1}(u), G^{-1}(v)) - H(F^{-1}(v), G^{-1}(u))|$$

$$= \sup_{x, y} |H(x, y) - H(F^{-1}G(y), G^{-1}F(x))|.$$

$\psi = 0$ if X and X^* are exchangeable. Small values of ψ thus indicate that X and Y can be fairly equated (although the resulting equating may not be very good).

Other metrics could also be used. For example,

$$\psi^* = \sum_{x, x^*} |h^*(x, x^*) - h^*(x^*, x)|$$

or

$$\psi^{**} = \tfrac{1}{2} \sum_{x, x^*} ([h^*(x, x^*)]^{1/2} - [h^*(x^*, x)]^{1/2})^2$$

$$= 1 - \sum_{x, x^*} [h^*(x, x^*)h^*(x^*, x)]^{1/2}.$$

Here h^* is the joint density of (X, X^*). ψ^* is a total variation distance and ψ^{**}, a Hellinger distance. There are other variations on the theme, such as replacing the absolute value by a square in defining ψ^* or taking expectations rather than just summing over x and x^*. For example, one could consider

$$\psi^{***} = E_{H^*}[H^*(X, X^*) - H^*(X^*, X)]^2.$$

There are similarly various possibilities for measuring concentration about the main diagonal. Rank correlations are particularly relevant because they allow one to work with the original scores X and Y just as well as with the equated scores. The rank correlations depend only on the joint distribution of $F(X)$ and $G(Y)$. In this way, exchangeability, if present, is automatically allowed for. The joint distribution of X and X^* or equivalently of $F(X)$ and $G(Y)$ concentrates *on* the main diagonal if

$$H(x, y) = F(x) \wedge G(y),$$

where $a \wedge b$ = minimum of a and b. In any case, $H(x, y) \le F(x) \wedge G(y)$, so one measure of concentration about the diagonal is

$$\Delta = \sup_{x,y} \{F(x) \wedge G(y) - H(x, y)\}.$$

Other such measures are the population Spearman correlation

$$\rho_S = \text{corr}[F(X), G(Y)],$$

which is equivalent to the dispersion quantity

$$\Delta^* = E_H[F(X) - G(Y)]^2,$$

and the population Kendall correlation (allowing for ties)

$$\rho_K = \frac{P((X_1 - X_2)(Y_1 - Y_2) > 0) - P((X_1 - X_2)(Y_1 - Y_2) < 0)}{P((X_1 - X_2)(Y_1 - Y_2) > 0) + P((X_1 - X_2)(Y_1 - Y_2) > 0)}.$$

Here (X_1, Y_1) and (X_2, Y_2) are two independent copies of (X, Y). For continuous random variables,

$$\rho_K = 4P(X_1 < X_2 \text{ and } Y_1 < Y_2) - 1 = 4 \int \int H(x, y) \, dH(x, y) - 1,$$

and this last quantity would probably serve equally well in the discrete case. (It may be viewed as a particular way of treating ties.) This last quantity also appears in an expectation version of Δ:

$$\Delta^* = \int \int \{F(x) \wedge G(y) - H(x, y)\} \, dH(x, y).$$

IRT suggests ways of estimating the foregoing quantities. As usual, one can obtain estimates of the item parameters for the two tests and then at least suitable approximations to the conditional distributions of X and Y given θ. Let these latter distributions be denoted by $\hat{F}(x|\theta)$ and $\hat{G}(y|\theta)$. Since X and Y are conditionally independent under IRT, one has

$$\hat{H}(x, y) = \int \hat{F}(x|\theta)\hat{G}(y|\theta) \, d\hat{K}(\theta)$$

as an estimate of H, where $\hat{K}(\theta)$ is an estimated distribution of ability for the calibrating group. Typically one would choose the calibrating group to be the pool of examinees that took either exam X or exam Y. Then

$$\hat{H}(x, y) = \frac{1}{n} \sum_{j=1}^{n} \hat{F}(x|\hat{\theta}_j)\hat{G}(y|\hat{\theta}_j),$$

where $\hat{\theta}_j$ is the estimated ability of the jth examinee in the combined group. However, it is not necessary to choose $\hat{K}$ in this way, and some imputed or "ideal" distribution may be used.*

Once $\hat{H}$ is available, it is clear, in principle, how to obtain estimates of the above quantities by substituting $\hat{H}$ for H. An alternative to this would be to generate scores by Monte Carlo techniques. One selects an examinee and, using his estimated ability $\hat{\theta}_j$, generates a simulated set of responses for the exam he did not take. In this way one obtains pairs of scores for each examinee of the form $(X, \hat{Y})$ or $(\hat{X}, Y)$, where the circumflex denotes a simulated score. One may then use the augmented data to form empirical estimates of H, G, and F. Of course, one could simulate both an X and a Y score for each examinee. It would seem reasonable to do this if for some reason one were interested in working with θ values not represented in the examinee pool.

A drawback to this way of estimating H is that it seems to depend heavily on the validity of the IRT assumptions. One way to at least partly check whether this seems tenable is to compare the IRT estimate of F for the X examinees with the empirical distribution of the observed X scores (and similarly for G). If there is an anchor test common to X and Y, it may be possible to get some reasonable information about H without using IRT. In this connection, see Rubin and Thayer (1978).

REFERENCE

Rubin, D. B., and Thayer, D. (1978). Relating tests given to different samples. *Psychometrika* *43*, 3–10.

---

* In fact, using the observed distribution of $\hat{\theta}$ as $\hat{K}$ is biased toward a larger dispersion than the actual dispersion of K.

On the Foundations of Test Equating

Carl N. Morris

1. Introduction

When the equating group external to ETS was convened, we were urged to think not only about practical issues, but also about foundations. Members of the ETS staff felt this would help by providing criteria for evaluating alternative equating methods. This paper attempts to respond to that request.

A conceptual basis for equating is based on the very old and simple idea of "statistical adjustment." Two test items are defined to be *equivalent* if and only if *every* member of the test population P has the same probability of correctly answering each item (the probabilities will depend on the examinee). An *equivalence class* of items is defined as a set of items containing only equivalent pairs of items. An examinee's probability of correct response or "true score" is introduced here to provide a system that makes equating meaningful, and thereby to provide criteria for judging equating methods. Existence of true scores is not postulated.

This development presupposes knowledge of the equivalence classes, i.e., which questions belong to the same equivalence classes (for P). Even in Section 3, when we parameterize equivalence classes functionally by their item characteristics and thereby move the theory closer to applications, we will continue to assume that the item characteristics are known. We gain

169

tremendous theoretical advantage in doing so because the equating of tests for one examinee will be independent of the other examinees' scores. Thus the form of test administration (spiraling, anchoring, etc.) can be ignored, and linear model methods will not be affected by the difficult-to-analyze "regression to the mean" phenomenon that often characterizes equating. These assumptions seem appropriate for developing a theory of equating.

Equating is thereby separated from the dual problem of evaluating item characteristics. Implementation of the theory must also deal with uncertainty and information about items, of course, perhaps derived from data involving many examinees and complex test administrations. But the equating problem, viewed in this manner, reduces to one of modeling and statistical theory. The role of the statistical theory becomes clearer by separating the problem of equating examinees' scores on tests from that of classifying items.

Section 2 defines "strong equating" and "weak equating" under idealized circumstances. "Idealized circumstances" pertain to the case just discussed, when the equivalence classes of items are known a priori. We also assume, unrealistically, that there are sufficiently few classes in relation to the test length for each examinee to be exposed to some items from every class. There then exists a unique way to equate "weakly," by which we mean that *every* examinee has the same *expected score* on both tests. Strong equating, meaning that the *score distribution* is identical for both tests for every examinee, then can be achieved only by the (unique) weak equating formula, a formula simply adjusting for the different proportions of items on the tests from each equivalence class. Using these results, strong equating, in the exact sense, is shown to be impossible under very general conditions.

Section 3 shows that strong equating can be approximated in practice if the test T being equated is not more reliable than the test T^* that T is being equated to. Next, a necessary condition for appropriateness of linear equating is examined and found wanting.

It is shown that no equating method from total scores to total scores works in the presence of more than one ability dimension. The dimension of the ability trait vector generally depends on the population P of examinees, i.e., success of an equating method is P-dependent. For one ability dimension, quantile equating is shown to be satisfactory for "highly reliable tests" (HRT), and a condition for correctness of linear equating in this case is given. Quantile equating, therefore, has wider applicability than linear equating, at least in the HRT case, although other considerations sometimes may make linear equating procedures preferable, for example, when there are few items.

The final section presents concepts for parametric equivalence classes [i.e., item response theory (IRT) models], which make the theory more practical. The need to equate before aggregating is reemphasized for linear equating in the separable case, "separable" meaning that each item involves only one ability parameter. Nonseparable cases are more complicated, although they are aided greatly by existence of a complete sufficient statistic, e.g., the logistic parameterization. This provides further justification for general IRT models.

When a complete sufficient statistic exists for the ability parameter, the weak equating method of Section 2 can be conditioned on this statistic to produce the minimum variance weak equating rule. A quadratic approximation to this rule is presented in Remark 7 and derived in Appendix I. The main body of the paper closes with an extended example comparing the equating method of Section 7 with a true-score equating method (Lord, 1977) and the rule of Remark 7. Conditions for preferring one method to the other are discussed.

Three appendixes include material related to the text.

2. Weak and Strong Equating under Idealized Circumstances

Frederic M. Lord has shown that two different tests generally cannot be equated. That is, no matter how one scores tests not equated by design (unless all information is discarded), some individuals will prefer one test to another because their score distributions on the two tests will differ. Here we define the term *strong equating* as follows:

Definition 1.   *Two tests are* strongly equated *if every individual in the test population has the same probability distribution for this score on both tests.*

An examinee would have no reason to prefer one of two strongly equated tests to another.

Although strong equating usually is not possible, we will show under an idealized model that weak equating is.

Definition 2.   *Two tests are* weakly equated *if each individual in the test population has the same expected score on both tests.*

"Strong" and "weak" are used properly here, for by these definitions strongly equated tests are weakly equated.

The initial discussion here takes place in a framework like Holland's (1979). Indeed, there and elsewhere Dr. Holland has favored developing foundations to provide criteria for answering such questions as whether to use equipercentile or linear equating. This seems completely appropriate, and should eventually provide practical benefits.

We follow Holland (1979) and Levine (1955) by partitioning the item pool Q into a priori strata

$$Q = Q_1 \cup \cdots \cup Q_m, \quad Q_i \cap Q_j = \phi, \quad i \neq j. \tag{1}$$

We further require that the Q_i be *equivalence classes* of items, in the sense that *every* examinee should be equally likely to answer any question in Q_i. This could happen because each examinee finds the items in Q_i equally difficult. This constrains the length, difficulty, and type of the question. The order of item presentation and standardizing of conditions Holland (1979) discusses present additional complications that are ignored here.

Requiring the $\{Q_i\}$ to be equivalence classes introduces strong restrictions. Verbal and quantitative questions cannot be grouped, at least for most examinee populations, because some examinees are more adept verbally than mathematically while others are not. Numerical problems cannot be grouped with mathematical word problems. Two numerical problems such as "Compute 73 + 49" and "Compute 67 + 54" both probably could be put in Q_1, but whether to include "Compute 273 + 449" also would depend on the examinee population. Third-graders would find the last one more difficult, so it could not be grouped with Q_1 for them, whereas it might be for a population of college graduates.

We extend Holland and Levine's definition of a test to be defined on equivalence classes.

Definition 3. *Let* $Q_1, \ldots, Q_m$ *be equivalence classes of items. A* test T *is a set of* specifications

$$\mathbf{q} = (q_1, q_2, \ldots, q_m) \tag{2}$$

with q_i *items chosen from the pool* Q_i.

A different test T^* with specification vector $\mathbf{q}^*$ is "L-parallel" to T if $q_i^* = q_i$ for all $i = 1(1)s$.

Remark 1. L-parallel tests are strongly equated if the $\{Q_i\}$ are equivalence classes.

Now consider a population P of examinees who will take test T. Let $j \in P$ be one of the examinees, and let

$$p_{ji} = P(\text{examinee } j \text{ answers an item in } Q_i \text{ correctly}). \tag{3}$$

The equivalence assumption permits p_{ji} to depend on the examinee and the equivalence class Q_i, but not on the item chosen from Q_i.

Let X_{ji} be the *fraction* of all items in Q_i answered correctly by j. Then j's score S_j on test T is

$$S_j = \sum_{i=1}^{m} q_i X_{ji}, \tag{4}$$

his total number of correct responses. (Take $X_{ji} = 0$ arbitrarily if $q_i = 0$.) Assume that X_{ji} is unbiased for p_{ji} whenever $q_i \geq 1$:

$$EX_{ji} = p_{ji}. \tag{5}$$

We do not require independence, although the binomial (q_i, p_{ji}) model for the total number correct, $q_i X_{ji}$, leads to this. Then the examinee's expected score ES_j is, from (4) and (5),

$$ES_j = \sum_{i=1}^{m} q_i p_{ji}. \tag{6}$$

What would j's score be on test T^*? Since the questions in different classes Q_i are not necessarily equally difficult for j, we would use

$$\hat{S}_j^* = \sum_{i=1}^{m} q_i^* X_{ji} \tag{7}$$

to (weakly) equate his scores from T to a score (total number right) on T^*, and call this "weak equating from T to T^*" by use of (7).

Theorem 1. *With $\hat{S}_j^*$ in* (7):

$$E\hat{S}_j^* = \sum_{i=1}^{m} q_i^* p_{ji} \tag{8}$$

is the score j would expect to get had he taken test T^* provided $q_i \geq 1$ whenever $q_i^* \geq 1$. Thus (7) *weakly equates from T to T^*.*

Note that we can weakly equate from T to T^* provided

$$q_i^*/q_i < \infty, \qquad i = 1(1)m, \tag{9}$$

interpreting $0/0$ as finite. Sometimes we can weakly equate from T to T^*, but not from T^* to T.

This weak equating idea is much used in statistics, called *direct standardization, stratified sampling estimation, statistical adjustment,* and (with linear model assumptions) *analysis of covariance.* Thus much statistical development can be used to implement these ideas for equating tests with just a few items.

The q_i^* are the only weights that weakly equate the expected scores from T to T^* without further modeling assumptions, as the next theorem shows.

Theorem 2. *The q_i^* weights in* (7) *are the unique set of weights that weakly equate from T to T*.*

Proof. The binomial distribution is complete. In that case, X_{ji} is the unique unbiased estimator of p_{ji}. Hence $\hat{S}_j^* = \sum_i q_i^* X_{ji}$ is the unique unbiased estimator of $\sum_i q_i^* p_{ji}$, the jth individual's expected score on T^*. Although the $\{X_{ji}: i = 1(1)s\}$ may be distributed according to any of a wide class of distributions, the fact that the binomials are a subset suffices to prove the result.

Theorem 3. *If it is possible to* strongly *equate from T to T*, this is done by*

$$\hat{S}_j^* = \sum_i^m q_i^* X_{ji}. \tag{10}$$

Proof. Strong equating implies weak equating, and (10) is the only way to weakly equate.   Q.E.D.

Lord has proved there is no function g such that $g(S_j)$ can be used to strongly equate from the score S_j for examinee j on T to T^*. His theorem still holds even when a wider class of functions is permitted.

Theorem 4. *There is no function $g: R^m \to R^1$ such that*

$$g(X_{j1}, \ldots, X_{jm}) \tag{11}$$

has the same distribution as $S_j^* = \sum_i^m q_i^* X_{ji}^*$ for all distributions on $\{X_{j1}, \ldots, X_{jm}\}$ parameterized by arbitrary vectors $(p_{j1}, p_{j2}, \ldots, p_{jm})$, and all $j \in C$, unless $q_i = q_i^*$ for every i. Thus two tests that are not L-parallel cannot be strongly equated, even by the very general transformations in (11).

Proof. Theorem 3 says g in (11) must have the form of (10). Since the $\{p_{ji}\}$ and the random variables $\{X_{ji}\}$ may have very general values and distributions ($EX_{ji} = p_{ji}$), let us assume that $q_i X_{ji} \sim$ binomial(q_i, p_{ji}) independently. Then, from (10)

$$\text{Var}(\hat{S}_j^*) = \sum_1^m (q_i^*)^2 p_{ji}(1 - p_{ji})/q_i, \tag{12}$$

while with $q_i^* X_{ji}^* \sim$ binomial(q_i^*, p_{ji})

$$\text{Var}(\sum_1^m q_i^* X_{ji}^*) = \sum_1^m q_i^* p_{ji}(1 - p_{ji}) \tag{13}$$

Since (12) and (13) must be equal for all $\{p_{ji}\}$, it follows that $(q_i^*)^2/q_i = q_i^*$ for all i, or $q_i^* = q_i$ for $i = 1(1)m$, unless $q_i^* = 0$.   Q.E.D.

By Lord's theorem (as restated in Theorem 4) one can never strongly equate, under the assumptions considered so far. However, we can *always* weakly equate (unless $q_i = 0$ and $q_i^* \geq 1$ for some i).

Section 2 shows that, while strong equating is not exactly possible, under certain assumptions weak equating *approximates* strong equating. Then we consider circumstances under which linear equating is justified.

3. Approximately Strong Equating and Linear Equating

Despite Theorem 4, there are instances when strong equating can be nearly accomplished. First consider formula (10) and ask if for individual j, the distribution of $\hat{S}_j^*$ in (10) is the same as

$$S_j^* = \sum_1^m q_i^* X_{ji}^*, \tag{14}$$

j's score on T^*. For a given individual we might expect both S_j^* and $\hat{S}_j^*$ to be nearly normally distributed provided (a) there are enough items on the tests, (b) the p_{ji} are not close to 0 or 1 very often, and (c) the $\{X_{ji}: i = 1(1)m\}$ and $\{X_{ji}^*: i = 1(1)s\}$ are at most weakly dependent. In that case T^* may be nearly strongly equated from T by (9) if S_j^* and $\hat{S}_j^*$ have approximately the same variances for every examinee $j \in C$, i.e., if

$$\mathrm{Var}(S_j^*) \doteq \mathrm{Var}(\hat{S}_j^*). \tag{15}$$

[Also note, as in Beaton (1976), if $\mathrm{Var}(S_j^*) < \mathrm{Var}(S_j)$, so that T^* is more reliable than T, the $\mathrm{Var}(S_j^*)$ may be increased by adding random noise with zero mean to S_j^*. Hence (2) can be achieved artificially in this manner, and approximate strong equating is still possible, as will be shown.]

Now assume independence and the binomial distribution, so that (15) is equivalent to

$$\sum_{i=1}^m q_i^* p_{ji}(1 - p_{ji}) \doteq \sum_{i=1}^m q_i^* \left(\frac{q_i^*}{q_i} \right) p_{ji}(1 - p_{ji}) \tag{16}$$

for all $j \in C$. While this cannot hold for all $\{p_{ji}\}$ (a fact used to prove Theorem 4), these two variances could be nearly equal in some instances, especially if $\sum_1^m q_i$ (the number of items on T) is about equal to (or perhaps slightly exceeds) $\sum_1^m q_i^*$ (since S_j^* may be more efficient than $\hat{S}_j^*$).

While (16) depends on j, it is common to speak as if it does not. When equating researchers refer to *the* "reliability" of a test, they do not refer to the "reliability for person $j \in P$". Thus the quantities in (16) usually are assumed (or known) to be relatively independent on j (the complement of reliability for person j is just $\text{Var}(S_j^*)/[\text{Var}(S_j^*) + VB]$, with VB defined to be the variance of true scores between examinees in P).

We have the following statement:

Weakly equated tests of equal length and reliability are likely to be nearly strongly equated.

We now consider a necessary and sufficient condition for the standard linear equating method to produce strong (and weak) equating under the assumptions of Section 1. It is the probability that examinee j answers a question in Q_i correctly can be expressed as

$$p_{ji} = a_i\theta_j + b_i. \tag{17}$$

Here a_i and b_i are constants so p_{ji} may depend on at most one latent trait θ_j characterizing the examinee.

Theorem 5.   *T and T* can be weakly equated by a linear transformation $S_j^* = C_0 + C_1 S_j$ for every j if and only if (17) holds for all j.*

Proof.   We suppress j, denote $p_i(\theta) = \theta'a_i + b_i$ permitting θ and a_i to be k-dimensional column vectors. Then

$$ES_\theta^* = C_0 + C_1 ES_\theta, \tag{18}$$

or because $ES_\theta^* = \sum_{i=1}^m q_i^* p_i(\theta)$ and $ES_\theta = \sum_{i=1}^m q_i p_i(\theta)$ from (14) and (6), then for all θ

$$\theta' \sum_{i=1}^m q_i^* a_i + \sum_{i=1}^m q_i^* b_i = C_0 + C_1 \theta' \sum_{i=1}^m q_i a_i + C_1 \sum_{i=1}^m q_i b_i. \tag{19}$$

Since θ is arbitrary, we must choose

$$C_1 = \sum q_i^* a_i / \sum q_i a_i \qquad \text{and} \qquad C_0 = \sum q_i^* b_i - C_1 \sum q_i b_i$$

to produce an identity.

Now if $p_{ji} - (a_i + \theta_j b_i) = \delta_{ji}$, say, and δ_{ji} is a nonlinear function of θ_j, then (18) leads to (19) with $\sum_i q_i^* \delta_{ji}$ added to the left-hand side and $C_1 \sum_i q_i \delta_{ji}$ added to the right of (19). Then C_0 and C_1 no longer can be chosen to give (18) for all values of θ and δ_{ji}.   Q.E.D.

Theorem 5 tells us that standard linear equating cannot weakly equate under any circumstances for *multi*factor ability models. Thus tests covering math and verbal abilities, or even different kinds of math abilities, cannot be linearly equated. We cannot even accommodate an additional nonlinear term in the single trait of θ_j itself. And of course the special linear form (17) is also required. Finally, even under the assumptions of Theorem 5, very special circumstances are required for the resulting weak equating to be strongly equated.

Despite these drawbacks, Theorem 5 offers a practical hint. There may be groups of Q_i for which only one trait is relevant for each examinee. And even though p_{ji} is nonlinear, it may be approximately linear, as in (17), or at least for some subset $\{Q_i : i \in G\}$

$$\sum_{i \in G} q_i p_{ji} \quad \text{and} \quad \sum_{i \in G} q_i^* p_{ji} \tag{20}$$

may be nearly linear functions of θ_j. An example is

$$p_{ji} = f_i(\theta_j) = g_i + (1 - g_i)\left(\frac{1}{1 + \exp[-a_i(\theta_j - d_i)]}\right), \tag{21}$$

the *three parameter logistic model* that has been used in the educational test literature. Then following (20), we would expand

$$\sum_{i \in G} q_i f_i(\theta_j) \tag{22}$$

around θ^0, say, and hope terms of order $(\theta_j - \theta^0)^2$ and higher would be negligible. Making such terms small might require further restriction of linear equating to homogeneous subgroups of examinees.

When these conditions hold, linear equating, at least within subgroups of items (and perhaps examinees) would be justifiable.

This idea also would work in the case of the *highly reliable test* (HRT) in which the variability of the total test score is small. If we also assume that θ_j is one-dimensional, then the weak equating formula (p_i not necessarily linear) gives

$$E\hat{S}_j^* = \sum_1 q_i^* p_i(\theta_j) - T^*(\theta_j)$$

(say), while

$$ES_j = \sum_1 q_i p_i(\theta_j) = T(\theta_j).$$

Theorem 6. *Percentile equating is appropriate for HRTs if only one ability dimension exists, i.e., in this case an examinee who scores s on test T would score*

$$s^* = F^{*-1}(F(s)) \tag{23}$$

with F^* and F being the distribution functions for T^* and T. This result is independent of the ability distribution of the $\{\theta_j\}$. This is linear equating if for some a, b

$$F^*(t) = F(a + bt) \tag{24}$$

for all t.

Proof. Let the set $\{\theta_j\}$ have distribution function G (assumed continuous). Then since $S_j \doteq T(\theta_j)$ for all j, the score S of a randomly chosen individual is

$$F(t) = P(S \leq t) \doteq P(T(\theta) \leq t)$$

$$= P(\theta \leq T^{-1}(t)) = G(T^{-1}(t)),$$

assuming $T(\theta) = ES|\theta$ is monotone increasing, and similarly

$$F^*(t) \doteq G(T^{*-1}(t)).$$

Thus $s^*(s) \doteq T^*(T^{-1}(s)) \doteq F^{*-1}(F(t))$ is the needed equating function. Q.E.D.

Theorem 6 justifies percentile equating in a wider range of cases than linear equating with high reliability and just one ability dimension. It is more nonparametric. The parametric assumption (24) justifies linear equating.

This relationship may change in practical cases, however, for in smaller samples with lower reliability, the substantial variance reduction achieved by linear models may far outweigh the loss from introduction of bias.

As in the discussion following Theorem 5, multifactor ability models require a higher dimensional statistic for proper equating of two tests. We define the *separable* case to mean that subsets of each test can be identified in which only one ability dimension operates. In separable cases, equating can be achieved by first equating the separate scores for each section, and then totaling the results. It is impossible to equate in multifactor situations by using the total score (or any one-dimensional statistic) on T.

Appendix II proposes a conceptual model for evaluating the relative efficiencies of linear and quantile equating.

4. The Transition to Statistical Models

The presentation thus far has been nonparametric. The number of sets Q_i has been assumed large so that items within Q_i are equivalent for all members of the test population. Unrealistically, many items would be required to use this method. The preceding only aims to establish guidelines. In practice, statistical models are used to estimate with moderate samples what one merely would compute with large samples.

How will modeling ideas be implemented here? Topological (distance) notions are needed to describe the items and the examinees. We cannot afford with typical sample sizes to distinguish between strata Q_i that are near to one another as we must for those that are very dissimilar. So we develop parametric models. It seems worth reviewing some models and statistical ideas suggested by the approach in this paper.

Examples of parametric equating models include the multiparameter latent trait models with items parameterized for item difficulty, guessing, item discrimination, etc. This permits (a) *prediction* of an individual's performance on items with given characteristics, followed by (b) *adjustment* of performance to account for different item characteristics on a second test. This sequence was used with the item strata in Section 2. Step a corresponds there to estimating response to Q_i by the observed proportion correct, step b adjusts to test T^* with different numbers q_i^* of items from Q_i. In that case the item parameters indicate the item strata themselves. Thus, using prediction models for adjustment, as Lord does, generalizes the method of Section 2. As is common in statistical modeling, a good model will reduce the number of parameters and thereby improve estimation by reducing variance. Skill is required in choosing an appropriate parameterization, but extensive data analysis can greatly enhance the required skill.

Remark 2.   Generalization of Section 2 to the continuous (parametric) case. The likelihood ratio

$$L(\gamma_1, \ldots, \gamma_k) = \frac{d\Pi^*(\gamma_1, \ldots, \gamma_k)}{d\Pi(\gamma_1, \ldots, \gamma_k)} \tag{25}$$

appears to be the appropriate generalization for moving from the $\{Q_i\}$ to item parameter vector $(\gamma_1, \gamma_2, \ldots, \gamma_k)'$, so $Q_i = (\gamma_{i1}, \gamma_{i2}, \ldots, \gamma_{ik})$. Here Π represents the probability distribution of item characteristics on T, effectively replacing the q_i of Section 2 (more precisely, the proportions $q_i/\sum_1^m q_j$), and the same for Π^* in relation to T^*. Then, if individual j has probability

$$f_j(\gamma_{i1}, \ldots, \gamma_{ik}) \tag{26}$$

of answering the ith item correctly, he is expected to answer the proportion

$$p_j = \int \cdots \int f_j(\gamma) \, d\Pi(\gamma) \tag{27}$$

correctly on T. We weakly equate from T to T^* by expecting him to answer the proportion

$$p_j^* = \int \cdots \int f_j(\gamma) L(\gamma) \, d\Pi(\gamma) \tag{28}$$

correctly on T^*, because

$$p_j^* = \int \cdots \int f_j(\gamma) \, d\Pi^*(\gamma).$$

This extends Section 2 to the continuous case with $L(\gamma)$ replacing q_i^*/q_i for Q_i. We note that if γ has dimension equal to the number of Q_i, and γ_i is chosen to be the indicator of Q_i, then (25) just reduces to q_i^*/q_i.

Remark 3.   The need to equate before aggregating (computing total score) when ability has more than one dimension. Separable items.   Let us consider the case of an underlying ability parameter θ with dimension $k \geq 2$. Let $p_i(\theta)$ be the probability that an individual with ability parameter θ answers item i correctly.

It is clear that any attempt to equate nonparallel tests from total scores in this situation must fail, for test T^* will emphasize some abilities more than T (otherwise the tests would be parallel) and then individuals with an abundance of that ability will do better on T^* than any one-dimensional equating procedure (e.g., linear or percentile equating) would predict.

The equating problem is *"separable"* if subsets of the tests T and T^* can be isolated that depend on just one ability dimension. That is, if for all items i in T and T^*, $p_i(\theta)$ depends on just one component of θ, then the problem of equating from T to T^* is "separable." Items meeting this same standard will be called "separable items."

To illustrate separability, suppose T and T^* involve just two abilities, quantitative and verbal. Suppose adequate linear equating functions exist for equating T to T^* on the verbal and quantitative sections separately:

$$X_v^* = a_v X_v - d_v \quad \text{and} \quad X_q^* = a_q X_q - d_q, \tag{29}$$

where we think of a_v and a_q as the item discrimination parameters for the two sections, d_v and d_q the difficulties, and X_v, $X_q(X_v^*, X_q^*)$ the proportions correct on $T(T^*)$. The predicted total score on T^* is

$$S^* = n_v^* X_v^* + n_q^* X_q^*, \tag{30}$$

which is *not* a linear function of the total score on T, $S = n_v X_v + n_q X_q$, unless it happens that

$$n_v^* a_v / n_v = n_q^* a_q / n_q. \tag{31}$$

Hence we cannot aggregate (compute total score) first and then equate. Therefore

If more than one ability dimension is involved, tests made up of separable items must be equated by first equating sections, then aggregating; not by aggregating, then equating.

This should be clear from Theorems 5 and 6 of Section 3. It suggests that equating based on total scores would generally be unfair to some examinees, when tests are not parallel.

To make this clearer suppose Mathew, a math major, answers 40% correctly on verbal questions and 80% on mathematical ones. Verna, a language major, is just the opposite: She gets 80% verbal correct, 40% math. If test T

has 75 verbal questions, 25 mathematical and T^* has 25 verbal, 75 mathematical, then Mathew will prefer T^* and Verna T, since they expect to get $0.80 \times 75 + 0.40 \times 25 = 70$ right on their preferred test, and $0.80 \times 25 + 0.40 \times 75 = 50$ on the other. The tests obviously are not parallel.

These tests can be equated, of course, but only by using both subsection scores. Standard equating methods based on total scores do not work in this case. For example, the distribution of total scores on both tests *could be the same*, in which case both standard equating methods would reduce to the identity function $S^* = S$ [this would happen, for example, if for every examinee with probabilities (p_1, p_2) of correctly answering verbal and math items, there is another with probabilities (p_2, p_1)]. Then standard equating methods based on aggregating before equating would indicate that Mathew, having scored 70 on T, would score 70 on T^*, and not the 50 he actually would expect on T^*.

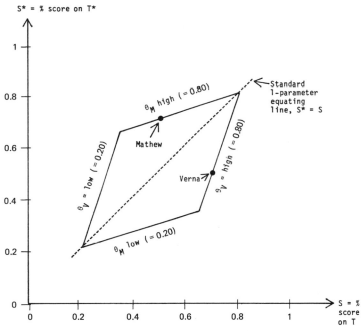

Fig. 1   Diamond example. Ability parameters are θ_V and θ_M (verbal and mathematical), uniformly and independently distributed, $0.2 \le \theta_V, \theta_M \le 0.8$. Test scores $S = 0.75\theta_V + 0.25\theta_M$ (mostly verbal questions) and $S^* = 0.25\theta_V + 0.75\theta_M$ (mostly mathematical). Mathew ($\theta_V = 0.4$, $\theta_M = 0.8$) and Verna ($\theta_V = 0.8$, $\theta_M = 0.4$) have symmetric abilities, but are greatly affected by which test they take. Linear, quantile, and IRT equating all yield $S^* = S$ as equating formula, but that is clearly wrong. The problem is that ability traits are two dimensional.

Figure 1 illustrates these points. Here we have two distinct, independent ability traits, distributed symmetrically in the population. The only reasonable one-dimensional equating method is $S^* = S$, but that works very poorly for individuals with θ_V substantially different from θ_M.

Remark 4.   The nonseparable case.   Items may not always be separable, and this substantially complicates matters. Logistic item response models, which admit sufficient statistics of dimension k (the dimension of θ), provide an appropriate generalization in such cases. For example, suppose an examinee with trait vector θ has probability of answering item i correctly given by

$$p_i(\theta) = [1 + \exp(d_i - a_i'\theta)]^{-1}, \qquad i = 1, \ldots, n \qquad (32)$$

with d_i the item difficulty and a_i the item discrimination vector (assumed known). Let **Y** be the n-dimensional vector of zeroes and ones indicating failure and success on the items:

$$P(y_i = 1) = p_i(\theta). \qquad (33)$$

Then a complete sufficient statistic for θ (and hence for any function of $\boldsymbol{\theta}$, in particular, the predicted score on another test) is the k-dimensional vector

$$\mathbf{z} = A\mathbf{y} \qquad (34)$$

with A the $k \times n$ matrix with ith column equal to a_i: $A = (a_i, \ldots, a_n)$.

These test items are separable if each a_i vector has $k - 1$ zero components. If not, then different components of **z** will involve the same item because some items depend on several ability dimensions.

Remark 5.   Utility of IRT models.   That models like (32) are valid is doubtful, but they can be useful. They are invalid since they do not involve guessing parameters, but including such parameters causes nonexistence of a sufficient statistic having the dimension of θ. Still, such models do capture the spirit of item dependence on more than one ability trait, thereby avoiding errors due to ignoring this point. Thus item response theory models, and logistic models in particular, promise to be valuable in practice.

Remark 6.   Rao–Blackwellization.   The complete sufficient statistic $\mathbf{z} = \mathbf{A}y$, when it exists, can be used to improve any unbiased estimate of score on T^* from the performance vector **y** on T via the Rao–Blackwell theory. For example, if the models of Section 2 apply, then they provide a formula for S^*, the predicted total on T^* based on observing score S on T. Here S^* must satisfy

$$E_\theta S^*(y) = E_\theta S^* = \text{expected score on } T^* \qquad (35)$$

for all θ (all examinees), S^* being a function of y, y observed from T. Then

$$E_\theta S^*(y)|z \qquad (36)$$

is independent of θ and is a better estimate of the expected score on T^* than is $S^*(y)$ itself [if the logistic model holds, as in (32)]. When (36) can be computed, it should provide a more accurate weak equating than true-score equating.

Remark 7. Example of Rao–Blackwellization—Theorem in Appendix I. Let

$$p_i(\theta) = e^\theta/(e^\theta + e^{d_i}) \qquad (37)$$

so that θ has dimension $k = 1$. Difficulties $\{d_i\}$ differ, but all items discriminate equally well [$a_i = 1$ for all i, a_i being the coefficient of θ in $p_i(\theta)$]. Items are scored 0 or 1, and we let $S = \sum n_i X_i$ be the total number correct on T, n_i items with difficulty d_i appear on T, X_i = fraction right on those n_i. Assume the item responses are independent (given θ).

Suppose on T^* we have n_i^* with difficulty d_i. Then by the theory of Braun and Holland (this volume):

$$F_0 \sum \frac{n_i^*}{n_i} X_i = E_\theta S^* \qquad (38)$$

since weak equating preserves expectations. Then by (36),

$$\hat{S}^* = E_\theta\left(\sum \frac{n_i^*}{n_i} X_i \Big| S\right) \qquad (39)$$

is a better estimate of $E_\theta S^*$, and we show in Appendix I for $N^* = \sum n_i^*$, $N = \sum n_i$, $r_i^* = n_i^*/N^*$, $r_i = n_i/N_i$ that

$$\frac{\hat{S}^*}{N^*} \doteq \frac{S}{N} + \frac{S}{N}\left(\frac{N-S}{N-1}\right)\sum (r_i - r_i^*)\, d_i. \qquad (40)$$

More simply, let $s^* = S^*/N$, $s = S/N$ be the fraction right on T^*, T, and $\bar{d}_{r*}$, $\bar{d}_r$ the average difficulties, $\bar{d}_{r*} = r_i^* d_i$, $\bar{d}_r = \sum r_i d_i$. Then

$$s^* \doteq s + s(1 - s)[\bar{d}_r - \bar{d}_{r*}]. \qquad (41)$$

Thus if $s = 0$ or $s = 1$, $s^* = s$, but otherwise we add an amount to the score on T^* if the average difficulty on T exceeds that on T^* (or subtract an equivalent amount in the opposite case). Thus quadratic equating is appropriate in this example, not linear equating.

Remark 8. Extended example comparing several equating methods with a one-dimensional ability parameter. We illustrate Lord's (1977) true-score equating method and the preceding results in the simple case of just two possible values of $p_i(\theta)$ (Q_1 and Q_2 are the only equivalence classes of items), with

$$p_i(\theta) = 1/[1 + \exp(-c(\theta - \tilde{d}_i))] = 2^\theta/(2^\theta + 2^{d_i}), \qquad c = \log(2), \quad (42)$$

$\tilde{d}_1 = -1, \tilde{d}_2 = 1, d_1 = -\log(2), d_2 = \log(2)$ (Q_2 has harder questions than Q_1). Both T and T^* have 50 items, but $n_1 = n_2 = 25$ on T while $n_1^* = 40$, $n_2^* = 10$ on T^*, so T^* is an easier test. We observe the fraction correct in Q_1, Q_2, X_1, X_2 from T, $n_1 X_1 \sim \text{Bin}[25, p_1(\theta)]$, $n_2 X_2 \sim \text{Bin}[25, p_2(\theta)]$ for an examinee, and wish to equate these results to an expected score on T^*. Methods are

(1)

$$\tilde{S}^* = 40X_1 + 10X_2, \tag{43}$$

based on Section 2 theory, ignoring the model.

(2)

$$\hat{S}^* = E(\tilde{S}^*|S) \doteq 50[s + s(1 - s)(0.6) \log(2)] \tag{44}$$

from the preceding example $(\bar{d}_r - \bar{d}_{r*} = 0 - \{0.8[-\log(2)] + 0.2[\log(2)\} = 0.6 \log(2), s = S/50)$.

(3)

$$\eta^*(S) = 40p_1(\hat{\theta}_S) + 10p_2(\hat{\theta}_S), \tag{45}$$

where $\hat{\theta}_S$ satisfies

$$S = 25p_1(\hat{\theta}_S) + 25p_2(\hat{\theta}_S), \tag{46}$$

which is Lord's true-score equating method.

Table 1 shows that $\hat{S}^*$ is fairly close to $\eta^*(S)$, Lord's true-score equating method. Lord's method will have some bias in that $\eta(\theta) = n_1^* p_1(\theta) + n_2^* p_2(\theta)$ is not generally a linear function of $\xi(\theta) = n_1 p_1(\theta) + n_2 p_2(\theta)$. This calculation shows that little bias may be introduced by Lord's method, which then gains considerably relative to Rao-Blackwellizing S^*, because Lord's method is much more easily extended to more complicated situations. (Actually $\hat{S}^*$ is also biased in that it approximates a very complicated unbiased estimate; see proof of theorem in Appendix I.)

Table 2 shows that the weak equating estimates $\tilde{S}^*$ vary with X_1 and X_2 separately, while $\hat{S}^*$ and $\eta^*(S)$ do not because S is, in these cases, a sufficient statistic under the model (48).

TABLE 1

Comparing $\hat{S}^*$ and $\eta^*(S)$

S	$\hat{S}^*$	$\hat{\theta}_S$	$\eta^*(S)$
0.0	0	$-\infty$	0
6.5	8.9	-3	8.6
11.1	14.8	-2	14.4
17.5	22.3	-1	22.0
25.0	30.3	0	30.0
32.5	37.3	1	37.0
38.9	42.6	2	42.2
43.5	45.9	3	45.6
50.0	50.0	∞	50.0

TABLE 2

Comparing $\hat{S}^*$ to $\tilde{S}^*$ and $\eta^*(S)$ When $X_1 + X_2$ Is Constant[a]

X_1	X_2	S	$\tilde{S}^*$	$\hat{S}^*$	$\eta^*(S)$
1.00	0.00	25	40.0	30.3	30
0.80	0.20	25	34.0	30.3	30
0.68	0.32	25	30.4	30.3	30
0.60	0.40	25	28.0	30.3	30
0.40	0.60	25	22.0	30.3	30
0.00	1.00	25	10.0	30.3	30
1.00	0.56	39	45.6	42.6	42.3
0.92	0.64	39	43.2	42.6	42.3
0.84	0.72	39	40.8	42.6	42.3
0.76	0.80	39	38.4	42.6	42.3
0.56	1.00	39	32.4	42.6	42.3

[a] Case 1, $\hat{\theta}_S$ (maximum likelihood estimate of θ) $= 0.00$ from $S = 25$.
Case 2, $\hat{\theta}_S = 2.02$ from $S = 39$.

Which method is better? Under the logistic model, $\log[p_i(\theta)/(1 - p_i(\theta)] = c(\theta - d_i)$, $\tilde{S}^*$ will be unbiased but inefficient, with variance about 35% greater than $\hat{S}^*$ or $\eta^*(S)$ when $\theta = 0$ (so $p_1 = p_1(0) = \frac{2}{3}$, $p_2 = p_2(0) = \frac{1}{3}$). However, if $p_1 = EX_1 \neq \frac{2}{3}$ but still $p_1 + p_2 = 1$, then the logistic model in the form (42) is invalid and $\hat{S}^*$ and $\eta^*(S)$ will be biased. In this case the bias is such that the mean-squared error of $\hat{S}^*$ for estimating $E_\theta S^* = \eta^*(\theta) = 40p_1 + 10p_2$ in this 50-item sample is

$$E_\theta(\hat{S}^* - (30p_1 + 10))^2 \tag{47}$$

which is less than

$$E_\theta(\tilde{S}^* - (30p_1 + 10))^2 = \text{Var}_\theta(\tilde{S}^*) \tag{48}$$

if and only if $0.59 < p_1 < 0.73$. If $p_1 < 0.51$ or $p_1 > 0.79$, the mean-squared error of $\hat{S}^*$ will more than double that of $\tilde{S}^*$.

Violation of the model (42) could occur because $\boldsymbol{\theta} = (\theta_1, \theta_2)$ with θ_1 the operative trait on Q_1, θ_2 on Q_2, and

$$p_1 = p_1(\theta_1), \quad p_2 = p_2(\theta_2), \quad p_1 + p_2 = 1. \tag{49}$$

$\tilde{S}^*$ will produce a proper equating in this case, even though this model is unknown. But $\hat{S}^*$ and $\eta^*(S)$ will equate incorrectly.

If, in fact, a more complicated logistic model involving discrimination parameters a_1, a_2 obtains in the form

$$\log[p_i(\theta)/(1 - p_i(\theta))] = a_i\theta - d_i, \tag{50}$$

then $a_1 X_1 + a_2 X_2$ is the complete sufficient statistic (css) for θ. It follows when $a_1 = 4a_2$ that $\tilde{S}^* = 40X_1 + 10X_2$ is the css for θ, and is the UMVUE of $\eta^*(\theta)$. Thus $\tilde{S}^*$ may be viewed as an inefficient equating method when $a_1 = a_2$, but with model uncertainty it looks much better and is the sufficient statistic under the model specification (50) if $a_1 = 4a_2$.

This example cannot show which method works best in practice. It only further illustrates a familiar theme in statistics, that more nonsampling information (better knowledge of the model) translates into better estimators. But if this information is substantially in error, then a less efficient procedure is preferred. A practical challenge for test equating is to learn more about appropriate models from test data.

5. Appendix I. Theorem of Remark 7: Approximating the Best Equating Method in a Case More General Than Model (37)

Let $Y_i^{\text{ind}} \sim \text{Bin}(n_i, p_i)$,

$$p_i = \frac{1}{1 + \exp(-a\theta + d_i)}, \quad i = 1(1)k.$$

Note $\sum Y_i$ is a css for θ, and let $A_t = \{\mathbf{Y}: \sum Y_i = t\}$. We make the definitions of Remark 7, $N = \sum n_i$, $N^* = \sum n_i^*$, $r_i = n_i/N$, $r_i^* = n_i^*/N^*$. Then

$$E[\sum n_i^*(Y_i/n_i) | \sum Y_i = t] \doteq \frac{N^*}{N}t + \frac{N^*}{N}\frac{t(N-t)}{N-1}\sum(r_i^* - r_i)\,d_i. \tag{51}$$

Proof.

$$\sum_1^k \frac{n_i^*}{n_i} \left[\frac{\sum_{y \in A_t} y_i \prod_1^k \binom{n_j}{y_j} p_j^{y_j} (1 - p_j)^{n_j - y_j}}{\sum_{y \in A_t} \prod_1^k \binom{n_j}{y_j} p_j^{y_j} (1 - p_j)^{n_j - y_j}} \right]$$

$$= \sum_1^k \frac{n_i^*}{n_i} \left[\frac{\sum_{y \in A_t} y_i \prod_1^k \binom{n_j}{y_j} \exp[(a\theta - d_j)y_j]}{\sum_{y \in A_t} \prod_1^k \binom{n_j}{y_j} \exp[(a\theta - d_j)y_j]} \right]$$

$$= - \sum_1^k \frac{n_i^*}{n_i} \frac{\partial}{\partial d_i} \log \left\{ \sum_{y \in A_t} \left[\prod_1^k \binom{n_j}{y_j} \right] \exp(-\sum d_j y_j) \right\}$$

$$= - \sum_1^k \frac{n_i^*}{n_i} \frac{\partial}{\partial d_i} \log[\tilde{E} \exp(-\sum d_j y_j)],$$

where $\tilde{E}$ requires the hypergeometric distribution for $\mathbf{y} \sim HG_k(t, \mathbf{r}; N)$. This has mean vector $t\mathbf{r}$ and covariance matrix

$$t \frac{N - t}{N - 1} (D_\mathbf{r} - \mathbf{rr}'),$$

with $D_\mathbf{r}$ a diagonal matrix with diagonal elements $\{r_i\}$.

Assuming approximate normality of $\sum d_j(y_j - tr_j)$ under the hypergeometric model (which may produce the best quadratic fit even if normality is not approximate),

$$E \left[\sum \frac{n_i^*}{n_i} Y_i \middle| \sum Y_i = t \right] = - \sum \frac{n_i^*}{n_i} \frac{\partial}{\partial d_i} \left[-\mathbf{d}'\mathbf{r}t + 1/2t \frac{N - t}{N - 1} [\sum d_j^2 r_j - (\mathbf{d}'\mathbf{r})^2] \right]$$

$$= \sum \frac{n^*}{n_i} \left[r_i t - \frac{t(N - t)}{N - 1} [r_i d_i - (\mathbf{d}'\mathbf{r})r_i] \right]$$

$$= \frac{N^*}{N} \left[t - \frac{t(N - t)}{N - 1} \sum_1^k (r_i^* - r_i)d_i \right] \quad \text{Q.E.D.}$$

6. Appendix II. Formulation of an Approach for Comparing the AREs of Linear and Quantile Equating

Questions have been asked about comparisons between linear and equipercentile equating. Neither of these can work if the test involves more than one ability dimension, but suppose there is just one and let S_j be the total score by j on test T, S_j^* the score on T^*. Suppose S_j has a distribution function, given θ_j, of

$$F_{\theta_j}(s)$$

and we assume the $\{\theta_j\} \sim G(\theta)$, the distribution of abilities in the population. Thus

$$H(s) = \int F_\theta(s)\, dG(\theta)$$

is the marginal distribution of test scores on T. Let H have mean and variance (μ_H, σ_H^2). If another random sample of examinees from the same population (or the same examinees could be asked to take the second test T^* too) is obtained, then their abilities also are distributed as $G(\theta)$, but their scores on T^*, given θ, have a different distribution $F_\theta^*(s)$, unless T^* is parallel to T. Thus

$$H^*(s) = \int F_\theta^*(s)\, dG(\theta)$$

is their marginal distribution function with first two moments $(\mu_{H^*}, \sigma_{H^*}^2)$.

Assuming large sample sizes (both the number of examinees n and the number k of items—it is enough to assume high reliability), the standard equating functions from $T \to T^*$ are

$$S_L^* = t_L(S) = \mu_{H^*} + \frac{\sigma_{H^*}}{\sigma_H}(S - \mu_H)$$

for linear equating and

$$S_Q^* = t_Q(S) = H^{*-1}(H(S))$$

for quantile equating.

Now we calculate the expected squared error

$$R_t(\theta) \equiv E_\theta(t(S) - \mu^*(\theta))^2$$

with $\mu^*(\theta) = \int s\, dF_\theta(s)$ the expected score, $t = t_L$, and $t = t_Q$. A summary of equating accuracy would be

$$r_t = \int R_t(\theta)\, dG(\theta).$$

Now assuming different models for F_θ and F_θ^* (normal, logistic, etc.) one can compare asymptotically the performance of t_L and t_Q in relation to each other and in relation to the optimal procedure. In particular, for situations making t_L optimal, how inefficient is t_Q? Generally t_Q is robust but t_L is not, so one also could construct examples to show how bad t_L can be.

We probably can learn more about the concepts of test equating by investigating the theory for highly reliable tests (HRT). This involves an asymptotic theory in which statistical issues are simplified substantially. Lord's true-score equating formula is in this spirit, as is the preceding example.

7. Appendix III. Additional Comments on Equating

Some comments related to the theory of this paper, but going beyond it are discussed here. These have to do with test design, randomization, and analysis issues that do not arise if data already are in hand and prior knowledge of either equivalence classes of item characteristics or the form of distribution of item responses is known.

A. Reducing the Number of Groups of Items and Examinees by Permitting Some Heterogenity

The number m of groups of equivalent items and of examinees assumed in Section 1 exceeds practical limitations. However, even $m = 10$ to 20 item strata may be quite useful for equating. We also might want to partition the set C of examinees as $C = C_1 \cup \cdots \cup C_r$ with each C_j a relatively homogeneous group of individuals. This could help analyses, perhaps even with r about 10. This latter decomposition might affect poststratification.

Randomization and controlled randomization based on item characteristics would help reduce problems due to the more hetergeneous Q_i. One then could define an examinee's "true score" on a test to be the probability he can correctly answer questions chosen at random from each of the Q_i with specs $(q_1, \ldots, q_m)$ for the Q_i. This concept avoids many difficulties, but may be difficult to implement (the set of all possible questions in Q_i would have to be constructed in advance), and the randomization, unless well controlled, may introduce unnecessary randomness.

B. Test Design and Construction

It is obvious to all of us concerned with equating that the more nearly parallel two tests are, the easier and more robust equating methods tend to be. "Parallel" means that the items in Q_i should be equivalent and that $q_i = q_i^*$. It is worth continuing effort in this direction.

C. Experimental Design

William Angoff discusses many kinds of designs used at ETS in his paper "Summary and Derivation of Equating Methods used at ETS" (1978); also see (Angoff, 1971). These ideas use randomization and/or some kind of common equating test. Obviously, equating is greatly aided by highly balanced and randomized designs such as Angoff's designs II (random groups, both forms administered, counterbalanced), and III (random groups, one test each plus common equating test). Design and randomization greatly diminish dependence on models, and so can be a great aid to equating. Further consideration of these matters might be very useful.

Although this fits better under topic B, randomization in the form of sampling items from strata to construct nearly parallel tests offers possibilities for parallel test construction. I am sure this has been done. New methods in design by computer, random sampling, but utilizing item characteristics also could be considered. Computer design has been successful in some applications in other areas.

D. Use of the Anchor Test

The anchor test, or W. Angoff's "equating test," is a test given to two groups to compare them as an interim step to equating two other tests taken separately by the groups. Angoff made many suggestions in his paper "Summary and Derivation of Equating Methods Used at ETS." David Hinkley reported on a nonparametric method, appropriate for use in the context of the assumptions of Section 2, during the March 1979 meeting at ETS. One problem will be to see how Hinkley's ideas can be adapted to parametric situations.

E. Poststratification

Section 2 assumes that the item strata, Q_i, are equivalence classes. Statistical evidence will be available to criticize this assumption after the test. If warranted, the Q_i could be redefined afterward. At the March 1979 meeting some of us suggested that the Q_i could be constructed after administration of the examination by letting Q_i be the set of all items for which a certain percentage were answered correctly by the entire group. It is clear now that while this is a necessary condition for equivalent strata, it is far from sufficient. Verbal items could be combined by this rule with quantitative items, but perhaps no individual examinee would be indifferent among items in Q_i. At the very least, the idea will have to be applied relative to homogeneous subgroups of examinees.

F. STATISTICAL METHODS FOR COMPROMISING BETWEEN DIFFERENT PROCEDURES

When faced with choices between different models for weak equating, for example, one unbiased but with large variance (as the method of Section 2 would be if put in practice), the other making more assumptions, and therefore probably biased, but with reasonably small variance (e.g. linear equating models and models based on latent traits and item characteristics), one sometimes can compromise by using a method that is better than either choice. The method averages the two kinds of equating results, with weights estimated by data. Appropriate methods for doing this include Bayesian and empirical Bayesian techniques. Another application would shrink quantile equating formulas toward linear equating formulas as far as the data permit.

Empirical Bayes methods also can be combined with analyses based on item similarity matrices (square matrices whose entries index the similarity of pairs of test items—analogous to the matrices used in multidimensional scaling). These matrices will be large but, presumably, sparse. When predicting an examinee's success probability on a T^* item, this method compromises between (a) estimates based only on equivalent items in T and (b) weighted averages of more items in T, each being less similar to the given item on T^*.

REFERENCES

Angoff, W. H. (1971). Scales, norms, and equivalent scores. In R. L. Thorndike (Ed.), *Educational Measurement* (2nd ed.). Washington, D. C.: American Council on Education, 508–600.

Angoff, W. H. (1978). *Summary and Derivation of Equating Methods Used at ETS.*

Beaton, A. E. (1976). Comments on equating on the Cohort Charge Study. *Educational Testing Service Memorandum*, December 14.

Holland, P. W. (1979). *Notes on Mathematical Foundations of Test Equating.*

Levine, R. (1955). *Equating the Score Scales of Alternative Forms Administered to Samples of Different Ability* (RB-55-23). Princeton, N. J.: Educational Testing Service.

Lord, F. M. (1977). Practical applications of item characteristic curve theory. *Journal of Educational Measurement*, *14*(2), 117–138.

Discussion of "On the Foundations of Test Equating"

David V. Hinkley

1. Introduction

As Carl Morris has pointed out, and ably illustrated, study of the foundations of test equating is important because one can investigate both the potential and limitations of equating in the absence of the complicating factors of particular applications. The essential questions include: What are the desirable objectives of equating? Which objectives are attainable, when, and how? Morris has used powerful, yet simple and understandable, ideas to obtain some interesting initial answers to some of the questions. In discussing the paper it would be appropriate to build on the paper with a view to existing problems of practical test equating. The final goal is to determine the extent of validity of existing equating methods, perhaps to devise new equating methods, and to describe ways in which the results of test equating can be checked for consistency.

2. Who or What Is Being Equated?

Here and elsewhere I shall try and follow the terminology of Morris's paper. Suppose that methods $M_1, \ldots, M_r$ are used to equate tests T and T^*,

193

method M_k translating subject j's scores $(X_{ji} : i = 1, \ldots, s)$ into $S^*_{j(k)}$. Suppose also that all r methods exactly equate the score distributions across a population P, meaning that for every value s

$$\text{frequency of } S^*_{j(k)} \leq s = \text{frequency of } S^*_j \leq s.$$

This is not sufficient, because a true equating method M_0 would have the stronger property that for *any* subpopulation P_j of P the distribution of $S^*_{(0)}$ is identical to that of S^*. Presumably, then, the best of $M_1, \ldots, M_r$ would be that for which an appropriate measure of distance between the distributions of $S^*_{(k)}$ and S^* within P_j is minimized for all relevant subpopulations P_j. What we would need to know is: What are the relevant subpopulations? Answers might include (i) demographically defined groups, (ii) ability groups, possibly defined objectively by appropriate surrogate traits, (iii) whatever the colleges and candidates view as peer groups.

3. Is the Latent Trait Model Appropriate?

Starting with the fundamental idea of equivalence classes of items—a useful notion brought to our attention by Paul Holland—both Morris and I came to view Lord's latent trait models as important vehicles for both the theory and practice of test equating. The natural question is: Is one factor adequate for the tests that must be equated? With the enormous data sets that are available to ETS and with their long experience in fitting these models, presumably one could determine whether or not this is so. But one should also ask: Are existing deviations from the single-trait model of any practical significance? One suggestion that I have may be outlined as follows.

Suppose that we are equating T to T^* as in Morris's paper, using the single-trait model to map $\xi \rightarrow \eta$ as described by Lord. The necessary ingredients are the subject trait θ and the item characteristics (a_i, d_i) of T and (a_i^*, d_i^*) of T^*. Scores on either test are stochastically increasing in θ. Associated with a particular score S_0 on T is a value θ_0 for which $S_0 = E(S|\theta_0)$, which then can be compared to the average of scores on T^* for subjects with θ's "very near" θ_0. But note that the averages may be equal under a multitrait model, as in Morris's "diamond example." What we should really look at are the theoretical frequency distribution of $S^*|\theta_0$ (according to the single-trait model) and the empirical frequency distribution of scores of T^* for subjects with θ^*s "very near" θ_0.

4. The Rao–Blackwellization Adjustment

Morris's development of Eq. (41) is elegant and instructive. It is also interesting statistically in that it relies on exact sufficiency, although an approximation is used at the final step. Yet clearly the result is qualitatively correct more generally, and deserves further study—perhaps using large-sample normal approximations for the linear combinations of the x_i at the first step. Unfortunately the general problem with unequal discrimination parameters a_i is not at all easy to deal with. First, Rao–Blackwellization conditions on the "wrong" statistic, namely, $\sum a_i X_i$ instead of the test-equater's $\sum X_i$. Then the partition of sample space effected by $\sum a_i X_i$ may not be fine enough; it could consist of a single point for certain a_is. (This is the point at which exact statistical theory becomes an embarrassment, since what is really of importance is conditioning on values "close to" the observed value of $\sum a_i X_i$.) Presumably a subtle use of approximations will lead to a solution analogous to (41), and in studying this problem we may learn something useful about equatability in general.

5. Smoothing the Tails of the Equating Curve

Most discussion of equating considers the entire range of scores, although from some practical points of view only the tails—particularly the upper one—are important. Note that relatively few testees determine the shape of the equating curve in the tails. It may then be worthwhile to consider methods of smoothing the tails.

Consider the scenario in Holland's paper in which the equating curve is

$$e(x) = F^{-1} \circ G(x)$$

with F and G the cumulative frequency distributions of S and S^*, respectively. For linear equating $G(x) = F(a + bx)$; cf. Morris's equation (24). Now suppose that $F \approx F_1$ and $G \approx G_1$ such that

$$e_1(x) = F_1^{-1} \circ G_1(x) = a + bx,$$

which then is a natural basis for $e(x)$. We may write

$$e(x) = (F^{-1} \circ F_1) \circ e_1 \circ (G_1^{-1} \circ G)(x),$$

which would be estimated by

$$\hat{e} = (\hat{F}^{-1} \circ F_1) \circ \hat{e}_1 \circ (G_1^{-1} \circ \hat{G}).$$

To see what might be done with this we recall a characterization of $\hat{F} \circ F^{-1}$ given by Durbin and Knott (1972), namely,

$$\hat{F} \circ F_1^{-1}(z) = z + \sum_1^\infty \hat{\beta}_j \sin(j\pi x),$$

where

$$\hat{\beta}_j = (2/j\pi) \int_0^1 \cos(j\pi x) \, d\hat{F}(x).$$

Truncation of the series at the last statistically significant term—or, better, use of empirical Bayes methods on the small coefficients $\hat{\beta}_j$—gives a smooth version of $\hat{F}$ by substitution of $z = F_1(x)$ in the modified series. A similar approach could be used with the above representation of $\hat{e}$.

Other smoothing methods are, of course, possible. The general point is that some compromise between "rough" equipercentile and "oversmooth" linear equating may well be useful, and empirical Bayes techniques may play a useful role as Morris suggests.

6. Use of the Anchor Test

Near the end of this paper Morris refers to earlier discussions about extending equations like (10) to the situation where an anchor test is available. In this situation one attaches common items to T and T^*. In a sense an anchor test is a surrogate for a linkage model such as the latent trait model discussed earlier. Let A denote the anchor test.

The group of examinees taking $T + A$ is E, with E^* taking $T^* + A$. A subject j^* in E^* may be said to be equivalent to a subject j in E if their frequencies of correct responses on A are the same. Thus equivalence classes of subjects may be defined, classes C_f and C_f^* consisting of all those subjects whose frequency correct on A was f. The equating curve $S \to S^*$ is then obtained by equating

$$\begin{array}{cc} \text{average } S_j & = \text{average } S_{j^*} \\ j \in C_f & j^* \in C_f^* \end{array}$$

for the series of values of f. Note that the resulting curve need not be monotone—if it were not, this might indicate a more fundamental difficulty for equating, such as existence of multiple traits.

A quite different approach is based on establishing equivalence classes of items, rather than subjects. Suppose that item k on A has $f = f_0$ in E and $f = f_0^*$ in E^*. Then the class of items I_0 in T with $f = f_0$ are said to be equivalent to the class of items I_0^* in T^* with $f = f_0^*$. Note that this is an empirical way of defining what Levene's model assumes. Given the sophistication of test construction, the empirical method might benefit from compromise with informed prior opinion as to which items are equivalent.

I must confess that I am not convinced of the utility of the ideas discussed here for construction of equating curves, but the suggestions may prove useful in attempting to assess consistency of given equating methods.

REFERENCE

Durbin, J., and Knott, M. (1972). Components of Cramér–von Mises Statistics I. *Journal of the Royal Statistical Society, B 34*, 290–307.

Part III

NEW METHODS FOR TEST EQUATING

Some Issues in Test Equating

Richard F. Potthoff

1. Introduction

The subject of test equating involves a number of different facets, many of which are related not only to test equating itself but also to other matters concerning testing. This paper has four main sections. Section 2 provides my general comments and my viewpoints on each of a number of issues pertaining to test equating. The remaining three sections are more technical, and present extended and detailed development of three of the issues that are included in Section 2. Section 3 covers a general mathematical formula for linear test equating that can be derived either from a model involving the conditional expectation of ability given observed score or by maximization of a particular correlation coefficient. Section 4 derives and illustrates a new method for determining differential scores for the different responses to multiple-choice items, for the case where no criterion variable is available and where omitted items may occur; such a method might produce several benefits, including sharply reduced bunching of test scores (which would make test equating easier). Section 5 deals with some anomalies concerning certain definitions of fairness, including fairness in test equating.

201

2. Overall Discussion of General Issues Pertaining to Equating

The following comments cover many general issues that are related to test equating in greater or lesser degree. In some cases, the comments are somewhat speculative in nature and are intended to suggest possible avenues of approach rather than to provide firm conclusions. In certain other cases, the points are rather obvious or already well established, but are listed anyway for sake of completeness.

2.1. Considerations Concerning Test Construction

The way that tests are constructed can have a substantial effect on how successfully they can be equated.

(a)   It is easier to try to be careful in constructing tests than it is to try to compensate for poor test construction afterward, at the time that equating is being done.

(b)   If two tests (or test forms) are to be equated, equating will generally be more accurate if average item difficulty and distribution of item difficulties do not differ greatly from one test to the other.

(c)   If tests are too easy for high-ability examinees, equating at the upper end of the score range will tend to be of unsatisfactory accuracy. In addition, though, discrimination among high-ability examinees will not be good even for each test individually.

(d)   Some tests contain small or large groups of linked items, i.e., items which are dependent on one another or on a common passage. The use of such groups of linked items restricts the options available to the test constructor; such reduction in flexibility makes it harder to meet test specifications, and thus may indirectly cause difficulties in equating later on. One might also question whether linked items would have other harmful effects, such as lower reliability and validity (stemming from positive correlation that could be expected to exist among the error scores for linked items) or a violation of an assumption of statistical independence required for certain models (e.g., latent-trait models). It may be wise to avoid the use of linked items wherever possible.

(e)   For some models (such as latent-trait models) that may be used in equating, omitted items may cause troubles, particularly in cases where examinees omit items because they have insufficient time to complete the test. This provides an argument in favor of going more toward power tests and away from speed tests (by allowing examinees greater average time per item). Actually, there may be a number of different advantages that would result from reducing the percentage of items that examinees omit because of insufficient time.

(f) If a test includes a nonoperational section that is to be used as an anchor test for equating, everything possible should be done to ensure that examinees will not be able to figure out which section is the nonoperational section.

2.2. General Formula for Linear Equating

Let x and y be two tests (or test forms) that are to be equated. For test x, let X denote an observed score, and suppose that, for a specified group or population of examinees, μ_x denotes mean of observed scores, σ_x^2 denotes variance of observed scores, $\sigma_{\tilde{x}}^2$ denotes variance of true scores, and $\rho_{xx} = \sigma_{\tilde{x}}^2/\sigma_x^2$ denotes reliability. Let Y, μ_y, σ_y^2, $\sigma_{\tilde{y}}^2$, and $\rho_{yy} = \sigma_{\tilde{y}}^2/\sigma_y^2$ have similar meanings for test y, where the parameters relate to the same specified population of examinees as for test x. To avoid extraneous complications, assume that the different parameters that were just defined are known (or can be estimated from available data with a high degree of accuracy).

If tests x and y have different reliabilities ($\rho_{xx} \neq \rho_{yy}$), they cannot be "equated" in the strict sense in which this word is often used: Since one test discriminates better than the other, there exists no transformation that can render scores on the two tests fully equivalent. Nevertheless, there may still be a practical need to "equate" tests x and y in some fashion even though $\rho_{xx} \neq \rho_{yy}$. For example, if x and y differ in length but are otherwise parallel, this alone would cause their reliabilities to differ. It is thus appropriate to try to find the least unsatisfactory solution to the equating problem for the case where reliabilities are unequal.

For the case where reliabilities may differ, a standard linear equating (or calibration) formula (see, e.g., Angoff, 1971, p. 571) is

$$(Y - \mu_y)/\sigma_{\tilde{y}} = (X - \mu_x)/\sigma_{\tilde{x}}. \tag{1}$$

Formula (1) is the same thing as

$$Y = \mu_y + \frac{\sigma_y \, \rho_{yy}^{1/2}}{\sigma_x \, \rho_{xx}^{1/2}} (X - \mu_x). \tag{2}$$

Section 3 develops an alternative linear equating formula,

$$Y = \mu_y + \frac{\sigma_y \, \rho_{xx}^{1/2}}{\sigma_x \, \rho_{yy}^{1/2}} (X - \mu_x), \tag{3}$$

which is the same as (2) except that the two reliability coefficients are traded. Formulas (2) and (3) will be equivalent if and only if the two reliabilities are equal.

Actually, formula (3) is a special case of more general linear formulas that are developed in Section 3. At the first level of generality, Section 3 considers the case of two tests x and y whose true scores are not necessarily perfectly correlated. Formula (3) is what results when this correlation is perfect, but it may also result in another way even when the correlation is less than perfect. At the second level of generality, Section 3 considers the case where there are three or more tests whose true scores are not necessarily perfectly correlated. Again, formula (3) is what results under perfect correlation of the true scores as well as under certain other conditions.

Section 3 derives the generalized versions of formula (3) in two different ways. In both derivations, ability (θ) is defined as a weighted average of the true scores from the different tests (where each true score is first transformed linearly so that it has mean 0 and variance 1). Then, in the first derivation, the equating formula is obtained by setting equal to each other the expressions for conditional expectation of ability (θ) given observed score on a test. It is assumed that, for each test, this conditional expectation is a linear function of the observed score on the test.

The second derivation starts by assuming that the equating among the tests is linear. An overall "test" (w) is defined in which each examinee takes one of the tests being equated, according to certain probabilities. The linear equating constants are then chosen so as to maximize the correlation between ability (θ, defined as before) and equated score from the "test" w. The equating formulas turn out to be the same as in the first derivation.

2.3. Scoring

The way that a test is scored can affect reliability and validity, but this is not all. In particular, the scoring scheme may have an impact on how accurately equating can be done. More specifically, if the total possible number of distinct scores is rather small, then there will naturally be heavy bunching of the test scores (i.e., there will be substantial "discreteness," as it is sometimes expressed), and heavy bunching will make equating more difficult.

Suppose, e.g., that there are six examinees who obtain the highest recorded score (not necessarily a perfect score) on test x, and likewise six examinees who receive the highest attained score on test y. Then, unless the highest scores on the two tests were exactly equated, any equating procedure would indicate that the six best examinees (from the two groups combined) are either all from the group that took test x or else all from the group that took test y—hardly a reasonable result. Of course, somewhat similar situations can occur at other points in the score distributions, where bunching may even be much heavier.

Certain difficulties involving smoothing are encountered in equipercentile equating. It would appear that these difficulties would be lessened if bunching of the test scores were reduced, since less bunching would mean a smoother score distribution to begin with.

Several possible approaches to scoring that would bring about reduced bunching of test scores will now be mentioned. The favorable impact on equating (resulting from less bunching) would probably not be the only benefit from some or all of these approaches; other possible benefits (perhaps even more important) might include higher reliability, higher validity, and lower test bias, although such additional benefits would not be universally ensured.

(a)   First, it should be noted that formula scoring can be expected to produce less bunching of test scores than number-right scoring, since the number of possible scores will be much greater with formula scoring. However, this is certainly not the most important reason for preferring formula scoring over number-right scoring.

(b)   The advantage just indicated for formula scoring will not be realized if formula scores are rounded to the nearest whole number. There would appear to be no justification for such rounding: All fractions should be retained in formula scoring.

(c)   Scoring based on latent-trait theory will cause bunching to be reduced. Of course, latent-trait theory can also provide a means for equating.

(d)   There would be less bunching of scores if examinees were allowed to mark more than one answer to an item. An item to which two or more responses were marked could then receive the average score of the marked responses. Under formula scoring for an item with five choices, e.g., an examinee who marked two of the choices including the correct one would receive a score of $\frac{1}{2}[1 + (-\frac{1}{4})] = \frac{3}{8}$, whereas an examinee who marked three or four choices including the correct one would score $\frac{1}{6}$ or $\frac{1}{16}$, respectively. (In any case where the correct choice was unmarked but one or more incorrect choices were marked, the score would be $-\frac{1}{4}$.) Under scoring schemes other than conventional formula scoring, multiple marking of responses could be handled in a similar way.

In theory, the concept just described could be generalized by allowing for a *weighted* average score of the marked responses, with the examinee providing the weights. Thus in the preceding example, an examinee who attached a weight of 0.7 to one response and a weight of 0.3 to another response would receive a score of $0.7(1) + 0.3(-\frac{1}{4}) = 0.625$ if the first response were the correct one. In practice, however, such a procedure would probably be far too complex for most tests.

(e)   There is no reason why all items need to receive the same weight in scoring. Different items could receive different scoring weights, which could be chosen in such a way as to try to maximize reliability or validity, or in some other fashion. Of course, the distribution of test scores would be considerably less bunched if each item had a different scoring weight.

(f)   Finally, there is no reason why each wrong response has to receive the same scoring weight ($-\frac{1}{4}$ in the case of conventional formula scoring for an item with five responses). Under a scoring scheme in which different wrong responses to an item receive different scoring weights, the different items in the test may each carry the same weight in some sense, or they may not. Section 4 will derive and illustrate the calculations for a scoring scheme for which different responses to an item receive different weights and for which (in addition) the different items are not weighted the same in any sense. The complete set of weights is chosen in such a way as to maximize the value of coefficient α (see, e.g., Lord and Novick, 1968, Section 4.4), which is, among other things, a lower bound on the reliability coefficient. Omitted items are handled in a special way.

A scoring scheme that assigns different weights to different incorrect responses will naturally result in less bunching of the test scores, although this is not generally the primary purpose for which it is intended. Previous empirical studies of such scoring schemes as applied to live test data (see, e.g., Cross and Frary, 1978; Thissen, 1976; and references contained therein) have reached mixed conclusions concerning what effects the schemes had on reliability and validity. However, none of these studies utilized a scoring scheme identical to that proposed in Section 4.

How well the method described in Section 4 would work in practice can be determined only by applying it to live data. Since the method tries to optimize reliability, one has to be on guard against the possibility that higher reliability may be achieved at the expense of lower validity. If there appear to be problems concerning validity, however, a remedy might be obtained by applying the method separately to different subgroups of items within a test, rather than to the test as a whole.

If adequate data are available on a suitable criterion variable and it is desired to use a scoring scheme with differential weighting, then any method that determines the weights so as to try to maximize reliability would probably not be the most desirable method in the first place. Instead, it would seem best to choose the scoring weights so as to maximize the validity with respect to the available criterion variable. This can be done through a simple application of multiple regression (see, e.g., Potthoff, 1966, p. 65).

In the construction of a test that is to be scored with differential weights for the different incorrect responses, it will be desirable to try to select these distractors in such a way that different distractors will tend to attract examinees of differing ability. In fact, with a test for which this was not done, one probably would not expect to approach the full realization of the potential benefits of differential weighting of distractors.

In some methods for differential scoring of distractors, including the method of Section 4, it is mathematically possible for a distractor to receive a higher score than the "correct" response. However, one might expect such occurrences to be confined mainly to poorly constructed items or to items for which there might be some debate as to which response is really "correct" in the first place.

Some of the scoring techniques that have been discussed here in Subsection 2.3 are rather complicated, both mathematically and computationally. Thanks to modern computers, however, scoring methods that would have been out of the question 25 years ago may be fully feasible today. As a general matter, it may be desirable to give serious consideration to scoring methods that are more sophisticated than the traditional ones.

2.4. Omitted Items and Guessing

Different scoring models and mathematical models of testing have been investigated for their possible use in connection with either basic or tangential aspects of the test equating problem. Unfortunately, many of these models assume a much simpler form if they ignore or inadequately address the complexities associated with omitted items and guessing. It would appear that serious difficulties from both ethical and practical standpoints would result from failure to give proper treatment to the matter of omitted items and guessing. Some of the relevant considerations are as follows.

(a)   The issue of number-right scoring versus formula scoring is a familiar one. As a matter of principle, I can see no justification for ever using number-right scoring rather than formula scoring on any multiple-choice test: Number-right scoring imposes a severe and improper penalty on those examinees who (for whatever reason) commit the simple offense of leaving items blank. In addition, number-right scoring increases the variance of error scores (and thus lowers reliability) to the extent that examinees mark answers randomly under compulsion.

(b)   More generally, any legitimate scoring system should observe the rule that an examinee who omits an item can expect to obtain a score on that item that is no worse than the average score of all the responses, i.e., no worse than the expected score that would result from randomly marking a response to the item.

(c)   The basis for this rule of (b) is simply as follows. First, most persons would probably consider it unethical *not* to inform the examinee that there is a penalty for leaving an item blank. Furthermore, if examinees were not informed of this penalty, probably some of them would find out about it anyway and some would not, in which case the test would not be measuring solely what it was supposed to measure but rather would also be measuring knowledge or ignorance of the penalty. (Examinees who knew about the penalty and its implications would presumably try to avoid omitting any items.)

Second, suppose that examinees *are* informed of the penalty for leaving an item blank. Then if there are still some examinees who leave some items blank, the test will again be measuring something besides what it is supposed to measure. Furthermore, since examinees are, in effect, forced to guess or mark at random, reliability will be reduced. Thus the penalty results in harm with no offsetting benefit.

For some purposes, it might be helpful to think of a test in terms of game theory, and to think about examinees' strategies in response to various possible approaches of the test-maker. It would not be surprising to find that an approach which is ethically somewhat questionable would also be rejected by game-theory considerations.

(d)   Conventional number-right scoring would clearly be barred by the rule given in (b). Also barred would be a somewhat more general scoring system, one in which the right responses to the various items would receive positive scoring weights that could differ from one another, but in which zero scores would still be assigned to all wrong responses and omitted items alike.

(e)   Certain scoring schemes that score the different distractors differently also provide for one additional scoring category that is to be applied if the item is omitted (e.g., see Cross and Frary, 1978, pp. 614–615, for a discussion of studies that have used this additional category). Thus for a five-choice item under such schemes there would be six scoring categories, one for each of the five choices plus one to be utilized when no choice is marked. Particularly in cases where the "omit" category produces a lower score than the average of the remaining categories for most or all items, it can be argued that this sort of scoring scheme essentially violates the rule given in (b). If the "omit" categories do not produce such poor scores overall, though, the issue becomes a bit fuzzier. In order to steer clear of various potential troubles, however, it would appear best to avoid using the separate "omit" category. In some situations a suitable alternative would be simply to set the "omit" score equal to the average score for the remaining responses; this is what is done in Section 4.

(f)   The matter of omitted items and guessing has been particularly troublesome with latent-trait models, but one recent development (Lord, 1974) has provided a substantial breakthrough. However, the technique of Lord (1974) makes use of a distinction between "omitted" and "not-reached" items: Those items after the last marked item are considered to be "not-reached." What possibility is there that this attempted distinction could lead to difficulties in connection with examinees who do not work the test items in sequence or who attempt but omit the very last one or more items? In applying the technique of Lord (1974), it might be best simply to drop this distinction and to treat *all* unmarked items as though they were attempted but omitted. Such a modification may be particularly suitable if a test is timed so that most examinees can comfortably finish it.

(g)   It is somewhat disconcerting that (see, e.g., Hambleton and Cook, 1977, p. 82) the "guessing" parameter in the three-parameter latent-trait model typically turns out to be less than the chance value of $\frac{1}{5}$ (where there are five choices). Might this imply that the test-maker is obligated, for ethical reasons or otherwise, to warn that an examinee who knows almost nothing about a question should omit (assuming formula scoring or other appropriate scoring) rather than try to guess? If so, this could change examinees' responses, and bring about more omissions.

2.5.   Linear Equating, and Possible Alternatives If It Does Not Work Well

It would be helpful if one could always equate linearly, since linear equating has the advantages of simplicity and objectivity. Unfortunately, however, there may be cases where linear equating does not provide a satisfactory fit to the data. Some observations concerning linear equating and possible alternatives to it follow.

(a)   The way that tests are scored could affect how well linear equating works. Subsection 2.3 discussed several scoring methods that bring about less bunching of test scores. Not only will such methods spread out the scores, but they might also tend to result in a better fit for linear equating formulas.

(b)   Since it is not always obvious whether or not linear equating provides a suitable fit, it would be desirable to have an objective means to try to determine this. In particular, if two tests are equated using linear equating with the assumption that the two reliabilities are equal, one might wish to test whether the two score distributions after equating are the same except for chance differences. The Kolmogorov–Smirnov test should provide a good means for testing this hypothesis. (The test will be slightly inaccurate—probably on the side of rejecting the null hypothesis too infrequently—because the data are used to estimate the two equating parameters.) If the null hypothesis cannot be rejected, there is probably no need to consider any type of nonlinear equating.

If two tests are unequal in reliability and if they are equated using either formula (2) or formula (3), then the two score distributions after equating will *not* be the same. Thus it would not be meaningful to run the Kolmogorov–Smirnov test in this case.

(c)   An important alternative to linear equating is equipercentile equating. My attitude toward equipercentile equating is less than fully enthusiastic, since I suspect that in many cases its advantages may be more cosmetic than real. There are certain possible problems with equipercentile equating:

(i)   Equipercentile equating customarily involves the use of subjective smoothing. This is a definite disadvantage. However, this disadvantage of subjectivity would be removed if an objective procedure for equipercentile equating could be found, accepted, and used. As noted earlier, smoothing problems with equipercentile equating will be reduced by the use of scoring methods which result in less bunching of test scores; on the other hand, though, such scoring methods may at the same time reduce the need for equipercentile equating in the first place, if they happen to bring about a better fit of the linear equating formula.

(ii)   Equipercentile equating procedures may be especially sensitive to sampling variability, particularly if sample sizes are small. However, this becomes less of a problem as sample sizes become larger.

(iii)   For the case where two tests differ in reliability, equipercentile equating is inconsistent with either of the formulas [(2) and (3)] that have been advanced for linear equating, since either one will produce two score distributions after equating that are *not* the same. This provides a strong conceptual objection to equipercentile equating.

(d)   Equating through latent-trait theory constitutes another important alternative to linear equating. It appears to me that latent-trait theory offers promise and should be pursued further, but that certain difficulties mentioned below should be resolved if possible.

(i)   Since the one-parameter and two-parameter latent-trait models seem to be inadequate, attention should be concentrated on the three-parameter model.

(ii)   Unfortunately, the three-parameter latent-trait model is complicated both mathematically and computationally. It would be desirable to try to resolve outstanding questions concerning the maximum-likelihood estimates of the parameters and the solution of the maximum-likelihood equations.

(iii)   As noted earlier, omitted items and related matters cause certain problems with latent-trait models. It may be possible to find improved ways of attacking these problems.

(iv)   Latent-trait models could be expected to provide a partial but not a complete solution to the problem of equating two tests with unequal reliabilities. There will not always be a cure for an unreliable test.

(e)   If linear equating or any other type of equating is known to be less than adequate, certain partial remedies can be applied in the reporting of the scores:

(i)   The identification of the test or form could be reported along with the equated score. A warning could then be included to the effect that equated scores from different tests or forms are not comparable to quite the full degree that one might like them to be.

(ii)   The report of a score could include not only the equated score itself, but also an appropriate type of confidence interval around this score. In case two tests had unequal reliabilities, the widths of their confidence intervals would also be different. If equating were based on formula (3), the 95% confidence interval could take the form

$$\rho_{xx}^{1/2}\left[(X - \mu_x)/\sigma_x\right] \pm 1.96\,(1 - \rho_{xx})^{1/2} \tag{4}$$

for test x, with an analogous formula for test y. For derivation of (4), see Subsection 3.5, which also provides some generalizations.

(f)   In some cases, linear equating may work well except at one or both ends of the score distribution. Problems at the ends might occur if, e.g., one or both tests do not have enough hard (or easy) items to discriminate well at the upper (or lower) end. Linear equating (and possibly other types of equating as well, for that matter) may then give good results in the middle but poor results at the ends. If such troubles cannot be largely avoided by widening the range of difficulty of the test items, another possible remedy consists in the various types of available schemes for giving tests of differing difficulty to different examinees, such that examinees of higher (lower) ability will take harder (easier) tests.

2.6.   *Estimation of Equating Parameters*

For certain situations it may be fruitful to investigate the possibility of using more sophisticated estimation techniques to estimate the parameters for linear equating.

(a)   In the equating of successive generations of tests as depicted on associated genealogical charts, it is apparently customary to equate a new form with only one other form in any single equating calculation. (A new form may actually be equated to *two* different previous forms, but in separate calculations whose results are then averaged.) In principle, one would think it ought to be possible to improve accuracy by somehow equating several different test forms simultaneously in a single equating calculation. Of course, details would have to be worked out concerning design of test administration, mathematical models, estimation of parameters (probably through maximum-likelihood estimation), and computation techniques. Also, it is certainly possible that the amount of computation would turn out to be prohibitively burdensome.

(b)   In those equating problems where two forms being equated are given to two separate groups and where furthermore both groups take a common equating test v, one might wish to consider an equating method that utilizes scores on component subtests within test v (or possibly even the individual item responses) rather than just the total score on test v. Theoretically, this ought to be able to improve the accuracy of equating, because more information would be utilized. However, any actual improvement might turn out to be minor. Also, the mathematical models and the computations (which would probably utilize maximum-likelihood estimation) would be more complex.

2.7. *The Matter of Fairness*

A certain degree of "unfairness" is inherent in any test, and is unavoidable. In particular, this is true in the sense that one might reasonably argue that a test is not "fair" to any examinee whose error score is negative.

There are other respects or situations in which it might be claimed that a test is not "fair." For example, suppose that examinees are divided into two groups in some nonrandom fashion, with each group receiving a different form of a test. If the two forms differ in reliability, then it is true that the test is not "fair" to the high-ability examinees who take the less reliable form, in the sense that these examinees have less chance to demonstrate their high ability levels.

It might then be further argued that equating through formula (3) above would be "unfair" to these same examinees, for much the same reason.

Now suppose that each examinee had been *randomly* assigned to one of the two forms or groups. Would there then be as strong a case for arguing that the test was "unfair" to the high-ability examinees who took the less reliable form? This is but one of many difficult questions that can crop up when issues involving fairness are considered.

For simplicity let us confine ourselves now to situations where the ability distributions of the two groups are the same. It might then be contended that a "fair" formula for equating the two unequally reliable test forms would be one for which the proportion of examinees receiving equated scores above any given value would be the same for each group. Both (2) and (3) would violate this requirement (in opposite directions). However, the equating formula

$$Y = \mu_y + (\sigma_y/\sigma_x)(X - \mu_x) \tag{5}$$

would satisfy the requirement and would also be equivalent to equipercentile equating. (This assumes that both score distributions are normal, or are otherwise of the same shape.) Formula (5) is "in between" (2) and (3).

Section 5 will consider what happens when the requirement of equal proportions that has just been described is imposed along with one additional requirement. The outcome is that, in order to meet the "fairness" requirements, it is apparently necessary to deliberately add a random-error term to the observed score of each examinee who took the more reliable test.

Section 5 will also consider another situation that concerns test fairness. This second illustration pertains not to equating, but to the test-bias question; however, there are interesting similarities to the first illustration. Specifically, the implications of a requirement for "fairness" (absence of bias) proposed in

Thorndike (1971) are examined, under a certain extended interpretation of this requirement. It again turns out to be necessary to deliberately add random-error terms to examinees' scores, in order to satisfy this extended "fairness" requirement.

In these two cases, "fairness" requirements that may initially sound entirely reasonable lead to conclusions that are difficult to defend. These results may simply reflect the great complexities of the fairness issue, and the hazards of oversimplifying it.

In view of the nebulousness that thus exists with respect to the overall issue of fairness, it does not appear that there is anything plainly "unfair" about formula (3). In fact, it is reasonable to argue that (3) is simply the least undesirable of the different possible formulas.

In the context of the test-bias question, Cole (1973) discusses six different definitions of "fairness." [The fifth definition is that of Thorndike (1971).] One might wonder at first whether the second, fifth, and sixth definitions of Cole (1973) would correspond, respectively, with the equations (3), (5), and (2) that have been given here. There are certainly some similarities, but the correspondence is not exact.

As a matter of general principle, even though some attempts to define "fairness" in testing may be seriously questioned, it would be a mistake to dismiss too quickly some of the underlying concerns that have motivated these attempts. For example, almost all controversies about test "fairness" have their roots in the fact that any test in the real world is neither perfectly reliable nor perfectly valid. How low would the reliability or validity of a test have to be before it would be considered improper or "unfair" to use the test? This is a question that as yet has no satisfactory answer.

As a final comment on fairness in testing, virtually all treatments of the subject consider it only within the context of a single test that is utilized for selection by a single educational institution or employer. In reality, a number of tests and/or other selection instruments will generally be applied to an individual by educational institutions and/or employers over a period of years during which the individual is considering different options concerning education and employment. An adverse or "unfair" result from one test or other selection instrument may be offset by an opposite result from another, but there will not be an exact balancing for everyone. A more realistic but far more difficult approach to the issue of fairness in tests and other selection instruments would take account of the broad picture just described, rather than restrict itself to the situation of a single test utilized by a single selecting organization. Furthermore, a more general approach of this type could probably not be confined within the realm of test theory and related statistical tools. In particular, the added perspective of economists specializing in labor economics might contribute significantly to the approach.

The remaining three sections of this paper are, of course, more mathematical than Section 2. At this juncture it should also be pointed out that there will be some overlap of the notational symbols that are used. In other words, if the same symbol or letter is used in two different developments, it will not necessarily have the same meaning. The notation of Section 3 is consistent with that of Section 2. However, Section 4 has its own notation. In Section 5, the notation of Subsection 5.1 is consistent with that of Sections 2 and 3, but Subsection 5.2 has its own notation.

3.   Derivations of General Formulas for Linear Equating

Presented here are the linear-equating derivations described in Subsection 2.2. These are derivations of generalized versions of formula (3). Subsections 3.1 and 3.2 will treat the case where just two tests are being equated, whereas Subsections 3.3 and 3.4 deal with the case where there may be more than two tests. (The only reason for giving separate derivations for the case of two tests is to make the derivations easier to follow.) Subsections 3.1 and 3.3 present the derivations based on the conditional-expectation model, whereas the derivations of Subsections 3.2 and 3.4 are based on the maximum-correlation model.

Finally, Subsection 3.5 will derive a general confidence-interval formula for use with linearly equated scores. Formula (4) (see Subsection 2.5) is a special case of it.

3.1.   Conditional-Expectation Derivation for Two Tests

Part of the notation required for the linear-equating derivation of this subsection was already introduced in Subsection 2.2. Additional notation is as follows. For test x let $\tilde{X}$ denote a true score and let $\mu_{\tilde{x}}$ denote mean of true scores; of course, $\mu_{\tilde{x}} = \mu_x$. Then define the variable

$$\theta_x = (\tilde{X} - \mu_{\tilde{x}})/\sigma_{\tilde{x}},$$

which has mean 0 and variance 1. Let e_x denote an error score for test x. The mean of e_x is 0; let σ_{ex}^2 denote its variance.

The analogous new notation for test y consists of $\tilde{Y}$, $\mu_{\tilde{y}}$, θ_y, e_y, and σ_{ey}^2.

Remember that all parameters (both those already introduced and those yet to be introduced) relate to a specified population of examinees—the same population for test x as for test y. Remember also that these parameters are all assumed to be known.

Since observed score is the sum of true score and error score, one can write

$$X = \tilde{X} + e_x = \mu_x + \sigma_{\tilde{x}}\theta_x + e_x, \qquad Y = \tilde{Y} + e_y = \mu_y + \sigma_{\tilde{y}}\theta_y + e_y. \quad (6)$$

Since true score and error score are assumed independent, it follows that

$$\sigma_x^2 = \sigma_{\tilde{x}}^2 + \sigma_{ex}^2, \qquad \sigma_y^2 = \sigma_{\tilde{y}}^2 + \sigma_{ey}^2, \tag{7}$$

which is another standard result.

Now let

$$\theta = P\theta_x + Q\theta_y \tag{8}$$

be the ability variable that it is desired to predict or estimate, where P and $Q = 1 - P$ represent weights whose values will be left open for the time being. If μ_θ and σ_θ^2 denote, respectively, the mean and variance of θ and if R denoted the correlation coefficient between θ_x and θ_y, then it follows from (8) that

$$\mu_\theta = 0 \tag{9}$$

and

$$\sigma_\theta^2 = P^2 + Q^2 + 2PQR. \tag{10}$$

Let $\sigma_{\theta x}$ denote the covariance of θ and X, and let $\sigma_{\theta y}$ denote the covariance of θ and Y. From (6) and (8) it follows that

$$\sigma_{\theta x} = \sigma_{\tilde{x}}(P + QR), \qquad \sigma_{\theta y} = \sigma_{\tilde{y}}(PR + Q). \tag{11}$$

As was noted earlier, it is assumed that the conditional expectation $E(\theta|X)$ is a linear function of X and that, similarly, $E(\theta|Y)$ is a linear function of Y. From this assumption, it easily follows (see, e.g., Cramér, 1951, pp. 273–274) that

$$E(\theta|X) = \mu_\theta + \frac{\sigma_{\theta x}}{\sigma_x^2}(X - \mu_x), \qquad E(\theta|Y) = \mu_\theta + \frac{\sigma_{\theta y}}{\sigma_y^2}(Y - \mu_y). \tag{12}$$

If (9) and (11) are substituted into (12), the result is

$$E(\theta|X) = (P + QR)\rho_{xx}^{1/2}\left(\frac{X - \mu_x}{\sigma_x}\right),$$

$$E(\theta|Y) = (PR + Q)\rho_{yy}^{1/2}\left(\frac{Y - \mu_y}{\sigma_y}\right). \tag{13}$$

Now $E(\theta|X)$ represents the best prediction or estimate of ability (θ) based on X, and $E(\theta|Y)$ similarly represents the best prediction of ability based on Y. It is reasonable to argue that two scores X and Y will be "equivalent" (at least to the extent that it is even possible to achieve "equivalence") if they give the same predicted value of θ. If this proposition is accepted, then it follows at once that the proper equating formula is

$$E(\theta|X) = E(\theta|Y), \tag{14}$$

which, upon substitution from (13), becomes

$$(P + QR)\rho_{xx}^{1/2}\left(\frac{X - \mu_x}{\sigma_x}\right) = (PR + Q)\rho_{yy}^{1/2}\left(\frac{Y - \mu_y}{\sigma_y}\right). \tag{15}$$

Of course, the "equivalence" which is achieved here is imperfect, in the sense that X and Y will not generally predict θ with the same accuracy. The difference in accuracy causes some problems (see Subsection 2.7). Nevertheless, there appears to exist no obvious alternative to (14) and (15) whose properties would be clearly more desirable.

In many cases where two forms or tests x and y are being equated, it will be reasonable to assume that θ_x and θ_y are the same variable, or virtually the same variable. If θ_x and θ_y are the same variable, then $R = 1$, and (15) reduces to (3).

Even if R is not 1, however, (15) will still reduce to (3) if $P = Q = \frac{1}{2}$, i.e., if the ability variable being predicted (θ) is considered to be just an unweighted average of θ_x and θ_y.

3.2. Maximum-Correlation Derivation for Two Tests

The second derivation of the general formula (15) uses the notation employed in Subsection 3.1 plus some additional notation that concerns a "test" w. This "test" w is defined in a fashion that is somewhat artificial but that also closely parallels what actually happens. In taking the "test" w, each examinee is considered to take test x with a probability p and test y with probability $q = 1 - p$. As indicated earlier, the equating formula linking tests x and y is assumed to be linear, i.e., of the form

$$Y = a + bX, \tag{16}$$

where a and b denote the equating parameters. Define a variable W (representing observed score on "test" w) to be $W = a + bX$ for an examinee who took test x and to be $W = Y$ for an examinee who took test y. Application of (6) then gives

$$W = a + b\mu_x + b\sigma_{\bar{x}}\theta_x + be_x \quad \text{if examinee took test } x \text{ (probability } p\text{)},$$
$$= \mu_y + \sigma_{\bar{y}}\theta_y + e_y \qquad \text{if examinee took test } y \text{ (probability } q\text{)}. \tag{17}$$

Now let $\rho_{\theta w}$ denote the correlation coefficient between the variables θ (8) and W (17). If $\sigma_{\theta w}$ denotes the covariance of θ and W and if σ_w^2 denotes the variance of W, then

$$\rho_{\theta w}^2 = \sigma_{\theta w}^2 / \sigma_\theta^2 \sigma_w^2. \tag{18}$$

The formula for σ_θ^2 has already been obtained (10). From (8), (9), and (17) it follows that

$$\sigma_{\theta w} = pb\sigma_{\bar{x}}(P + QR) + q\sigma_{\bar{y}}(PR + Q). \tag{19}$$

To find the formula for σ_w^2, note first from (17) that

$$E(W) = p(a + b\mu_x) + q\mu_y,$$

and from (17) and (7) that

$$E(W^2) = p(a + b\mu_x)^2 + pb^2\sigma_x^2 + q\mu_y^2 + q\sigma_y^2.$$

Then

$$\sigma_w^2 = E(W^2) - [E(W)]^2 = pb^2\sigma_x^2 + q\sigma_y^2 + pq(a + b\mu_x - \mu_y)^2. \tag{20}$$

Now suppose that the equating parameters a and b are chosen in such a way as to maximize (18), the square of the correlation coefficient between ability and the "test" score W. This would appear to be a reasonable way to select a and b. It will turn out that, when the resulting values of a and b are substituted into the equating formula (16), the latter will become equivalent to (15), the equating formula derived in Subsection 3.1.

To prove this, note first that the parameter a appears in (20) but not in (19) or (10). Hence this parameter needs to be chosen so as to minimize (20), which implies that

$$a = \mu_y - b\mu_x. \tag{21}$$

Combining (18), (19), (20), and (21) then gives

$$\rho_{\theta w}^2 = \frac{[bp\sigma_{\bar{x}}(P + QR) + q\sigma_{\bar{y}}(PR + Q)]^2}{(b^2p\sigma_x^2 + q\sigma_y^2)\sigma_\theta^2}. \tag{22}$$

Note that σ_θ^2 (10) does not depend on b. It now remains only to select b so as to maximize (22).

Let C, D, E, and F denote positive constants, and consider a function of the form

$$\varphi(b) = (bC + D)^2/(b^2E + F).$$

Define $b^* = CF/DE$. It is easily shown that $\varphi(b)$ is maximized with respect to b at $b = b^*$ and that

$$\varphi(b^*) = (C^2/E) + (D^2/F).$$

It thus follows at once that $\rho_{\theta w}^2$ (22) will be maximized with respect to b if

$$b = \frac{p\sigma_{\bar{x}}(P + QR)q\sigma_y^2\sigma_\theta^2}{q\sigma_{\bar{y}}(PR + Q)p\sigma_x^2\sigma_\theta^2} = \frac{(P + QR)\rho_{xx}^{1/2}/\sigma_x}{(PR + Q)\rho_{yy}^{1/2}/\sigma_y} \tag{23}$$

and that, for this value of b,

$$\rho_{\theta w}^2 = \frac{p\rho_{xx}(P + QR)^2 + q\rho_{yy}(PR + Q)^2}{P^2 + Q^2 + 2PQR}, \tag{24}$$

where the denominator of the right side of (24) is the expression for σ_θ^2 from (10). Finally, if a (21) and b (23) are substituted into the linear equating relation (16), the result will be equivalent to (15). Note that the formulas for a, b are not dependent on what p and q are.

In the case where $R = 1$, (24) reduces to

$$\rho_{\theta w}^2 = p\rho_{xx} + q\rho_{yy},$$

which is a weighted average of the two reliability coefficients. Especially for the case where $R = 1$, $\rho_{\theta w}^2$ could be interpreted more or less as a quasi-reliability-coefficient of the "test" w.

3.3.   Conditional-Expectation Derivation for Any Number of Tests

This section generalizes Subsection 3.1. Let there be k forms or tests $x_1, x_2, \ldots, x_k$ that are to be equated. For test x_h ($h = 1, 2, \ldots, k$), use X_h to denote an observed score, $\tilde{X}_h$ to denote a true score, e_h to denote an error score, μ_h to denote mean of observed scores, $\mu_{\tilde{h}}(= \mu_h)$ to denote mean of true scores, $\sigma_{\tilde{h}}^2$ to denote variance of true scores, σ_{eh}^2 to denote variance of error scores, $\sigma_h^2(= \sigma_{\tilde{h}}^2 + \sigma_{eh}^2)$ to denote variance of observed scores, and $\rho_{hh}(= \sigma_{\tilde{h}}^2/\sigma_h^2)$ to denote reliability. Define

$$\theta_h = (\tilde{X}_h - \mu_{\tilde{h}})/\sigma_{\tilde{h}},$$

which has mean 0 and variance 1. Then

$$X_h = \tilde{X}_h + e_h = \mu_h + \sigma_{\tilde{h}}\theta_h + e_h. \tag{25}$$

The conditional-expectation derivation for the case of k tests follows the same lines as the derivation of Subsection 3.1. Let

$$\theta = \sum_{h=1}^{k} P_h\theta_h \tag{26}$$

denote the ability variable to be predicted or estimated, where the P_h's are weights such that

$$\sum_{h=1}^{k} P_h = 1. \tag{27}$$

The mean and variance of θ are, respectively, $\mu_\theta = 0$ and

$$\sigma_\theta^2 = \sum_{h=1}^{k} \sum_{H=1}^{k} P_h P_H R_{hH}, \qquad (28)$$

where R_{hH} denotes the correlation coefficient between θ_h and θ_H. If $\sigma_{\theta h}$ denotes the covariance of θ and X_h, then it follows from (25) and (26) that

$$\sigma_{\theta h} = \sigma_{\tilde{h}} \sum_{H=1}^{k} P_H R_{hH}. \qquad (29)$$

It may be noted in passing that, for $h \neq H$,

$$R_{hH} = R_{hH}^* / \rho_{hh}^{1/2} \rho_{HH}^{1/2},$$

where R_{hH}^* denotes the correlation coefficient between X_h and X_H. This formula might be of some help if it is ever desired to estimate the R_{hH}'s.

The assumption is made that $E(\theta|X_h)$ is a linear function of X_h. It then follows that

$$E(\theta|X_h) = \mu_\theta + \frac{\sigma_{\theta h}}{\sigma_h^2}(X_h - \mu_h),$$

which becomes

$$E(\theta|X_h) = \left(\sum_{H=1}^{k} P_H R_{hH}\right) \rho_{hh}^{1/2} \left(\frac{X_h - \mu_h}{\sigma_h}\right) \qquad (30)$$

upon application of (29). Thus (30) would be used as the equated score corresponding to an observed score of X_h on test h, for $h = 1, 2, \ldots, k$. Of course, (30) is a generalization of (13).

If the k different θ_h's are all the same variable, then every R_{hH} will be 1. For this case, it is clear that equating based on (30) becomes identical with equating based on (3), for any pair of tests.

Even if some R_{hH}'s are not 1, however, equating based on (30) will still reduce to equating based on (3) for every pair of tests if

$$\sum_{H=1}^{k} P_H R_{hH} = \lambda \qquad (h = 1, 2, \ldots, k) \qquad (31)$$

holds [along with (27)]. Here λ denotes a constant and, because of (28), its value must be equal to σ_θ^2.

Observe that the P_h's will obey (31) and (27) if they are chosen in such a way as to minimize σ_θ^2 (28) subject to (27); this result can easily be established through differentiation with the use of a Lagrange multiplier. Thus if θ (26) is chosen so that its variance is minimized, then the equating formula of the form (3) will be applicable for any two of the k tests.

3.4. Maximum-Correlation Derivation for Any Number of Tests

By proceeding similarly to the development in Subsection 3.2, this subsection will furnish a second derivation of the k-test equating formula based on (30). The notation of Subsection 3.3 will be used, except that test x_k will be referred to as test y, and certain other analogous changes will be made: Y will be substituted for X_k, μ_y for μ_k, σ_y^2 for σ_k^2, $\sigma_{\tilde{y}}^2$ for $\sigma_{\tilde{k}}^2$, e_y for e_k, σ_{ey}^2 for σ_{ek}^2, θ_y for θ_k, and ρ_{yy} for ρ_{kk}.

Additionally, the "test" w is defined so that each examinee is considered to take test x_h with probability $p_h (h = 1, 2, \ldots, k - 1)$ and test y with probability $q = p_k$, where

$$q = 1 - \sum_{h=1}^{k-1} p_h.$$

The equating formula linking test x_h with test y is

$$Y = a_h + b_h X_h \qquad (h = 1, 2, \ldots, k - 1), \tag{32}$$

where a_h and b_h denote the equating parameters to be determined. The observed score on the "test" w is then defined to be $W = a_h + b_h X_h$ for an examinee who took test $x_h (h = 1, 2, \ldots, k - 1)$ and $W = Y$ for an examinee who took test y. Application of (25) now gives

$$
\begin{aligned}
W &= a_h + b_h \mu_h + b_h \sigma_{\tilde{h}} \theta_h + b_h e_h && \text{if examinee took test } x_h \text{ (probability } p_h), \\
&= \mu_y + \sigma_{\tilde{y}} \theta_y + e_y && \text{if examinee took test } y \text{ (probability } q);
\end{aligned}
\tag{33}
$$

the first of the two lines is for $h = 1, 2, \ldots, k - 1$.

Let $\rho_{\theta w}$, $\sigma_{\theta w}$, and σ_w^2 have meanings like what they had in Subsection 3.2. The objective will be to choose values for the a_h's and b_h's so as to maximize $\rho_{\theta w}^2 = \sigma_{\theta w}^2 / \sigma_\theta^2 \sigma_w^2$ [σ_θ^2 is given by (28)]. First, it follows from (26) and (33) that

$$\sigma_{\theta w} - \sum_{h=1}^{k-1} p_h b_h \sigma_h \sum_{H=1}^{k} P_H R_{hH} + q \sigma_{\tilde{y}} \sum_{H=1}^{k} P_H R_{kH}. \tag{34}$$

To obtain an expression for σ_w^2, start by noting from (33) that

$$E(W) = \sum_{h=1}^{k-1} p_h (a_h + b_h \mu_h) + q \mu_y$$

and

$$E(W^2) = \sum_{h=1}^{k-1} p_h \left[(a_h + b_h \mu_h)^2 + b_h^2 \sigma_h^2 \right] + q(\mu_y^2 + \sigma_y^2).$$

Then

$$\sigma_w^2 = E(W^2) - [E(W)]^2$$

$$= \sum_{h=1}^{k-1} p_h b_h^2 \sigma_h^2 + q\sigma_y^2 + \frac{1}{2} \sum_{h=1}^{k} \sum_{H=1}^{k} p_h p_H (M_h - M_H)^2, \quad (35)$$

where

$$M_h = a_h + b_h \mu_h \quad (h = 1, 2, \ldots, k - 1), \qquad M_k = \mu_y.$$

Since σ_w^2 (35) depends on the a_h's but $\sigma_{\theta w}$ (34) and σ_θ^2 (28) do not, the a_h's will need to be chosen so as to minimize (35). It is thus required that $M_1 = M_2 = \cdots = M_k$, or

$$a_h = \mu_y - b_h \mu_h \quad (h = 1, 2, \ldots, k - 1). \quad (36)$$

From (34), (35), and (36) one then obtains

$$\rho_{\theta w}^2 = \frac{\sigma_{\theta w}^2}{\sigma_\theta^2 \sigma_w^2} = \frac{(\sum_{h=1}^{k-1} b_h p_h \sigma_{\tilde{h}} \sum_{H=1}^{k} P_H R_{hH} + q\sigma_{\tilde{y}} \sum_{H=1}^{k} P_H R_{kH})^2}{(\sum_{h=1}^{k-1} b_h^2 p_h \sigma_h^2 + q\sigma_y^2)\sigma_\theta^2}, \quad (37)$$

which is to be maximized with respect to the b_h's.

It is not hard to prove (by mathematical induction on k, e.g.) that if C_h, D, E_h, and F are positive constants and $b_h^* - C_h F / D E_h$, then the function

$$\varphi(b_1, b_2, \ldots, b_{k-1}) = \left(\sum_{h=1}^{k-1} b_h C_h + D\right)^2 \Bigg/ \left(\sum_{h=1}^{k-1} b_h^2 E_h + F\right)$$

is maximized with respect to the b_h's at $b_h = b_h^*$ ($h = 1, 2, \ldots, k - 1$), where it assumes the value

$$\varphi(b_1^*, b_2^*, \ldots, b_{k-1}^*) = \sum_{h=1}^{k-1} (C_h^2 / E_h) + (D^2 / F).$$

It thus follows that $\rho_{\theta w}^2$ (37) will be maximized with respect to the b_h's if

$$b_h = \frac{(\sum_{H=1}^{k} P_H R_{hH})\rho_{hh}^{1/2}/\sigma_h}{(\sum_{H=1}^{k} P_H R_{kH})\rho_{yy}^{1/2}/\sigma_y} \quad (h = 1, 2, \ldots, k - 1), \quad (38)$$

for which (37) becomes

$$\rho_{\theta w}^2 = \frac{\sum_{h=1}^{k-1} p_h \rho_{hh}(\sum_{H=1}^{k} P_H R_{hH})^2 + q\rho_{yy}(\sum_{H=1}^{k} P_H R_{kH})^2}{\sum_{h=1}^{k} \sum_{H=1}^{k} p_h P_H R_{hH}}. \quad (39)$$

The denominator of the right side of (39) is, of course, σ_θ^2 from (28). If all R_{hH}'s are 1, then (39) reduces to

$$\rho_{\theta w}^2 = \sum_{h=1}^{k-1} p_h \rho_{hh} + q\rho_{yy}.$$

Finally, the substitution of a_h (36) and b_h (38) into the linear equating relation (32) produces a set of equating formulas that is equivalent to (30). Note that formulas (36) and (38) are a generalization of formulas (21) and (23).

3.5. Derivation of Confidence-Interval Formula

This subsection derives a general confidence-interval formula for use in connection with the equated scores of (30). Formula (4) of Subsection 2.5 will be a special case of this general formula.

Notation and assumptions will be the same as what were used earlier here in Section 3, except that two additional assumptions will be introduced. First, let $\text{var}(\theta|X_h)$ denote the conditional variance of θ given X_h; it will be assumed that $\text{var}(\theta|X_h)$ is constant (i.e., is independent of the value of X_h). Second, in the last stage of the derivation it will be assumed that the conditional distribution of θ given X_h is normal.

From the first assumption and from Eq. (21.9.1) on p. 280 of Cramér (1951) it follows that

$$\sigma_\theta^2 = \text{var}(\theta|X_h) + E\{[E(\theta|X_h) - \mu_\theta]^2\}. \tag{40}$$

On substitution of (28) and (30), one obtains

$$\text{var}(\theta|X_h) = \sum_{H=1}^{k} \sum_{H'=1}^{k} P_H P_{H'} R_{HH'} - \rho_{hh}\left(\sum_{H=1}^{k} P_H R_{hH}\right)^2 \tag{41}$$

from (40). Finally, from the normality assumption it now follows that

$$E(\theta|X_h) \pm 1.96[\text{var}(\theta|X_h)]^{1/2} \tag{42}$$

will provide a 95% confidence interval for θ (for a given value of X_h), where the formulas for $E(\theta|X_h)$ and var $(\theta|X_h)$ are furnished by (30) and (41), respectively.

Two special cases may be considered. First, suppose that the k different θ_h's are all the same variable, so that every R_{hH} is 1. If $R_{hH} = 1$ is substituted into (30) and (41) and the resulting formulas are next substituted into (42), then (42) will be the same thing as (4). Thus one concludes that formula (4) is applicable if every R_{hH} is 1.

Second, consider the special case where (31) holds. For this case (42) becomes

$$\lambda\rho_{hh}^{1/2}[(X_h - \mu_h)/\sigma_h] \pm 1.96(\lambda - \rho_{hh}\lambda^2)^{1/2}. \tag{43}$$

Using (43) is equivalent to using

$$\rho_{hh}^{1/2}[(X_h - \mu_h)/\sigma_h] \pm 1.96[(1/\lambda) - \rho_{hh}]^{1/2}, \tag{44}$$

which is a simpler formula. Of course, λ in (44) is the value of (28) that is obtained when the P_h's are chosen so as to minimize (28).

Note that (44) is the same as (4), except that the 1 in (4) has been replaced by $1/\lambda$ in (44). Since λ can never exceed 1, confidence intervals calculated from (44) could never be narrower than confidence intervals calculated from (4).

4.   A Technique for Differential Weighting of Distractors

As was indicated in (f) of Subsection 2.3, Section 4 provides details concerning a possible scoring scheme that attaches different scoring weights to the different responses to multiple-choice items. First, Subsection 4.1 will introduce the basic notation. Then Subsection 4.2 furnishes the mathematical derivation of the method for determining the set of scoring weights so as to maximize the value of coefficient α. Finally, Subsection 4.3 illustrates the use of the method, by showing the calculation steps to obtain the entire set of weights for a short numerical example based on artificial data. The reader who is interested in application but not theory might read Subsection 4.1, skim over Subsection 4.2, and then read Subsection 4.3.

The scoring scheme that is developed in what follows bears a close resemblance to that of Guttman (1941), even though its mathematical structure is somewhat different. However, the most important difference between these two schemes concerns the handling of omitted items. On the one hand, the method of Guttman (1941) is not set up to handle omitted items, except by the rather questionable procedure of creating a separate "omit" category [see earlier discussion in (e) of Subsection 2.4]. On the other hand, the scheme that is presented here in Section 4 is more general in that it enables one to set the score for an omitted item equal to the average of the scores for all the responses to the item, thus providing an appropriate means of treating omitted items. In Reilly (1975) the notion is advanced that the use of the separate "omit" category in conjunction with the scheme of Guttman (1941) may be responsible for lowered validities reported by investigators who tried this type of scoring. As an alternative, in fact, Reilly (1975) proposes a particular scoring scheme (different from the one to be presented here) in which an omitted item receives the average score of the different responses. Besides Guttman (1941) and Reilly (1975), references that provide some background material for what is developed below include Lord (1958) and Chapter 12 of Torgerson (1958).

4.1.   Basic Notation and Preliminaries

Necessary preliminaries and all notation that is needed to understand the method of calculating the scoring weights will be provided in this subsection. Subsection 4.2 will introduce some limited additional notation that is needed only for the derivation given in that subsection.

The test that is to be scored is assumed to consist of m multiple-choice items. Each item is assumed to have n^* responses from which the examinee chooses. Actually, the development that follows can easily be generalized in an obvious manner to cover the situation where the number of possible responses is not the same (n^*) for all items, but such a generalized treatment has been avoided in order to keep the notation simpler.

Suppose that data are available for N individuals who have taken the test. Define a variable x_{ijk} $(i = 1, 2, \ldots, N; j = 1, 2, \ldots, m; k = 1, 2, \ldots, n^*)$ that is 1 if the ith individual has marked the kth response to the jth item, 0 if she or he has marked some other response to the jth item, and $1/n^*$ if she or he has left the jth item blank.

This definition of x_{ijk} assumes that an examinee is allowed to mark no more than one response to an item. However, the definition is easily generalized to handle a situation where the examinee is permitted to mark more than one response: If the ith individual marks n'_{ij} (> 1) responses to the jth item, then set $x_{ijk} = 1/n'_{ij}$ for each k-value for which the response is marked and $x_{ijk} = 0$ for all other k-values. For example, if the examinee marks responses 1 and 3, then x_{ij1} and x_{ij3} will each be $\frac{1}{2}$, and the remaining x_{ijk}'s will be 0.

Now define $n = n^* - 1$, and let $w_{jk}(j = 1, 2, \ldots, m; k = 1, 2, \ldots, n)$ denote the scoring weight for the kth response to the jth item. Then the score for the ith individual on the jth item is

$$S_{ij} = \sum_{k=1}^{n} w_{jk} x_{ijk} = \mathbf{w}'_j \mathbf{x}_{ij},\qquad(45)$$

where

$$\mathbf{w}_j(n \times 1) = (w_{j1}, w_{j2}, \ldots, w_{jn})'$$

and

$$\mathbf{x}_{ij}(n \times 1) = (x_{ij1}, x_{ij2}, \ldots, x_{ijn})',$$

and the score for the ith individual on the entire test is

$$S_{i.} = \sum_{j=1}^{m} S_{ij} = \sum_{j=1}^{m} \sum_{k=1}^{n} w_{jk} x_{ijk} = \mathbf{w}' \mathbf{x}_i,\qquad(46)$$

where

$$\mathbf{w}(mn \times 1) = (\mathbf{w}'_1; \mathbf{w}'_2; \ldots; \mathbf{w}'_m)'$$

and

$$\mathbf{x}_i(mn \times 1) = (\mathbf{x}'_{i1}; \mathbf{x}'_{i2}; \ldots; \mathbf{x}'_{im})'.$$

Note that a scoring weight is *not* being assigned to the n^*th response; in effect, w_{jn^*} is taken to be equal to 0. The reason for this needs to be explained.

In the first place, no generality is lost by arbitrarily setting $w_{jn^*}' = 0$. More specifically, for any set of weights that attaches a score to each of the n^* responses to each item, there exists an equivalent set of weights for which $w_{jn^*} = 0$ for each item. To see that this is so, note first that the preceding definition of x_{ijk} implies the linear restriction

$$\sum_{k=1}^{n^*} x_{ijk} = 1, \quad \text{i.e.,} \quad x_{ijn^*} = 1 - \sum_{k=1}^{n} x_{ijk}. \tag{47}$$

Now let w_{jk}^0 $(j = 1, 2, \ldots, m; k = 1, 2, \ldots, n^*)$ be any set of weights that assigns a value to each of the n^* responses to each item. For this set of weights, the total score on the test for the ith individual would be

$$S_{i.}^0 = \sum_{j=1}^{m} \sum_{k=1}^{n^*} w_{jk}^0 x_{ijk}. \tag{48}$$

On application of (47), (48) becomes

$$S_{i.}^0 = \sum_{j=1}^{m} \sum_{k=1}^{n} w_{jk} x_{ijk} + w_{.n^*}^0, \tag{49}$$

where

$$w_{jk} = w_{jk}^0 - w_{jn^*}^0 \quad (j = 1, 2, \ldots, m; k = 1, 2, \ldots, n)$$

and

$$w_{.n^*}^0 = \sum_{j=1}^{m} w_{jn^*}^0.$$

Thus when (48) is reexpressed in the form (49), a scoring weight for the n^*th response is no longer used. One concludes that w_{jn^*} may be taken to be 0 without loss of generality.

Setting $w_{jn^*} = 0$ not only results in no loss of generality, but, as an important second consideration, also causes both the derivation and the calculations that are developed later to be definitely simpler than they would otherwise be. In fact, certain $n \times n$ matrices that are inverted later would become singular if they were each augmented with an n^*th row and an n^*th column.

For each item, one of the responses will have to be identified as the n^*th one, i.e., as the response that is to receive a scoring weight of 0. Mathematically, it makes no difference which response is so chosen. From a practical standpoint, however, there is one particular choice that appears to be best: The response that is judged to be the "correct" one should be considered to be the n^*th response. If, for each item, the "correct" response is thus chosen to be the one that receives no weight, then all the scoring weights in **w** will turn out to be of the same sign—assuming that there is no case of a distractor that receives a better score than the "correct" response. This should make it possible to avoid some confusion in interpreting the elements of **w** that will be obtained as the outcome of the calculations to be described later.

The test-score definition (46) may be put into a clearer perspective if its relationship to ordinary formula scoring is explained. This relationship is a simple one. Suppose that the correct response is identified as the n*th response, for each item. Suppose that $w_{jk} = -n^*/n$ ($j = 1, 2, \ldots, m; k = 1, 2, \ldots, n$) is substituted into (46). Then it is not hard to show that the resulting value of $S_{i.}$ will be equal to the formula score minus m. Thus, using the same value for all w_{jk}'s in (46) is equivalent to formula scoring.

Some further notation is required for the calculations that will be specified in what follows. Define

$$\mathbf{x}_{.j}(n \times 1) = \sum_{i=1}^{N} \mathbf{x}_{ij} = \left(\sum_{i=1}^{N} x_{ij1}, \sum_{i=1}^{N} x_{ij2}, \ldots, \sum_{i=1}^{N} x_{ijn} \right)'$$

and

$$\mathbf{x}_{.}(mn \times 1) = \sum_{i=1}^{N} \mathbf{x}_i = (\mathbf{x}'_{.1} ; \mathbf{x}'_{.2} ; \ldots ; \mathbf{x}'_{.m})'.$$

Next, define the matrix

$$\mathbf{G}(mn \times mn) = \sum_{i=1}^{N} \mathbf{x}_i \mathbf{x}'_i - \frac{1}{N} \mathbf{x}_{.} \mathbf{x}'_{.}. \tag{50}$$

Define also the matrices

$$\mathbf{H}_j(n \times n) = \sum_{i=1}^{N} \mathbf{x}_{ij} \mathbf{x}'_{ij} - \frac{1}{N} \mathbf{x}_{.j} \mathbf{x}'_{.j}, \tag{51}$$

each of which will be assumed to be nonsingular. Then let $\mathbf{0}_{nn}$ denote the $n \times n$ null matrix, and define

$$\mathbf{H}(mm \times mn) = \begin{bmatrix} \mathbf{H}_1 & \mathbf{0}_{nn} & \cdots & \mathbf{0}_{nn} \\ \mathbf{0}_{nn} & \mathbf{H}_2 & \cdots & \mathbf{0}_{nn} \\ \vdots & \vdots & \ddots & \vdots \\ \mathbf{0}_{nn} & \mathbf{0}_{nn} & \cdots & \mathbf{H}_m \end{bmatrix}. \tag{52}$$

Finally, the largest characteristic root of the matrix $\mathbf{H}^{-1}\mathbf{G}$ will be denoted by λ^*.

For purposes of obtaining **H**, it will be helpful to note that each nonzero element of **H** is the same as the corresponding element of **G**.

In the calculations, the matrix $\mathbf{H}$ has to be inverted. Note that this does not imply that a large $(mn \times mn)$ matrix will have to be inverted as a whole, since the inversion of $\mathbf{H}$ can be accomplished simply through inverting each of the m individual matrices $\mathbf{H}_j$. Of course, each $\mathbf{H}_j$ is just $n \times n$.

4.2. *Derivation*

The vector of scoring weights, which is being denoted by $\mathbf{w}$ $(mn \times 1)$, will be chosen in such a way as to maximize coefficient α. This subsection provides the derivation of the method for determining $\mathbf{w}$.

Coefficient α is discussed at length in Section 4.4 of Lord and Novick (1968). It is closely related to the reliability coefficient: Bringing about an increase in coefficient α could generally be expected to bring about an increase in reliability also.

The formula for coefficient α, as given, e.g., by equation (4.4.8) of Lord and Novick (1968), is

$$\frac{m}{m-1}\left[1 - \frac{\sum_{j=1}^{m} \text{var}(S_{ij})}{\text{var}(S_{i.})}\right] \tag{53}$$

when expressed in terms of the notation that was introduced in Subsection 4.1 of the present paper. If the population values $\text{var}(S_{ij})$ and $\text{var}(S_{i.})$ are replaced by their respective estimates from the sample of N examinees, then (53) changes to

$$\alpha = \frac{m}{m-1}\left(1 - \frac{\sum_{j=1}^{m} v_j}{v}\right), \tag{54}$$

where

$$v_j = V_j/(N-1), \qquad v = V/(N-1), \tag{55}$$

$$V_j = \sum_{i=1}^{N} S_{ij}^2 - \frac{1}{N} S_{.j}^2, \tag{56}$$

$$V = \sum_{i=1}^{N} S_{i.}^2 - \frac{1}{N} S_{..}^2, \tag{57}$$

$$S_{.j} = \sum_{i=1}^{N} S_{ij} = \mathbf{w}_j' \mathbf{x}_{.j}, \tag{58}$$

and

$$S_{..} = \sum_{i=1}^{N} S_{i.} = \sum_{i=1}^{N} \sum_{j=1}^{m} S_{ij} = \mathbf{w}' \mathbf{x}_{..}. \tag{59}$$

From (45), (58), (56), and (51) it follows that

$$V_j = \mathbf{w}_j' \mathbf{H}_j \mathbf{w}_j. \tag{60}$$

Then (52) and (60) lead at once to the equation

$$\sum_{j=1}^{m} V_j = \mathbf{w}' \mathbf{H} \mathbf{w}. \tag{61}$$

In addition, the equation

$$V = \mathbf{w}' \mathbf{G} \mathbf{w} \tag{62}$$

is obtained by using (46), (59), (57), and (50).

The objective, of course, is to choose $\mathbf{w}$ so as to maximize (54). On combining (54), (55), (61), and (62), one finds that α (54) will be maximized if $\mathbf{w}$ is chosen so as to maximize

$$V / \sum_{j=1}^{m} V_j = \mathbf{w}' \mathbf{G} \mathbf{w} / \mathbf{w}' \mathbf{H} \mathbf{w}. \tag{63}$$

The derivative of $\mathbf{w}' \mathbf{G} \mathbf{w} / \mathbf{w}' \mathbf{H} \mathbf{w}$ with respect to $\mathbf{w}$ is the vector

$$[2/(\mathbf{w}' \mathbf{H} \mathbf{w})] \mathbf{G} \mathbf{w} - [2(\mathbf{w}' \mathbf{G} \mathbf{w})/(\mathbf{w}' \mathbf{H} \mathbf{w})^2] \mathbf{H} \mathbf{w}. \tag{64}$$

Let $\mathbf{0}_{mn,1}$ denote the $mn \times 1$ null vector. If the vector (64) is set equal to $\mathbf{0}_{mn,1}$ and the resulting set of equations is simplified, then one obtains

$$\left(\mathbf{G} - \frac{\mathbf{w}' \mathbf{G} \mathbf{w}}{\mathbf{w}' \mathbf{H} \mathbf{w}} \mathbf{H} \right) \mathbf{w} = \mathbf{0}_{mn,1}. \tag{65}$$

It is not hard to show that $\mathbf{w}$ will satisfy this set of equations (65) if and only if $\mathbf{w}$ is a characteristic vector of the matrix $\mathbf{H}^{-1} \mathbf{G}$. Also, in order to maximize $\mathbf{w}' \mathbf{G} \mathbf{w} / \mathbf{w}' \mathbf{H} \mathbf{w}$ as desired, $\mathbf{w}$ must be a characteristic vector corresponding to λ^*, the *largest* characteristic root of $\mathbf{H}^{-1} \mathbf{G}$.

Furthermore, for this optimal choice of $\mathbf{w}$, (63) will simply be equal to λ^*. One thus concludes that, if $\mathbf{w}$ is chosen so as to maximize α (54), then (54) will assume the value

$$\alpha = \frac{m}{m-1} \left(1 - \frac{1}{\lambda^*} \right). \tag{66}$$

4.3. Numerical Example

Subsection 4.2 showed that coefficient α will be maximized [and will be equal to the value indicated by (66)] if the weight vector $\mathbf{w}$ ($mn \times 1$) is chosen to be a characteristic vector of $\mathbf{H}^{-1} \mathbf{G}$ that corresponds to λ^*, the largest characteristic root of $\mathbf{H}^{-1}\mathbf{G}$. Here in Subsection 4.3, the calculation of the optimal $\mathbf{w}$ will be illustrated by means of a short numerical example that uses artificial test results.

This example is based on a test consisting of just $m = 2$ items. Each item has just $n^* = 3$ responses to choose from, so that $n = n^* - 1 = 2$. For each item the responses are numbered (or renumbered) so that the n^*th or third response ($k = 3$) is the "correct" one.

There are data for $N = 100$ examinees (a much smaller value of N than would normally be sufficient if the data were not artificial). No examinee left either item blank. The way that the ith examinee answered the two items is, of course, indicated by $\mathbf{x}_i = (x_{i11}, x_{i12}; x_{i21}, x_{i22})'$. The 100 examinees were distributed in the following manner with respect to their $\mathbf{x}_i$ vectors (or, equivalently, their respective responses to items 1 and 2):

9 examinees had $\mathbf{x}_i = (1, 0; 1, 0)'$ (i.e., they marked responses 1, 1);
4 examinees had $\mathbf{x}_i = (1, 0; 0, 1)'$ (i.e., they marked responses 1, 2);
11 examinees had $\mathbf{x}_i = (1, 0; 0, 0)'$ (i.e., they marked responses 1, 3);
5 examinees had $\mathbf{x}_i = (0, 1; 1, 0)'$ (i.e., they marked responses 2, 1);
4 examinees had $\mathbf{x}_i = (0, 1; 0, 1)'$ (i.e., they marked responses 2, 2);
5 examinees had $\mathbf{x}_i = (0, 1; 0, 0)'$ (i.e., they marked responses 2, 3);
24 examinees had $\mathbf{x}_i = (0, 0; 1, 0)'$ (i.e., they marked responses 3, 1);
6 examinees had $\mathbf{x}_i = (0, 0; 0, 1)'$ (i.e., they marked responses 3, 2);
32 examinees had $\mathbf{x}_i = (0, 0; 0, 0)'$ (i.e., they marked responses 3, 3).

To obtain $\mathbf{G}$ (50), note first that

$$\sum_{i=1}^{N} \mathbf{x}_i \mathbf{x}_i' = 9 \begin{bmatrix} 1 & 0 & 1 & 0 \\ 0 & 0 & 0 & 0 \\ 1 & 0 & 1 & 0 \\ 0 & 0 & 0 & 0 \end{bmatrix} + 4 \begin{bmatrix} 1 & 0 & 0 & 1 \\ 0 & 0 & 0 & 0 \\ 0 & 0 & 0 & 0 \\ 1 & 0 & 0 & 1 \end{bmatrix}$$

$$+ \cdots + 32 \begin{bmatrix} 0 & 0 & 0 & 0 \\ 0 & 0 & 0 & 0 \\ 0 & 0 & 0 & 0 \\ 0 & 0 & 0 & 0 \end{bmatrix} = \begin{bmatrix} 24 & 0 & 9 & 4 \\ 0 & 14 & 5 & 4 \\ 9 & 5 & 38 & 0 \\ 4 & 4 & 0 & 14 \end{bmatrix},$$

$$\mathbf{x}_. = 9 \begin{bmatrix} 1 \\ 0 \\ 1 \\ 0 \end{bmatrix} + 4 \begin{bmatrix} 1 \\ 0 \\ 0 \\ 1 \end{bmatrix} + \cdots + 32 \begin{bmatrix} 0 \\ 0 \\ 0 \\ 0 \end{bmatrix} = \begin{bmatrix} 24 \\ 14 \\ 38 \\ 14 \end{bmatrix},$$

and

$$\frac{1}{N}\mathbf{x}.\mathbf{x'} = \begin{bmatrix} 5.76 & 3.36 & 9.12 & 3.36 \\ 3.36 & 1.96 & 5.32 & 1.96 \\ 9.12 & 5.32 & 14.44 & 5.32 \\ 3.36 & 1.96 & 5.32 & 1.96 \end{bmatrix}.$$

Formula (50) now gives

$$\mathbf{G} = \begin{bmatrix} 18.24 & -3.36 & -0.12 & 0.64 \\ -3.36 & 12.04 & -0.32 & 2.04 \\ -0.12 & -0.32 & 23.56 & -5.32 \\ 0.64 & 2.04 & -5.32 & 12.04 \end{bmatrix}.$$

As noted earlier, the nonzero elements of $\mathbf{H}(52)$ can simply be obtained from $\mathbf{G}$. Thus

$$\mathbf{H} = \begin{bmatrix} 18.24 & -3.36 & 0 & 0 \\ -3.36 & 12.04 & 0 & 0 \\ 0 & 0 & 23.56 & -5.32 \\ 0 & 0 & -5.32 & 12.04 \end{bmatrix}.$$

Then

$$\mathbf{H}^{-1} = \begin{bmatrix} 12.04/208.32 & 3.36/208.32 & 0 & 0 \\ 3.36/208.32 & 18.24/208.32 & 0 & 0 \\ 0 & 0 & 12.04/255.36 & 5.32/255.36 \\ 0 & 0 & 5.32/255.36 & 23.56/255.36 \end{bmatrix}$$

and

$$\mathbf{H}^{-1}\mathbf{G} = \begin{bmatrix} 1 & 0 & -2.52/208.32 & 14.56/208.32 \\ 0 & 1 & -6.24/208.32 & 39.36/208.32 \\ 1.96/255.36 & 7.00/255.36 & 1 & 0 \\ 14.44/255.36 & 46.36/255.36 & 0 & 1 \end{bmatrix}.$$

The characteristic roots of $\mathbf{H}^{-1}\mathbf{G}$ are the values of λ that satisfy the equation

$$|\mathbf{H}^{-1}\mathbf{G} - \lambda\mathbf{I}| = 0,$$

where $\mathbf{I}$ denotes the $mn \times mn$ identity matrix. This equation becomes

$$(1 - \lambda)^4 - 0.03733992(1 - \lambda)^2 + (3.007711 \times 10^{-8}) = 0,$$

and its four roots are

$$\lambda = 0.8067667, 0.9991025, 1.0008975, 1.1932333.$$

Thus the largest characteristic root of $\mathbf{H}^{-1}\mathbf{G}$ is

$$\lambda^* = 1.1932333,$$

for which the resulting value of α (66) is

$$\alpha = \frac{2}{2-1}\left(1 - \frac{1}{1.1932333}\right) = 0.32388.$$

The weight vector $\mathbf{w} = (w_{11}, w_{12}; w_{21}, w_{22})'$ is now obtained by solving the system

$$(\mathbf{H}^{-1}\mathbf{G} - \lambda^*\mathbf{I})\mathbf{w} = \mathbf{O}_{41}$$

for $\mathbf{w}$. Of course, this system does not have a unique solution for $\mathbf{w}$. In order to obtain a specific solution for $\mathbf{w}$, one of the elements of $\mathbf{w}$ has to be chosen arbitrarily. If w_{22} is arbitrarily taken to be -1, then the weight vector $\mathbf{w}$ turns out to be

$$\mathbf{w} = (-0.352346, -0.954620; -0.149419, -1)'.$$

With this weight vector, all test scores will obviously be either zero (perfect score) or negative, but this need not cause any difficulty.

It has not been mentioned previously, but the methodology developed here in Section 4 can also be applied to obtain the scoring weights for the type of scoring scheme that weights different items differentially but does not weight individual distractors differentially. Such application is accomplished simply by replacing $\mathbf{x}_{ij}$ $(n \times 1) = (x_{ij1}, x_{ij2}, \ldots, x_{ijn})'$ with the single value $x_{ij} = \sum_{k=1}^{n} x_{ijk}$, and replacing n with 1. (It is assumed here that the correct response was taken to be the n^*th response, for each item.)

For the data of the numerical example just presented, the calculations would proceed as follows. Note first that the original test-result data need to be reexpressed in terms of the new $\mathbf{x}_i$, which is $\mathbf{x}_i = (x_{i1}; x_{i2})'$. Specifically, observe that the 100 examinees were distributed in the following manner with respect to this new $\mathbf{x}_i$:

22 examinees had $\mathbf{x}_i = (1; 1)'$ (i.e., they marked both items wrong);
16 examinees had $\mathbf{x}_i = (1; 0)'$ (i.e., they marked item 1 wrong, item 2 right);
30 examinees had $\mathbf{x}_i = (0; 1)'$ (i.e., they marked item 1 right, item 2 wrong);
32 examinees had $\mathbf{x}_i = (0; 0)'$ (i.e., they marked both items right).

Using the new $\mathbf{x}_i$'s, one now obtains

$$\sum_{i=1}^{N} \mathbf{x}_i\mathbf{x}_i' = 22\begin{bmatrix} 1 & 1 \\ 1 & 1 \end{bmatrix} + 16\begin{bmatrix} 1 & 0 \\ 0 & 0 \end{bmatrix} + 30\begin{bmatrix} 0 & 0 \\ 0 & 1 \end{bmatrix} + 32\begin{bmatrix} 0 & 0 \\ 0 & 0 \end{bmatrix} = \begin{bmatrix} 38 & 22 \\ 22 & 52 \end{bmatrix},$$

$$\mathbf{x}_. = 22\begin{bmatrix} 1 \\ 1 \end{bmatrix} + 16\begin{bmatrix} 1 \\ 0 \end{bmatrix} + 30\begin{bmatrix} 0 \\ 1 \end{bmatrix} + 32\begin{bmatrix} 0 \\ 0 \end{bmatrix} = \begin{bmatrix} 38 \\ 52 \end{bmatrix},$$

and

$$\frac{1}{N}\mathbf{X}.\mathbf{X}'. = \begin{bmatrix} 14.44 & 19.76 \\ 19.76 & 27.04 \end{bmatrix}.$$

Equation (50) then gives

$$\mathbf{G} = \begin{bmatrix} 23.56 & 2.24 \\ 2.24 & 24.96 \end{bmatrix}.$$

Also,

$$\mathbf{H} = \begin{bmatrix} 23.56 & 0 \\ 0 & 24.96 \end{bmatrix}, \quad \mathbf{H}^{-1} = \begin{bmatrix} 1/23.56 & 0 \\ 0 & 1/24.96 \end{bmatrix},$$

$$\mathbf{H}^{-1}\mathbf{G} = \begin{bmatrix} 1 & 2.24/23.56 \\ 2.24/24.96 & 1 \end{bmatrix}.$$

The characteristic equation to be solved is

$$|\mathbf{H}^{-1}\mathbf{G} - \lambda\mathbf{I}| = (1 - \lambda)^2 - 2.24^2/(23.56 \times 24.96) = 0,$$

whose roots are easily found to be $\lambda = 0.9076285, 1.0923715$. Hence the larger characteristic root of $\mathbf{H}^{-1}\mathbf{G}$ is $\lambda^* = 1.0923715$, so that the resulting value of α (66) is

$$\alpha = \frac{2}{2 - 1}\left(1 - \frac{1}{1.0923715}\right) = 0.16912.$$

A solution for the weight vector $\mathbf{w} = (w_{11}; w_{21})'$ can now be obtained by arbitrarily setting $w_{21} = -1$ and then solving the system

$$(\mathbf{H}^{-1}\mathbf{G} - \lambda^*\mathbf{I})\mathbf{w} = \mathbf{0}_{21}.$$

The result is

$$\mathbf{w} = (-1.029283; -1)'.$$

At this point, it should be noted in passing that the value of coefficient α for *formula scoring* can easily be obtained from the $\mathbf{G}$ and $\mathbf{H}$ matrices. The original $mn \times mn$ $\mathbf{G}$ and $\mathbf{H}$ matrices may be used, but the result will be the same if one calculates it using the smaller $m \times m$ $\mathbf{G}$ and $\mathbf{H}$ matrices (like those just exhibited) that treat all distractors alike. (It is again assumed that the correct response is identified as the n^*th response, for each item.) To obtain the value of coefficient α for formula scoring, simply divide the sum of all the elements of $\mathbf{H}$ by the sum of all the elements of $\mathbf{G}$, subtract the quotient from 1, and multiply by $m/(m - 1)$. For the data of the above numerical example, the value of coefficient α for formula scoring then turns out to be

$$\alpha = \frac{2}{2-1}\left(1 - \frac{48.52}{53.00}\right) = 0.16906.$$

Thus, for the data just examined, coefficient α for formula scoring is only slightly lower than coefficient α for differential weighting of different items (0.16906 versus 0.16912), whereas coefficient α for differential item weighting is substantially lower than coefficient α for differential weighting of distractors (0.16912 versus 0.32388). However, one should certainly avoid trying to draw any generalizations just from this example, since the data for the example are artificial and the test contains only two items. On the other hand, for the scoring methodology proposed here in Section 4 it will always be true that coefficient α for formula scoring will not exceed coefficient α for differential item weighting, and that, in turn, coefficient α for differential item weighting will not exceed coefficient α for differential weighting of distractors.

As noted earlier, the literature contains the results of different empirical studies of differential weighting of distractors. These studies utilized a variety of methods for determining the scoring weights. In some cases, the definition of the method may be somewhat fuzzy, or there may be uncertainty as to what some of the mathematical properties of the method are. In any event, though, none of these methods described in the literature can produce a value of coefficient α greater than what will be obtained through the method presented here in Section 4, simply because the method proposed here produces the maximum possible value of coefficient α. (To make this statement, one has to assume, of course, that omitted items are handled in equivalent fashion for different methods.)

Even though the method proposed here is optimal with respect to coefficient α, there are two other points that also need to be considered. First, coefficient α is not the only yardstick that can be used to assess the desirability of different methods of determining the scoring weights. Second, it is possible that, at least some of the time if not consistently, an alternative method may produce a value of coefficient α that is close to the maximum attainable even if it does not hit the maximum exactly.

The preceding numerical example was constructed with no omitted items, in order to keep it simpler. Earlier discussion had suggested, however, that the method developed here in Section 4 may be beneficial particularly because of the way that it handles omitted items. Thus it appears desirable at this point to outline an extension of this example that does include some examinees who left items blank.

This extended example will be constructed by using the same 100 examinees as before but adding 25 more examinees to give a total of $N = 125$. The 25 additional examinees were distributed as follows, with respect to their vectors $\mathbf{x}_i = (x_{i11}, \ x_{i12}; \ x_{i21}, \ x_{i22})'$ (or, equivalently, their respective responses to items 1 and 2):

5 examinees had $\mathbf{x}_i = (1, 0; \frac{1}{3}, \frac{1}{3})'$ [i.e., they responded (1, omit)];
3 examinees had $\mathbf{x}_i = (0, 1; \frac{1}{3}, \frac{1}{3})'$ [i.e., they responded (2, omit)];
6 examinees had $\mathbf{x}_i = (0, 0; \frac{1}{3}, \frac{1}{3})'$ [i.e., they responded (3, omit)];
4 examinees had $\mathbf{x}_i = (\frac{1}{3}, \frac{1}{3}; 1, 0)'$ [i.e., they responded (omit, 1)];
2 examinees had $\mathbf{x}_i = (\frac{1}{3}, \frac{1}{3}; 0, 1)'$ [i.e., they responded (omit, 2)];
4 examinees had $\mathbf{x}_i = (\frac{1}{3}, \frac{1}{3}; 0, 0)'$ [i.e., they responded (omit, 3)];
1 examinee had $\mathbf{x}_i = (\frac{1}{3}, \frac{1}{3}; \frac{1}{3}, \frac{1}{3})'$ [i.e., response was (omit, omit)].

To find $\mathbf{G}$ (50), first obtain

$$
\sum_{i=1}^{N} \mathbf{x}_i \mathbf{x}_i' =
\begin{bmatrix} 24 & 0 & 9 & 4 \\ 0 & 14 & 5 & 4 \\ 9 & 5 & 38 & 0 \\ 4 & 4 & 0 & 14 \end{bmatrix}
+ 5 \begin{bmatrix} 1 & 0 & \frac{1}{3} & \frac{1}{3} \\ 0 & 0 & 0 & 0 \\ \frac{1}{3} & 0 & \frac{1}{9} & \frac{1}{9} \\ \frac{1}{3} & 0 & \frac{1}{9} & \frac{1}{9} \end{bmatrix}
+ 3 \begin{bmatrix} 0 & 0 & 0 & 0 \\ 0 & 1 & \frac{1}{3} & \frac{1}{3} \\ 0 & \frac{1}{3} & \frac{1}{9} & \frac{1}{9} \\ 0 & \frac{1}{3} & \frac{1}{9} & \frac{1}{9} \end{bmatrix}
+ \cdots
$$

$$
+ 1 \begin{bmatrix} \frac{1}{9} & \frac{1}{9} & \frac{1}{9} & \frac{1}{9} \\ \frac{1}{9} & \frac{1}{9} & \frac{1}{9} & \frac{1}{9} \\ \frac{1}{9} & \frac{1}{9} & \frac{1}{9} & \frac{1}{9} \\ \frac{1}{9} & \frac{1}{9} & \frac{1}{9} & \frac{1}{9} \end{bmatrix}
= \frac{1}{9} \begin{bmatrix} 272 & 11 & 109 & 58 \\ 11 & 164 & 67 & 52 \\ 109 & 67 & 393 & 15 \\ 58 & 52 & 15 & 159 \end{bmatrix},
$$

$$
\mathbf{x}_. =
\begin{bmatrix} 24 \\ 14 \\ 38 \\ 14 \end{bmatrix}
+ 5 \begin{bmatrix} 1 \\ 0 \\ \frac{1}{3} \\ \frac{1}{3} \end{bmatrix}
+ 3 \begin{bmatrix} 0 \\ 1 \\ \frac{1}{3} \\ \frac{1}{3} \end{bmatrix}
+ \cdots + 1 \begin{bmatrix} \frac{1}{3} \\ \frac{1}{3} \\ \frac{1}{3} \\ \frac{1}{3} \end{bmatrix}
= \frac{1}{3} \begin{bmatrix} 98 \\ 62 \\ 141 \\ 63 \end{bmatrix},
$$

and

$$
\frac{1}{N} \mathbf{x}_. \mathbf{x}_.' = \frac{1}{9} \begin{bmatrix} 76.832 & 48.608 & 110.544 & 49.392 \\ 48.608 & 30.752 & 69.936 & 31.248 \\ 110.544 & 69.936 & 159.048 & 71.064 \\ 49.392 & 31.248 & 71.064 & 31.752 \end{bmatrix}.
$$

Then

$$
\mathbf{G} = \frac{1}{9} \begin{bmatrix} 195.168 & -37.608 & -1.544 & 8.608 \\ -37.608 & 133.248 & -2.936 & 20.752 \\ -1.544 & -2.936 & 233.952 & -56.064 \\ 8.608 & 20.752 & -56.064 & 127.248 \end{bmatrix}.
$$

Also,

$$\mathbf{H}^{-1} =$$

$$9 \begin{bmatrix} 133.248/24591.384 & 37.608/24591.384 & 0 & 0 \\ 37.608/24591.384 & 195.168/24591.384 & 0 & 0 \\ 0 & 0 & 127.248/26626.752 & 56.064/26626.752 \\ 0 & 0 & 56.064/26626.752 & 233.952/26626.752 \end{bmatrix}$$

and

$$\mathbf{H}^{-1}\mathbf{G} =$$

$$\begin{bmatrix} 1 & 0 & -316.152/24591.384 & 1927.440/24591.384 \\ 0 & 1 & -631.080/24591.384 & 4373.856/24591.384 \\ 286.128/26626.752 & 789.840/26626.752 & 1 & 0 \\ 1927.296/26626.752 & 4690.368/26626.752 & 0 & 1 \end{bmatrix}.$$

The largest characteristic root of $\mathbf{H}^{-1}\mathbf{G}$ is found to be $\lambda^* = 1.1900068$, which, on substitution into (66), gives

$$\alpha = \frac{2}{2-1}\left(1 - \frac{1}{1.1900068}\right) = 0.31934$$

as the value of coefficient α. The weight vector $\mathbf{w} = (w_{11}, w_{12}; w_{21}, w_{22})'$, with w_{22} taken to be -1, turns out to be

$$\mathbf{w} = (-0.401317, -0.913747; -0.165348, -1)'.$$

For the scoring scheme that weights different items differentially but does not weight individual distractors differentially, the calculations are, of course, based on the modified $\mathbf{x}_i$ vector of the form $\mathbf{x}_i = (x_{i1}; x_{i2})'$. It is readily seen that the 25 additional examinees were distributed as follows with respect to this $\mathbf{x}_i$:

8 examinees had $\mathbf{x}_i = (1; \frac{2}{3})'$ [i.e., they responded (wrong, omit)];
6 examinees had $\mathbf{x}_i = (0; \frac{2}{3})'$ [i.e., they responded (right, omit)];
6 examinees had $\mathbf{x}_i = (\frac{2}{3}; 1)'$ [i.e., they responded (omit, wrong)];
4 examinees had $\mathbf{x}_i = (\frac{2}{3}; 0)'$ [i.e., they responded (omit, right)];
1 examinee had $\mathbf{x}_i = (\frac{2}{3}; \frac{2}{3})'$ [i.e., response was (omit, omit)].

One then obtains

$$\mathbf{G} = \frac{1}{9}\begin{bmatrix} 253.200 & 24.880 \\ 24.880 & 249.072 \end{bmatrix},$$

$$\mathbf{H}^{-1}\mathbf{G} = \begin{bmatrix} 1 & 24.880/253.200 \\ 24.880/249.072 & 1 \end{bmatrix}.$$

Next, the larger characteristic root of $\mathbf{H}^{-1}\mathbf{G}$ is easily determined to be $\lambda^* = 1.0990732$. This gives

$$\alpha = \frac{2}{2-1}\left(1 - \frac{1}{1.0990732}\right) = 0.180285$$

for coefficient α, and, with w_{21} taken to be -1, gives

$$\mathbf{w} = (-0.991815; -1)'$$

as the weight vector $\mathbf{w} = (w_{11}; w_{21})'$.

For formula scoring, the value of coefficient α is

$$\alpha = \frac{2}{2-1}\left(1 - \frac{502.272/9}{552.032/9}\right) = 0.180279.$$

5. Two Cases of Unexpected Implications of "Fairness" Definitions

As was indicated in Subsection 2.7, Section 5 explores the implications of certain definitions or requirements of "fairness" in testing, for two different situations. The first situation, which concerns the equating of two test forms with unequal reliabilities, is covered in Subsection 5.1. Subsection 5.2 deals with the second situation, which concerns the test-bias issue. In each case, what may sound like a reasonable definition or requirement of "fairness" apparently leads to a necessity for random-error terms to be deliberately added to examinees' scores.

5.1. Equating Two Test Forms with Unequal Reliabilities

As in Subsections 2.2 and 2.7, consider two tests or test forms x and y that are to be equated and that have respective observed scores X and Y and reliabilities ρ_{xx} and ρ_{yy}. The means of the observed scores are μ_x and μ_y, and the variances of the observed scores are σ_x^2 and σ_y^2.

Assume that X and Y are each normally distributed. Assume also that the two reliabilities are unequal and that ρ_{xx} is the larger, so that $\rho_{xx} > \rho_{yy}$.

As was indicated in Subsection 2.7, equating of the two forms through the equating formula (5) will satisfy the "fairness" requirement that the proportion of examinees receiving equated scores above any given value is to be the same for each test form. Suppose now, however, that this "fairness" requirement of equal proportions is deemed to be inadequate to achieve its purpose, and that a further "fairness" requirement is therefore deemed to be necessary.

Specifically, it can be contended that a selecting organization could circumvent the equated scores obtained from (5), if it knew which form each examinee took and also knew the values of certain parameters including ρ_{xx} and ρ_{yy}. In other words, with just this limited knowledge the selecting organization could make its own calculations to convert equated scores based on (5) to equated scores based on (3), if it preferred to use the latter in order to obtain better prediction.

It might thus be contended that an additional "fairness" requirement should be imposed that will be sufficient to prevent this from happening, since the equated scores based on (3) do not satisfy the "fairness" condition of equal proportions. Such an additional requirement could stipulate that reported scores are to be determined in such a way that the user would not be able to improve the predictive ability of these scores by running recalculations of the type described in the previous paragraph. With this second "fairness" requirement in effect, the user would have no incentive to violate the purpose of the first "fairness" requirement (the requirement of equal proportions) by circumventing and modifying the equating in a manner like that just described.

Satisfying both the first and second "fairness" requirements is not easy. Only one way of doing it comes to mind. This involves adding random-error terms to examinees' test scores. Let Z denote a random variable that is normally distributed with mean 0 and variance 1. For each examinee who took form x (the form with the higher reliability), draw a value of Z, and then calculate a modified score X^* from the formula

$$X^* = X + \sigma_x[(\rho_{xx}/\rho_{yy}) - 1]^{1/2}Z. \tag{67}$$

For each examinee who took form y, the original observed score Y is retained, and no random-error term is added.

Now let the X^* scores be used for equating and then reporting for the examinees who took form x, and let the Y scores be used for equating and then reporting for the examinees who took form y. The scores defined by X^* have the same reliability as the Y scores, viz., ρ_{yy}. To see that this is so, let $\sigma_{\tilde{x}*}^2$, σ_{x*}^2, and ρ_{x*x*} denote, respectively, the variance of true scores, variance of observed scores, and reliability for the "test" associated with X^*. Note first that $\sigma_{\tilde{x}*}^2$ is equal to $\sigma_{\tilde{x}}^2$ (variance of true scores associated with X), since true score is not changed in going from X to X^*. Then observe from (67) that

$$\sigma_{x*}^2 = \sigma_x^2 + \sigma_x^2[(\rho_{xx}/\rho_{yy}) - 1] = \rho_{xx}\sigma_x^2/\rho_{yy} = \sigma_{\tilde{x}}^2/\rho_{yy} = \sigma_{\tilde{x}*}^2/\rho_{yy}.$$

Since $\rho_{x*x*} = \sigma_{\tilde{x}*}^2/\sigma_{x*}^2$, it follows at once that $\rho_{x*x*} = \rho_{yy}$.

Because reliability is thus the same for X^* as for Y, the two equating formulas (5) and (3) (if applied to X^* and Y rather than to X and Y) will yield exactly the same results as each other. One therefore concludes that *both of* the "fairness" requirements will be satisfied if equating is based on X^* instead of on X. The price that had to be paid to achieve such "fairness" was to add a random-error term to each of the X scores so as to reduce their reliability to that of the Y scores.

5.2. The Test-Bias Problem

This subsection examines the implications of a particular definition of "fairness" that has been proposed for the test-bias problem. In certain respects the results will turn out to be similar to those of Subsection 5.1.

Notation in this subsection bears no relation to the notation of Subsection 5.1. Here let X denote score on a test that is being used for selection and let Y denote score on a criterion variable that X is intended to predict. The essence of the test-bias problem is, of course, to assess under what conditions the test (with score X) may be considered to be biased against one group or another when it is used as a basis for selection for an educational or employment situation in which success can be measured by the indicated criterion variable (with score Y). The development that follows will be confined to the case where it is assumed that, for any group G being considered, the conditional distribution of Y given X is normal with variance σ^2 (for all X) and with conditional mean given by

$$E(Y|X) = \alpha + \beta X \qquad \text{(regardless of } G\text{)}. \qquad (68)$$

Under one widely used definition of bias that is challenged in Thorndike (1971), a test with score X will be free of bias for predicting a criterion with score Y if $E(Y|X)$ is, as in (68), the same function of X for each group G being considered. Thus, with this definition, the test discussed in the previous paragraph is free of bias, or "fair," with respect to the indicated criterion variable and the specified groups.

The definition of "fairness" proposed in Thorndike (1971) is, however, entirely different. Briefly, it specifies (see Thorndike, 1971, p. 68, last paragraph) that the test is considered to be used "fairly" in connection with the criterion if, for each group G being considered, the probability of selection based on the test is equal to the probability of success on the criterion.

This "fairness" definition of Thorndike (1971) may be expressed in more mathematical terms. Suppose that there are n possible scores on the test, to be denoted by $X_j (j = 1, 2, \ldots, n)$. Assume at this point that the population of examinees has been classified into m mutually exclusive groups for purposes of trying to assess and achieve "fairness," and let these m groups be indexed by i ($i = 1, 2, \ldots, m$). Let f_{ij} denote the fraction of the entire examinee population that belongs to group i and obtains the score X_j on the test. Then

$$f_{i.} = \sum_{j=1}^{n} f_{ij}$$

will denote the fraction of the population which belongs to group i, and

$$h_{ij} = f_{ij}/f_{i.}$$

will denote the fraction of group i that receives score X_j.

Now let S_{ij} denote the probability of selection for an individual in group i whose score is X_j. The values of the S_{ij}'s are yet to be chosen—chosen in such a way as to satisfy the "fairness" definition. Conventionally, each S_{ij} would be either 0 or 1. However, no such restriction will be imposed here.

Let Y_0 denote a score on the criterion variable such that an individual with criterion score Y is considered to be successful on the criterion if $Y \geq Y_0$ but unsuccessful if $Y < Y_0$. Let P_j denote the probability of success on the criterion for an individual whose score on the test is X_j. Note that this probability does not depend on i (i.e., on the group the examinee belongs to), since it was assumed in the beginning that the conditional distribution of Y given X is the same for every group. The formula for P_j is, of course,

$$P_j = \Phi\left(\frac{\alpha + \beta X_j - Y_0}{\sigma}\right), \tag{69}$$

where $\Phi(\cdot)$ denotes the cumulative distribution function of the normal distribution with zero mean and unit variance.

The probability of selection for members of group i (i.e., the fraction of group i that will be selected) is $\sum_{j=1}^{n} h_{ij}S_{ij}$. The probability of success on the criterion for group i (i.e., the fraction of the entire group that would score no lower than Y_0 on the criterion variable if selected) is $\sum_{j=1}^{n} h_{ij}P_j$. Thus the "fairness" definition of Thorndike (1971) is equivalent to stipulating that

$$\sum_{j=1}^{n} h_{ij}S_{ij} = \sum_{j=1}^{n} h_{ij}P_j \qquad \text{for each } i\ (i = 1, 2, \ldots, m). \tag{70}$$

The development of Thorndike (1971) seems to be oriented essentially toward a situation where (70) is stipulated with just $m = 2$ mutually exclusive groups. In actuality, however, one could define a large number of mutually exclusive groups with respect to which the "fairness" requirement (70) might be imposed. Such groups could be based simultaneously on a variety of factors, such as race, sex, national origin, religion, age, geographic location, family income, and others, including even the test score X. The m mutually exclusive groups would then correspond to all possible combinations of the designated levels or categories of the designated factors.

Consider now an extended interpretation of the "fairness" requirement (70). Suppose first that S_{ij}'s are to be chosen so that (70) will hold for a particular set of m mutually exclusive groups that is based on designated levels for a designated list of factors like those just mentioned. Then suppose it is further stipulated that this list of factors is to include the test score X, whose designated levels are to be $X_1, X_2, \ldots, X_n$. Thus the "fairness" requirement is made to apply to each of a number of groups that are distinguished in part by test score itself.

Under this extended interpretation of the "fairness" requirement of Thorndike (1971), each group i for which (70) has to hold will be composed of individuals who all have the same test score, i.e., the same value of X. Thus if X_J [where $J = J(i)$] is the test score that is common to all the individuals in a particular group i, then, for this group i,

$$
\begin{aligned}
h_{ij} &= 1 && \text{for} && j = J(i), \\
h_{ij} &= 0 && \text{for all} && j \neq J(i).
\end{aligned}
\tag{71}
$$

Substitution of (71) into (70) then yields at once the result

$$
S_{i,J(i)} = P_{J(i)} \qquad (i = 1, 2, \ldots, m).
\tag{72}
$$

Of course, there is no need to try to determine S_{ij} for any $j \neq J(i)$, since there are no individuals in group i whose test scores are other than X_J.

Thus, under the extended interpretation of the "fairness" requirement (70), the first implication is that the probability of selection for all individuals with test score X_j must be equal to P_j (69). This proposition follows at once from (72): For each group i whose common test score is X_J, probability of selection must be P_J. Obviously, the proposition holds regardless of what factors besides test score were used in defining the m groups.

What is essentially a converse of this proposition may be noted in passing. Regardless of how the m groups are set up or what factors they involve (and regardless of whether or not test score is one of these factors), the "fairness" requirement (70) will clearly be satisfied if $S_{ij} = P_j$, i.e., if the probability of selection for anyone with test score X_j is made equal to P_j.

Now P_j (69) is neither 0 nor 1 for any j, but rather is in between. Hence, some sort of randomization device will have to be employed if each person with test score X_j is to be selected with probability P_j. One way of achieving the randomization is as follows. Let Z denote a random variable drawn from a normal distribution with mean 0 and variance 1. For each examinee, draw a value of Z, and then calculate

$$
X^* = X + (\sigma/\beta)Z,
\tag{73}
$$

which is the original test score with a random-error term added to it. Now use a selection rule that selects all those examinees for whom

$$\alpha + \beta X^* \geq Y_0, \quad \text{i.e.,} \quad X^* \geq (Y_0 - \alpha)/\beta. \tag{74}$$

From (73) and (74) it follows that the probability of selection for any examinee with test score X_j (for any given value of X_j) will be

$$Pr\{X_j + (\sigma/\beta)Z \geq (Y_0 - \alpha)/\beta\} = Pr\{-Z \leq (\alpha + \beta X_j - Y_0)/\sigma\},$$

which is simply P_j (69), as desired.

Note that it does not matter what Y_0 is in order for the variable X^* (73) to be used for selection in the manner just indicated. In other words, regardless of what Y_0 is, the examinees who would be selected would be the ones with the highest X^* scores.

The results of Subsection 5.2 may now be summarized as follows. This subsection considered an extended interpretation of the "fairness" definition of Thorndike (1971), an interpretation in which it is stipulated that probability of selection based on the test is to be equal to probability of success on the criterion for each of a number of groups that are distinguished from one another by a set of variables *including* test score. Roughly speaking, the essence of this extension of the "fairness" concept of Thorndike (1971) was to require that probability of selection be equal to probability of success both for individuals with low test scores and for individuals with high test scores. Under this extended interpretation of the "fairness" requirement of Thorndike (1971), the final implication is that, in effect, a random-error term must be added to the test score in order for the use of the test to be considered "fair."

Thus Subsections 5.1 and 5.2 followed similar paths. Each subsection examined a particular definition of test "fairness" for a particular situation, and each concluded that it was necessary to add random-error terms to test scores in order to satisfy the definition of "fairness." Since this adulteration of the test scores with random-error terms would generally be regarded as undesirable, one is inclined to question in each case whether the definition of "fairness" was appropriate in the first place, even though originally each definition may have appeared (at least on the surface) to be reasonable.

Of course, the results of Section 5 do not imply that questions of test fairness are unimportant or should be disregarded. Rather, the purpose of Section 5 has been to suggest that trying to define test "fairness" may be a hazardous undertaking. It would appear that some aspects of trying to define test "fairness" may involve difficult if not intractable problems.

REFERENCES

Angoff, W. H. (1971). Scales, norms, and equivalent scores. In R. L. Thorndike (Ed.), *Educational Measurement* (2nd ed.). Washington, D. C.: American Council on Education, 508–600.

Cole, N. S. (1973). Bias in selection. *Journal of Educational Measurement, 10,* 237–255.

Cramér, H. (1951). *Mathematical Methods of Statistics.* Princeton, N.J.: Princeton University Press.

Cross, L. H., and Frary, R. B. (1978). Empirical choice weighting under "guess" and "do not guess" directions. *Educational and Psychological Measurement, 38,* 613–620.

Guttman, L. (1941). The quantification of a class of attributes: A theory and method of scale construction. In P. Horst (Ed.), *The Prediction of Personal Adjustment.* New York: Social Science Research Council, 321–347.

Hambleton, R. K., and Cook, L. L. (1977). Latent trait models and their use in the analysis of educational test data. *Journal of Educational Measurement, 14,* 75–96.

Lord, F. M. (1958). Some relations between Guttman's principal components of scale analysis and other psychometric theory. *Psychometrika, 23,* 291–296.

Lord, F. M. (1974). Estimation of latent ability and item parameters when there are omitted responses. *Psychometrika, 39,* 247–264.

Lord, F. M., and Novick, M. R. (1968). *Statistical Theories of Mental Test Scores.* Reading, Mass.: Addison-Wesley.

Potthoff, R. F. (1966). Statistical aspects of the problem of biases in psychological tests (Institute of Statistics Mimeo Series No. 479). Chapel Hill, N. C.: Department of Statistics, University of North Carolina.

Reilly, R. R. (1975). Empirical option weighting with a correction for guessing. *Educational and Psychological Measurement, 35,* 613–619.

Thissen, D. M. (1976). Information in wrong responses to the Raven Progressive Matrices. *Journal of Educational Measurement, 13,* 201–214.

Thorndike, R. L. (1971). Concepts of culture-fairness. *Journal of Educational Measurement, 8,* 63–70.

Torgerson, W. S. (1958). *Theory and Methods of Scaling.* New York: Wiley.

Discussion of "Some Issues in Test Equating"

John C. Flanagan

My remarks concerning Dr. Potthoff's paper entitled "Some Issues in Test Equating" need to be viewed in a historical setting since my active work in the field was done a number of years ago. While studying with Truman Lee Kelley at Harvard University in the early 1930s I became acquainted with Arthur S. Otis's 1922 article on test equating. In the formal courses with Dr. Kelley and my work as a research assistant with him I became acquainted with his extensive use and development of test-equating procedures in connection with the standardization of the Stanford Achievement Tests and his writings on this topic.

In 1935 on the recommendation of Dr. Kelley I joined the staff of the Cooperative Test Service as statistician under the direct supervision of Dr. Ben Wood. In the 6 years I was a member of the staff of the Cooperative Test Service, I equated the new annual forms of many tests. At that time we also developed the system of scaled scores to provide a greater degree of comparability for the scores from tests in various fields. The bulletin on *Scaled Scores* was published in 1939 and acknowledged the use of concepts developed by Dr. L. L. Thurstone, Dr. T. L. Kelley, Dr. W. A. McCall, Dr. A. S. Otis, Dr. P. J. Rulon, and Sir Francis Galton.

Because of the relative newness of the field of educational and psychological measurement at that time, a fairly extended treatment of the problems of obtaining comparable scores was presented. This included a discussion of

243

the weaknesses of some of the practices such as constructing parallel forms using comparable distributions of item difficulties, standard scores, and "best estimates" of true scores. It was suggested that the most generally useful procedure was to assume that "true" scores on the two tests being compared that are exceeded by equal proportions of the group are "comparable." In this discussion "true" scores are defined as the mean score the individual would obtain if that person took an indefinitely large number of tests with the same characteristics. This, of course, is somewhat different from the "best estimate" of the person's "true" score knowing the reliability coefficient of the test, the mean and standard deviation, and the obtained score. Of course, if the reliability coefficients for the two forms of the test being equated are quite similar and high, very little error will be introduced by using the distributions of obtained scores rather than the theoretical distributions of "true" scores.

These concepts proved very useful in equating new forms of specific aptitude tests and in obtaining comparable scores for various types of aptitudes in the Aviation Psychology Program in the United States Army Air Forces where some of the staff from the Cooperative Test Service and many other psychologists and measurement professionals spent the next five years—1941–1946. In this program we simplified the handling of scores by reducing them to a single digit (with a mid-point of five and an interval of a half standard deviation). These scores were called stanines rather than scaled scores.

In working with practical testing programs such as those mentioned above there are two distinct but closely related problems. One is the equivalence of scores for parallel forms of the same test. The other is the comparability of scores from tests intended to measure various characteristics of the individual's performance. In considering problems of test equating it is important to keep two factors in mind. The first consideration is the degree of similarity or identity in the content included in the two measures being equated. The second consideration is the extent of similarity or identity of the population samples used. If the content is practically identical in the two forms, the scores indicated to represent comparable performance on them will be the same no matter what population is used for equating them. Conversely if the sample is large and representative of either the defined national population or some other defined reference group then the scores will be comparable for that reference group regardless of differences in the content of the tests.

In practice neither of these conditions is fully met and this is why test equating requires careful attention and treatment of all the factors in the specific situation. Potthoff's paper includes some very constructive suggestions on how to design tests to avoid problems in test equating. These include suggestions on ways of scoring and weighting items and responses to smooth out distributions. The discussion of the best ways to handle omitted items and guessing is useful.

Potthoff's section on linear equating, and possible alternatives if it does not work well, contains a variety of suggestions. He points out the subjectivity usually involved in smoothing the equipercentile points to obtain a curve to indicate comparable scores for two test forms. This problem seems of special concern to mathematical statisticians because it introduces modifications that cannot be well described mathematically. The problem can be overcome to a considerable extent by using moving averages to avoid abrupt deviations in the trend. The special problems introduced when the two tests differ in reliability are also mentioned. However, it seems that this can be as readily handled by using distributions of "true" scores as by methods available through attempting linear equating.

In discussing the use of linear equating in cases where it does not work well, Potthoff suggests, "A warning could then be included to the effect that equated scores from different tests or forms are not comparable to quite the full degree that one might like them to be." Another problem he discusses is introduced with the statement, "In some cases, linear equating may work well except at one or both ends of the score distribution." He suggests as a possible remedy for this problem that high and low scores on the test be avoided by giving a different form to persons who might make such scores.

These suggestions on avoiding the use of equipercentile equating procedures in favor of linear equating seem somewhat labored. Linear equating by its very nature does a good job at the center of the score distribution in many if not most situations. It seems likely that it rarely does a satisfactory job for the easy and difficult scores. In examining the many hundreds or perhaps thousands of sets of equating points that I drew curves for between 1935 and 1951 when I wrote the discussion of equating in E. F. Lindquist's *Educational Measurement*, I cannot recall even once when I looked at the points and decided that the line of best fit was a straight line.

In closing the chapter, Potthoff discusses estimation of equating parameters and the matter of fairness. My own view is that the argument for equating subscores separately should be made even more strongly and that wherever possible an independent equating should be done to check the equating. This is especially true where a series of equatings may produce trends or where the anchor may differ somewhat in content from the latest forms. On the matter of fairness Potthoff's viewpoint appears to be somewhat similar to mine. Although he indicates "it is apparently necessary to deliberately add a random-error term to the observed score of each examinee who took the more reliable test" when equating tests with unequal reliability coefficients, he later states that these conclusions are "difficult to defend" and goes on to recommend other solutions. It should be made clear that adding "a random-error term" is not recommended under any circumstances. In my opinion the fairest procedure in all cases is to provide for each candidate the best estimate of that candidate's "true" score. If one test is highly reliable, the candidate's score should be approximately the obtained score. If the other test has zero reliability, the candidate's score should be reported at the mean of the group. Any attempt to "add a random-error term to the observed score of each examinee who took the more reliable test" appears to be definitely unfair to the candidates. Probably the "fairest" procedure is to report the obtained score or the "best estimate" of the "true" score with the expected standard error range appropriate to the form taken and the position of that score in the score distribution.

Equating Using the Confirmatory
Factor Analysis Model

Donald A. Rock

1. Introduction

The equating procedures proposed here use maximum-likelihood factor analysis procedures to estimate the equating parameters. If the data fit the factor model (either by design or good fortune), the maximum-likelihood estimate (MLE) uses all the information in the data and provides large sample statistical tests of a number of psychometric equalities that are assumed to hold but are seldom subject to appropriate tests. In its most general form, the MLE procedure outlined here would allow for (1) the simultaneous equating of several different test forms simultaneously, (2) equating based on component subtests within the common equating tests, (3) efficient and consistent estimates of the equating parameters since it uses all the available information, and (4) large sample likelihood ratio tests of the equality of certain equating parameters across both measures and populations.

247

2. Relationship between the Factor Model and Calibration of Tests

Let $\mathbf{x}$ be a vector of order p representing p observed test scores that are assumed to be all measuring the same construct. To keep things simple, we are assuming that a single common factor (f) is sufficient to explain the observed covariances. This is a narrow definition of congeneric measures (the lowest level of test equivalence) that in its broader meaning assumes that the tests have the same factor makeup but not necessarily single-factoredness. The assumption of single-factoredness should not be a limiting factor in equating tests since the methods to be described here calibrate components or subsections of tests (e.g., verbal analogies), which are more likely to meet this assumption. This is a testable assumption within the factor model and if found inconsistent with the data, then equating cannot be done using the proposed factor methods.

Classical test theory [see, e.g., Lord and Novick, 1968] decomposes a test score x_i into a true score t_i and an error e_i, i.e.,

$$x_i = t_i + e_i, \qquad i = 1, p, \tag{1}$$

where $\mathrm{cov}(t_i e_i) = 0$ and $\mathrm{cov}(e_i, e_j) = 0$, $i \neq j$.

Following Jöreskog (1971, p. 111), if the test scores are congeneric, then any pair of true scores t_i and t_j correlate unity and therefore there exists a random variable f such that t_i is linearly related to f, i.e.,

$$t_i = v_i + \Lambda_i f. \tag{2}$$

Now substituting (2) into (1) gives

$$x_i = v_i + \Lambda_i f + e_i. \tag{3}$$

Then let Σ be the variance covariance matrix of $\mathbf{x}$ and ψ be the diagonal matrix with error variances on the diagonal. Then

$$\Sigma = \Lambda \Phi \Lambda' + \Psi, \tag{4}$$

where Φ is the variance of f. Equations (3) and (4) are the basic equations of the factor analysis model with one common factor. If it is reasonable to assume that the vector $\mathbf{x}$ of test scores has a multivariate normal distribution, maximum likelihood estimates of the unknown parameters $\mathbf{v}$, Λ, Φ, and Ψ as well as their standard errors can be estimated.

It can further be shown that the following levels of test form equivalence are special cases of the preceding factor model:

(1) Parallel tests

$$\Lambda_1 = \Lambda_2, \ldots, = \Lambda_p,$$

$$\Psi_1 = \Psi_2, \ldots, = \Psi_p,$$

$$v_1 = v_2, \ldots, = v_p.$$

Thus parallel tests have equal scaling units Λ's, error variances Ψ's, and means v. In factor analysis terminology the corresponding factor loadings and uniqueness are equal.

(2) Tau-equivalent tests

$$\Lambda_1 = \Lambda_2, \ldots, = \Lambda_p,$$

$$v_1 = v_2, \ldots, = v_p,$$

i.e., the tests have equal scaling units Λ's and means v.

(3) Essentially tau-equivalent $\Lambda_1 = \Lambda_2, \ldots, = \Lambda_p$, i.e., tests have equal scaling units.

(4) Congeneric tests. The tests are all measures of the same underlying construct. COFAMM, a confirmatory factor analysis program (Sörbom and Jöreskog, 1976), allows one to test the preceding assumptions about form and test equivalence both within and across calibrating populations by systematically constraining certain of the above parameters to be equal (depending on which of the above models is assumed both within and across calibrating populations) and subsequently forming likelihood ratio tests. Under the hypothesis H_0 implied by the constrained model, the maximum-likelihood estimate of the population covariance matrix Σ is $\hat{\Sigma} = \hat{\Lambda}\hat{\Phi}\hat{\Lambda} + \hat{\Psi}$. To test the overall goodness of fit of the model under H_0 we compare it to the less restrictive hypothesis H_1 that Σ is any positive definite matrix of similar order. H_1 is the least restrictive hypothesis possible for the model and hence provides a standard for a perfect fit. The test consists of comparing the likelihood of H_0 generating the observed data (L_0) to the likelihood of H_1 generating the observed data (L_1). It is known that $-2 \log(L_0/L_1)$ is approximately distributed as χ^2 when the null is true and N is large. The degrees of freedom associated with the test are equal to the total number of independent parameters under H_1 minus the number of independent parameters under H_0.

Constraints available under H_0 include (1) fixed parameters, which have been assigned given values, (2) constrained parameters, which are unknown but yet equal to one or more other parameters, and (3) free parameters, which are unknown and not constrained to be equal to any other parameters. Equality constraints may be used both within and across populations. If the constrained model under H_0 is identified, standard errors are computed for each of the unknown parameters.

3. Equating Two Forms Using an Anchor Test

Figure 1 shows a design for equating two operational forms of a verbal test. This anchor test design leads to a two-factor model. The circles indicate factors and the boxes indicate observed variables. Spiraling is carried out so that sample 1 examinees receive x_1, x_2, x_3, and x_4 while sample 2 examinees receive x_1, x_2', x_3, and x_4. Although x_2 is constructed to be parallel to x_2', it need only be congeneric for the proposed calibration to take place. As pointed out earlier, congeneric measures will in general have different units of measurement, (unequal, Λ's) dissimilar scale origins (unequal v's) and different error variances (unequal e's). From Fig. 1, the sample 1 factor equations defining two sets of congeneric measures (i.e., two verbal congeneric measures and the math split halved congeneric measures) are

$$
\begin{bmatrix} x_1 \\ x_2 \\ x_3 \\ x_4 \end{bmatrix} = \begin{bmatrix} v_1 \\ v_2 \\ v_3 \\ v_4 \end{bmatrix} + \begin{bmatrix} 1.0 & 0 \\ \Lambda_{21} & 0 \\ 0 & 1.0 \\ 0 & \Lambda_{42} \end{bmatrix} \begin{bmatrix} f_1 \\ f_2 \end{bmatrix} + \begin{bmatrix} e_1 \\ e_2 \\ e_3 \\ e_4 \end{bmatrix} \tag{5}
$$

and

$$
\Phi_1 = \begin{bmatrix} v(f_1) & \\ v(f_1 f_2) & v(f_2) \end{bmatrix}, \qquad \Psi_1 = \begin{bmatrix} v(e_1) & & & \\ 0 & v(e_2) & & \\ 0 & 0 & v(e_3) & \\ 0 & 0 & 0 & v(e_4) \end{bmatrix},
$$

and sample 2 equations are

$$
\begin{bmatrix} x_1' \\ x_2' \\ x_3' \\ x_4' \end{bmatrix} = \begin{bmatrix} v_1' \\ v_2' \\ v_3' \\ v_4' \end{bmatrix} + \begin{bmatrix} 1.0 & 1 \\ \Lambda_{21}' & 0 \\ 0 & 1.0 \\ 0 & \Lambda_{42}' \end{bmatrix} \begin{bmatrix} f_1 \\ f_2 \end{bmatrix} + \begin{bmatrix} e_1' \\ e_2' \\ e_3' \\ e_4' \end{bmatrix} \tag{6}
$$

and

$$
\Phi_2 = \begin{bmatrix} v'(f_1) & \\ v'(f_1 f_2) & v'(f_2) \end{bmatrix}, \qquad \Psi_2 = \begin{bmatrix} v(e_1') & & & \\ 0 & v(e_2') & & \\ 0 & 0 & v(e_3') & \\ 0 & 0 & 0 & v(e_4') \end{bmatrix}.
$$

A necessary and sufficient condition for calibrating x_2 and x_2' is that the above two factor congeneric models be consistent with the data. That is, the above two-factor pattern "fits" in both populations. If the data are not consistent with this congeneric specification, then the factor model should not be used for calibrating. The reader will note that Λ_{11} and Λ_{32} are fixed at unity; this arbitrarily scales f_1 and f_2 in the same units as the x_1 and x_3 observed scores.

One might want to test other interesting assumptions at this point. For example, it may well be that the anchor test (x_1) is a "scaled-down" version

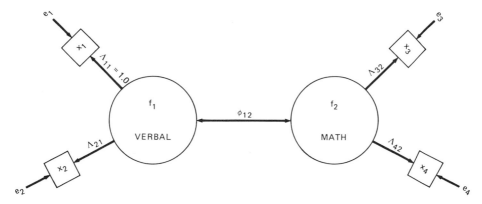

Fig. 1 Anchor test model. Sample 1: x_1, anchor test score (verbal); x_2, operational verbal score; x_3, split halved math score; x_4, split halved math score. Sample 2: same as sample 1, but we have x_2' assumed to be parallel to x_2 in sample 1.

of the operational verbal test (x_2). If the anchor test is a scaled-down parallel version of the operational test, then the scaling units Λ_{11} and Λ_{21} will be exactly proportional to their respective test lengths. (Lord and Novick, 1968). That is, if x_1 has 25 items and x_2 has 30 items, then a test of the critical assumption of parallelness (equality of scaling units) is simply to fix $\Lambda_{21} = 1.25$ and observe the increment in χ^2 remembering that Λ_{11} is already fixed at 1.0. The ability to carry out such tests is particularly important if one is vertically equating. For example, the anchor test could have been an operational test from a previous administration, in which case assumptions for both vertical and cross-sectional equating can be tested by this procedure.

The most commonly used technique for calibrating two measures is a linear procedure (Angoff, 1971; Werts et al., 1980) in which standard score deviates are set equal for the two tests. This observed-score procedure assumes that the two measures x_2 and x_2' have the same reliability, which in turn implies that the ratio of the two observed variances is equal to the ratio of true variances. In this proposed confirmatory factor approach the equal reliability assumption can be tested by changing equation (3) to $x_i = v_i + \Lambda_i f + \Lambda_i e_i$ (Werts et al., 1980). That is, the error variance is rescaled by the scaling units (factor loading) for the corresponding measure. For sample 1, the following parameter specifications would be appropriate for testing the equality of the reliabilities of x_2 and x_2'.

$$\begin{bmatrix} x_1 \\ x_2 \\ x_3 \\ x_4 \end{bmatrix} = \begin{bmatrix} v_1 \\ v_2 \\ v_3 \\ v_4 \end{bmatrix} + \begin{bmatrix} \Lambda_{11} & 0 & 0 \\ \Lambda_{21} & \Lambda_{21} & 0 \\ 0 & 0 & \Lambda_{32} \\ 0 & 0 & \Lambda_{42} \end{bmatrix} \begin{bmatrix} f_1 \\ e_2 \\ f_2 \end{bmatrix} + \begin{bmatrix} e_1 \\ 0 \\ e_3 \\ e_4 \end{bmatrix} \qquad (7)$$

and

$$\Phi = \begin{bmatrix} 1.0 & & \\ 0 & v(e_2) & \\ r(f_1 f_2) & 0 & 1.0 \end{bmatrix}, \quad \Psi = \begin{bmatrix} v(e_1) & & & \\ 0 & 0 & & \\ 0 & 0 & v(e_3) & \\ 0 & 0 & 0 & v(e_4) \end{bmatrix}.$$

Similar design matrices would be posed for sample 2. The above model is first run without and then with the constraint that $v(e_2) = v(e_2')$. The increase in χ^2 with one degree of freedom is a test of the hypothesis that the reliability of x_2 equals that of x_2'. If $v(e_2)$ is not equal to $v(e_2')$, the appropriate MLE estimate of the reliabilities are $1 - v(e_2)/\sigma_{x_2}^2$ and $1 - v(e_2')/\sigma_{x_2'}^2$. The reader should note that we are using information from the mathematics measure as well as the anchor test in estimating the reliability parameters.

If the reliabilities are equal, then the model specification in (7) can be rerun with the additional constraint that $v_2 = v_2'$. This specification provides a χ^2 test with 1 df of the equality of the intercepts of x_2 and x_2' conditional on equal reliabilities. If the data are consistent with this constrained model, then one might test the parallelness of the two forms by constraining $\Lambda_{21} = \Lambda_{21}'$, $v_2 = v_2'$ and $e_2 = e_2'$ in model specifications (5) and (6). Clearly if the data are consistent with this parallel model, one does not need to equate.

The weakest form of equating would be the case where only the congeneric model specified in (5) and (6) was found to be consistent with the data. That is, none of the more constrained models fit the data. Since in this more general case the two measures would have different units of measurement ($\Lambda_{21} \neq \Lambda_{21}'$), dissimilar scale orgins ($v_2 \neq v_2'$), different error variances ($e_2 \neq e_2'$), and unequal reliabilities, the two forms would not be interchangeable. As Lord points out (Angoff, 1971), there is no single transformation of units for unequally reliable tests that will render scores interchangeable. However, in practice the need to make congeneric measures comparable may arise. This type of equating we shall simply refer to as calibrating. Calibrating two or more measures that have been shown to be congeneric [i.e., models (5) and (6) hold] can be accomplished by subtracting the v_i constants from each x_i scale and then dividing each x_i by its respective factor loading (Λ_i). This can be shown to be equivalent to true-score equating as described by Potthoff (this volume). This follows since the standardized factor loadings are equivalent to the true-score standard deviations in Potthoff's true-score model (Potthoff's paper, this volume). The difference here being that these true-score standard deviation estimates are MLE estimates that use all available information. Thus they have the properties of being both efficient and consistent estimates.

4. Section Pre-Equating

We will now consider an operational test x with, for example, three operational verbal section scores x_1, x_2, and x_3 and three new yet-to-be-scaled sections x_1', x_2', and x_3'. We are restricting this example to just three sections but the method easily generalizes to K sections. Corresponding new and old sections are theoretically constructed to be parallel but only need be congeneric from the viewpoint of the factor approach to calibration. The three different tests and at least one of the new sections are spiraled so that three random samples generate the following data patterns.

	x_1	x_2	x_3	x_1'	x_2'	x_3'
S_{1a}	*	*	*	*	—	—
S_{2b}	*	*	*	—	*	—
S_{3c}	*	*	*	—	—	*

Note: * is observed data; — is missing data.

The problem here is, of course, to provide an equated total score $x' = x_1' + x_2' + x_3'$. A pictorial representation of the factor models to be fitted in each of the three samples is shown in Fig. 2. As before the circles indicate factors and the boxes indicate observed variables. The arrows indicate that the observed variables are regressed on their respective factors; thus in sample 1, Λ_{1a}' is the regression of the new section x_{1a}' on the first factor. Since the regression coefficients (Λ's) are ratios of the scale units of the x_i to those of the factors (f_i), we have to fix the units of each f. In sample a, we will fix Λ_{1a} at unity; then the units of f_{1a} will be identical to those of x_{1a}. Consequently, if x_{1a}' (the new section) is at least essentially tau equivalent to x_{1a}, then $\Lambda_{1a} = \Lambda_{1a}' = 1.0$ within sampling error. To test the essentially tau equivalence of each of the new sections with their corresponding old sections we would set

$$\Lambda_{1a} = \Lambda_{1a}' = \Lambda_{2b} = \Lambda_{2b}' = \Lambda_{3c} = \Lambda_{3c}' = 1.0.$$
$$\text{Sample } a \qquad \text{Sample } b \qquad \text{Sample } c$$

One would get the maximum-likelihood estimate of Σ conditional on this constrained model. If the likelihood ratio suggests that the data are not inconsistent with the constrained model, we have the enviable situation that the corresponding new and old sections appear to at least have the same scale units.

In general, we would not immediately test for essentially tau equivalence but would first test whether the corresponding sections were congeneric. If

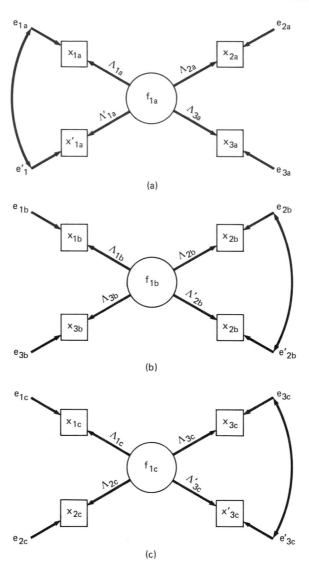

Fig. 2 (a) Sample a, (b) Sample b, (c) Sample c.

they are found not to be, then we cannot equate within the factor model. The following parameter matrices provide tests of the preceding congeneric single-factor model.

Ignoring for the moment the vector of intercepts we have

$$\text{Sample } a$$

$$
\begin{bmatrix} x_{1a} \\ x'_{1a} \\ x_{2a} \\ x_{3a} \end{bmatrix} = \begin{bmatrix} 1.0 \\ \Lambda'_{1a} \\ \Lambda_{2a} \\ \Lambda_{3a} \end{bmatrix} [f_1] + \begin{bmatrix} e_{1a} \\ e'_{1a} \\ e_{2a} \\ e_{3a} \end{bmatrix} \quad \text{and} \quad \Phi_a = v(f)
$$

$$
\text{and} \quad \Psi_a = \begin{bmatrix} v(e_{1a}) & & & \\ v(e_1 e'_1) & v(e'_{1a}) & & \\ 0 & 0 & v(e_{2a}) & \\ 0 & 0 & 0 & v(e_{3a}) \end{bmatrix}. \quad (8)
$$

$$\text{Sample } b$$

$$
\begin{bmatrix} x_{1b} \\ x_{2b} \\ x'_{2b} \\ x_{3b} \end{bmatrix} = \begin{bmatrix} 1.0 \\ \Lambda_{2b} \\ \Lambda'_{2b} \\ \Lambda_{3b} \end{bmatrix} [f] + \begin{bmatrix} e_{1b} \\ e_{2b} \\ e'_{2b} \\ e_{3b} \end{bmatrix} \quad \text{and} \quad \Phi_b = v(f)
$$

$$
\text{and} \quad \Psi_b = \begin{bmatrix} v(e_{1b}) & & & \\ 0 & V(e_{2b}) & & \\ 0 & V(e_2 e'_2) & V(e'_{2b}) & \\ 0 & 0 & 0 & V(e_{3b}) \end{bmatrix}. \quad (9)
$$

$$\text{Sample } c$$

$$
\begin{bmatrix} x_{1c} \\ x_{2c} \\ x_{3c} \\ x'_{3c} \end{bmatrix} = \begin{bmatrix} 1.0 \\ \Lambda_{2c} \\ \Lambda_{3c} \\ \Lambda'_{3c} \end{bmatrix} [f] + \begin{bmatrix} e_{1c} \\ e_{2c} \\ e_{3c} \\ e'_{3c} \end{bmatrix} \quad \text{and} \quad \Phi_c = V(f)
$$

$$
\text{and} \quad \Psi_c = \begin{bmatrix} V(e_{1c}) & & & \\ 0 & V(e_{2c}) & & \\ 0 & 0 & V(e_{3c}) & \\ 0 & 0 & V(e_3 e'_3) & V(e'_{3c}) \end{bmatrix}. \quad (10)
$$

The tau model specification not only estimates the scaling units (Λ's) and error variances $V(e$'s) but also the possible covariance between correlated "method" errors. That is x_{1a} and x'_{1a} covary because they share common trait variance (e.g., both are measures of verbal ability) but they also may share method variance (e.g., they could both be reading item formats). The remaining measures in sample a, x_{2a} and x_{3a}, share common trait variance but are characterized by differing item types and thus do not share method variance with either each other or x_{1a} or x'_{1a}. There is one degree of freedom to test the goodness of fit of these models within each sample. If the correlated errors are small compared to their standard errors, then the model can be rerun with the Ψ matrices constrained to be diagonal. This flexibility allows one to calibrate two measures on the basis of shared trait variance while shared variance due to common methods may be excluded in the calibrating process.

Assuming that the tau equivalent models are consistent with the data (with or without the specification of correlated errors due to method variance), the next step would be to test the equality of the reliabilities of the corresponding sections within each sample.

Similar to the anchor-test model, the following parameter matrices would be specified in sample a

$$
\begin{bmatrix} x_{1a} \\ x'_{1a} \\ x_{2a} \\ x_{3a} \end{bmatrix} = \begin{bmatrix} \Lambda_{11} & \Lambda_{11} & 0 \\ \Lambda'_{11} & 0 & \Lambda'_{11} \\ \Lambda_{21} & 0 & 0 \\ \Lambda_{31} & 0 & 0 \end{bmatrix} \begin{bmatrix} f \\ e_1 \\ e'_1 \end{bmatrix} + \begin{bmatrix} 0 \\ 0 \\ e_2 \\ e_3 \end{bmatrix},
$$

$$
\Phi = \begin{bmatrix} 1.0 \\ 0 & V(e_1) \\ 0 & 0 & V(e'_1) \end{bmatrix}, \qquad \Psi = \begin{bmatrix} 0 \\ 0 & 0 \\ 0 & 0 & v(e_2) \\ 0 & 0 & 0 & v(e_3) \end{bmatrix}.
$$

(11)

Similar parameter matrices would be posed within samples b and c. The test of equal reliabilities for x_{1a} and x'_{1a} based on the preceding model would be to estimate the unknown parameters with and without the constraint $V(e_1) = V(e'_1)$. In sample b the parallel test would, of course, be $V(e_2) = V(e'_2)$ and similarly in the third sample $V(e_3) = V(e'_3)$. Stronger models such as parallelness could also be tested assuming that equal reliabilities hold. Assuming that at least the congeneric model holds, the next step is the estimation of statistical characteristics of the new composite test x'.

5. Estimating the Statistical Characteristics of x'

To calibrate the mean $\bar{x}'$ we have $\bar{x}' = \bar{x}'_1 + \bar{x}'_2 + \bar{x}'_3$ where the $\bar{x}'_i$'s are the maximum-likelihood estimates conditional on the highest-order equivalence model that was fitted (e.g., essentially tau equivalent). The variance $\sigma^2_{x'}$ is

$$
\sigma^2_{x'} = \sum_{i=1}^{3} \sigma^2_{x_i} + 2 \sum \mathrm{cov}(x'_i x'_j).
$$

The covariances are estimated as

e.g.,
$$
\mathrm{cov}(x'_i x'_j) = \Lambda'_i \Phi \Lambda'_j,
$$
$$
\mathrm{cov}(x'_1 x'_2) = \Lambda'_1 \Phi \Lambda'_2,
$$

where Λ'_1 is estimated in the first sample and Λ'_2 in the second sample. Φ is estimated from data from all three samples, conditional on a well-fitting model that constrains Φ to be unknown but equal across the three populations.

The estimate of reliability of x'

$$r_{x'x'} = 1 - \left[\Sigma V(e_i')\right]/\sigma_{x'}^2$$

where the error-variance Ψ's are taken from the appropriate sample estimates.

6. Summary

The accuracy of equating parameters depends to a certain extent on how well the measures being calibrated meet certain psychometric assumptions. In the past, systematic procedures for testing the adequacy of the assumptions and the implications for violations were not readily available. However, given certain equating designs MLE factor analysis procedures can be used to test these assumptions as well as arrive at consistent estimates of the equating parameters. This paper outlines parameter estimation procedures as well as tests of the equating assumptions for both an anchor-test design and a section pre-equating scheme. The more crucial advantages of this approach over the traditional calibrating methods are as follows:

(a)   Given appropriate data, an explicit test is made of the assumption that the measures are indicators of the same trait.

(b)   The scaling constants used for calibration are specially relevant to calibration and are generated in the population in which the data arise.

(c)   Any assumptions about equal reliabilities are tested for fit to the data. If justified, the calibration constants can be estimated with this constraint in the model.

(d)   All available information is used in arriving at consistent estimates of the scaling parameters.

REFERENCES

Angoff, W. H. (1971). Scales, norms and equivalent scores. In R. L. Thorndike (Ed.), *Educational Measurement*. Washington, D. C.: American Council on Education, 508–600.

Jöreskog, K. G. (1971). Statistical analysis of sets of congeneric tests. *Psychometrika, 36,* 109–133.

Lord, F. M., and Novick, M. R. (1968). *Statistical Theories of Mental Tests Scores*. Reading, Mass.: Addison-Wesley.

Sörbom, D., and Jöreskog, K. G. (1976). *COFAMM-Confirmatory Analysis with Model Modification*. Chicago, Ill.: National Educational Resources, Inc.

Werts, E. E., Grandy, J., and Schabacker, W. H. (1980). A confirmatory approach to calibrating congeneric measures. *Multivariate Behavioral Research, 15,* 109–122.

Discussion of "Equating Using the Confirmatory Factor Analysis Model"

Leon Jay Gleser

The strategy proposed by Rock (this volume) is to perform preliminary likelihood ratio tests on equating data in order to choose the simplest model, in a hierarchy of factor-analytic models, that fits the data. Once the model is chosen, equating formulas are to be based on maximum-likelihood estimators of the parameters of that model. Because their homogeneity of content makes it reasonable to assume that their scores are generated by a single dimension (factor) of ability, subsections of tests are to be equated or "calibrated," rather than the tests as a whole. The use of subsections has the added advantage of allowing a greater variety of equating designs to be utilized.

Unfortunately, as noted in Section 1 of this discussion, the reduced number of items included in subsection scores may result in a poor fit to the factor-analytic models utilized by Rock. Guidelines for choosing items to improve the fit of factor-analytic models to subsection scores over a wide range of individual abilities are suggested. It is also noted that if these guidelines are followed, increased reliability of subsection scores can result.

In Section 2, we point out some potential weaknesses of preliminary testing as part of a methodology for estimation, note that use of factor-analytic models in mental testing contexts implies certain inequality restraints on factor loadings not accounted for in Rock's methodology, and emphasize the central importance of checking for equal subsection reliabilities before attempting to equate subsection scores.

259

Finally, in Section 3, we comment on an alternative strategy for equating tests that is based on choosing appropriate weighted sums of subsection scores.

1.   Item Design for Subsection Scores

The typical mental ability test utilizes items (such as multiple-choice items) with item scores Y_i having one of two possible values: $Y_i = u$ for a correct response and $Y_i = w$ for an incorrect response. (A third possible value, assigned to omitted items, will not be discussed here.) An *item characteristic curve model* for Y_i based on a single underlying measure (factor) of ability T for an individual is

$$P\{Y_i = u|T\} = p_i(T),$$
$$P\{Y_i = w|T\} = 1 - p_i(T) \equiv q_i(T),$$

(1)

where $p_i(T)$ is a cumulative distribution function (cdf). For a test (or subsection) of n items with scores $Y_1, Y_2, \ldots, Y_n$, it is also assumed that

given T, the Y_i's are mutually statistically independent.   (2)

Any population of individuals is assumed to have abilities T ranging over some interval of the real line, with distribution governed by the density function $f(T)$. Finally, the summary score X for an individual on the test (or subsection) is defined by the total, $X = \sum_{i=1}^{n} Y_i$, of the item scores.

Using (1) and (2), it is easily seen that

$$E(X|T) = nw + (u - w) \sum_{i=1}^{n} p_i(T) \equiv nw + (u - w)p.(T), \qquad (3)$$

$$\text{Var}(X|T) = (u - w)^2 \sum_{i=1}^{n} p_i(T)q_i(T). \qquad (4)$$

On the other hand, a factor-analytic model for X based on the factor (or *true score*) T requires that for some constants a, b, σ^2,

$$E(X|T) = a + bT, \qquad (5a)$$

$$\text{Var}(X|T) = \sigma^2 \qquad \text{(independent of } T), \qquad (5b)$$

given T, $e = X - E(X|T)$ has a normal distribution.   (5c)

When $n = 1$ (so that $X = Y_1$), the equations (5a) and (5b) can hold as approximations to (3) and (4) only over a very narrow interval of T values, while (5c) cannot hold at all. When n is large, it is usually asserted that (5a)–(5c) provide a close approximation to the distribution of X "because of the central limit theorem." For many mental ability *tests*, experience does show that the approximation is usually fairly good, although equating formulas based on the model (5a)–(5c) do not do particularly well at extremes of ability. Even though the central limit theorem may help to explain why (5c) holds to an approximation, the fit of (5a) and (5b), as will be seen, is less due to the large numbers of items in such tests than to their diversity in terms of difficulty.

Since subsection scores X are based on a smaller number n of items, and also involve items of less diversity, one must be cautious about assuming, as Rock (this volume) does, that (5a)–(5c) provide a good fit to the distribution of X. Since equating formulas based on factor-analytic models are functions only of first and second sample moments, it is worth considering how best to *design* subsections by choosing items that permit (5a) and (5b) to hold approximately over as wide an interval of individual abilities T as possible.

Let $p_i^{(r)}(T)$ be the rth derivative of $p_i(T)$ with respect to T, $r = 1, 2$, $1 \le i \le n$. The closeness to linearity [see (5a)] of (3) depends on the closeness of

$$p^{(2)}(T) = \sum_{i=1}^{n} p_i^{(2)}(T) \tag{6}$$

to zero. If all items have identical item characteristic curves (*ICCs*) $p_i(T) \equiv p(T)$, then $p^{(2)}(T) = p^{(2)}(T)$ and can be approximately zero only over a narrow range of T values. However, suppose that the ICCs $p_i(T)$ are *S-shaped*:

$$p_i^{(2)}(T) \text{ is } \begin{cases} >0, & T < T_i^*, \\ =0, & T = T_i^*, \\ <0, & T > T_i^*, \end{cases}$$

$1 \le i \le n$. Then, by choosing items such that the *item modal points* T_i^* are dispersed over the real line, one can ensure that for a wide range of values of T, $p^{(2)}(T)$ is the sum of both negative and positive terms, thus broadening the range of T values for which $p^{(2)}(T) \approx 0$ and (3) is approximately linear.

In particular, suppose that

$$p_i(T) = H(c_i(T - d_i)), \qquad i = 1, 2, \ldots, n, \tag{7}$$

where $H(u)$ is an S-shaped cdf with modal point $u = 0$. In this case the item modal points $T_i^* = d_i$, $1 \le i \le n$, are called *item difficulties*, and the preceding discussion leads to the recommendation that items should be chosen with item difficulties spread out over the real line.

The closeness to constancy [see (5b)] of (4) depends on the closeness of

$$\frac{d}{dT} \sum_{i=1}^{n} p_i(T)q_i(T) = \sum_{i=1}^{n} \left[p_i^{(1)}(T)(1 - 2p_i(T)) \right] \qquad (8)$$

to zero. Noting that $p_i^{(1)}(T) \geq 0$, $1 \leq i \leq n$, all T, the range of T values at which (8) is close to zero can be extended by making sure that for every fixed T, $p_i(T) \leq \frac{1}{2}$ for some items i, and $p_i(T) \geq \frac{1}{2}$ for other items. Put another way, one should choose items with a restricted range of "difficulties" so that for any individual it is *not* the case that all items are too hard $[p_i(T) < \frac{1}{2}$, all $i]$, or too easy $[p_i(T) > \frac{1}{2}$, all $i]$.

Optimal designs for items to make (3) as linear as possible and (4) as constant as possible over the widest range of T values is a worthy subject for analytical investigation. The argument here, however, suggests two guidelines for this choice:

(a) Items should have varying difficulties d_i (or item modal points T_i^*);
(b) Items should not all be too hard, or all too easy, for any subgroup of individuals.

Note that these guidelines are somewhat antithetical to each other, in that (a) says to spread item difficulties as widely as possible, but (b) says not to move item difficulties to the extremes.

Note that $E(X|T)$ is always linear in $p.(T) = \sum_{i=1}^{n} p_i(T)$ and that in typical situations the transformation from T to $F = p.(T)$ is strictly mono-tone increasing. In this case F could be used as the single factor in a factor-analytic model for X. It is not hard to show that following guideline (b) helps to make $\mathrm{Var}(X|F)$ constant over as wide a range of F values as possible, and also helps to force the distribution of $e = X - E(X|F) = X - E(X|T)$ to be nearly symmetric about zero (thus yielding a closer fit to normality even when the number n of items is not large). However, choosing F as the single factor eliminates the need to make $E(X|F)$ linear in F, and thus seems to make guideline (a) unnecessary. Nevertheless, even in this situation (and also in the previous situation where T is the factor), guideline (a) helps to indicate a way of increasing the reliability

$$R(X) = \frac{E_F[(E(X|F))^2]}{E_F[(E(X|F))^2] + E_F[\mathrm{Var}(X|F)]} \qquad (9)$$

of X. The basis for this assertion can be found in results of Gleser (1975), which show that under the model (1) and (2), $\mathrm{Var}(X|F)$ is minimized, for fixed $F = \sum_{i=1}^{n} p_i(T)$, by choosing items for which the $p_i(T)$ values are as widely dispersed as possible. Therefore, in either case, guideline (a) modified by guideline (b) gives test designers a way of choosing items so as both to achieve a good fit to a single-factor model (5a)–(5c) for X and to maximize the reliability of X.

2. Preliminary Testing

Although estimation procedures using preliminary tests of significance may not have been previously explored in the literature on test equating, "inference procedures using preliminary tests of significance" have been of considerable interest to statisticians and econometricians (Bancroft, 1944; Kitagawa, 1963; Bock et al., 1973; Sclove et al., 1973). It is clear from the logic of significance tests (in which the null hypothesis is favored by bounding the risk of false rejection) that use of the usual levels of significance ($\alpha = 0.05$, 0.01) in preliminary tests leads to a bias in favor of choosing overly simple models (and thus to biased terminal inference procedures), particularly in samples of small or moderate size. In very large samples, preliminary tests at $\alpha = 0.01$ or $\alpha = 0.05$ can be overly precise in that they tend to reject null hypotheses because of deviations from such hypotheses that are irrelevant to the accuracy of the terminal inference desired (in the present case, estimation of means, factor loadings, and error variances). Statisticians who have studied preliminary testing methodology thus recommend using levels of significance for preliminary tests that are sample size dependent (Bancroft, 1968; Cohen, 1974).

In the case of preliminary tests leading to terminal estimation procedures, there is evidence in regression and ANOVA contexts that estimators which smoothly incorporate the preliminary test statistics themselves (rather than indicator functions specifying the model chosen by these tests) give superior performance when compared to preliminary test estimators (Bock et al., 1973; Sclove et al., 1973). Because of the similarities between true-score models and ANOVA models, similar results should hold in the context of estimating equating formulas.

In contrast to classical equating formula estimators, which are based on method-of-moments estimators of the parameters of a true-score (congeneric single factor) model for scores X, and thus only involve first and second moments of the distribution of X, Rock's (this volume) preliminary-testing methodology requires accurate knowledge of the shape of the distribution of X in order to set rejection regions for the various preliminary tests. Although Section 1 suggests that items in subsections can be designed to give good first and second moment fits to a single-factor model for X [assumptions (5a) and (5b)], it is not clear that one can design items so as to guarantee a good fit to assumption (5c), normality of the distributions of the errors e, particularly in the tails of the distribution. Thus Rock's claims of efficiency for his methods may not hold when his methods are applied to actual mental test data. In particular, inaccuracies in controlling the significance levels of the repeated tests required by his approach may result in errors that overwhelm any gains that may be achieved by reducing parameterization.

It should also be noted that the models which Rock proposes to utilize do not take full advantage of prior knowledge about the context in which test scores appear. Suppose that two subtest scores X and X^*, which are to be equated, obey a single-factor model

$$\begin{pmatrix} X \\ X^* \end{pmatrix} = \begin{pmatrix} v \\ v^* \end{pmatrix} + \begin{pmatrix} \Lambda \\ \Lambda^* \end{pmatrix} F + \begin{pmatrix} e \\ e^* \end{pmatrix}, \tag{10}$$

where $F \sim N(0, \phi)$,

$$\begin{pmatrix} e \\ e^* \end{pmatrix} \sim \text{BVN}\left(\begin{pmatrix} 0 \\ 0 \end{pmatrix}, \begin{pmatrix} \psi & 0 \\ 0 & \psi^* \end{pmatrix} \right),$$

F and $(e, e^*)'$ are statistically independent, and $\Lambda = 1$. In this model F serves to measure the common dimension of ability tested, and the two subtests would be poorly designed indeed if X and X^* did not both vary in the same manner with F. That is, since $\Lambda = 1$, it can be assumed in advance that

$$\Lambda^* \geq 0. \tag{11}$$

Rock's (this volume) methodology does not take the restriction (11) into account, either when searching for a model or when fitting the model eventually chosen by preliminary testing. [In contrast, classical equating formulas based on true-score models implicitly account for (11) in their formulas.] The computer program COFAMM advocated by Rock for use in his methodology neither imposes restriction (11) nor allows it to be imposed by users of the program. Thus Rock's methodology does not achieve maximum efficiency in test-equating contexts, even when model (10) holds. [Ignoring (11) in equating contexts should result in the loss of about the same amount of information from the data as would be lost by ignoring the fact that the intraclass correlation ρ must be nonnegative in components of variance ANOVA contexts.]

Actually, the Fletcher–Powell optimization method used by COFAMM to fit models such as (10) should be easy to modify to take account of inequality restraints on factor loadings. Such a modified program may already exist—in which case, it should be used in place of COFAMM to implement Rock's methodology. At the very least, if COFAMM is used for finding equating formulas, users should check the fitted value of Λ^* before accepting the results obtained.

It is not at all clear that Rock's attempts to choose the simplest factor model compatible with the data can produce a gain in efficiency of estimation large enough to justify the trouble that must be taken to test the hierarchy of models listed by Rock. On the other hand, the tests of significance given by Rock can be useful *after* equating the data to tell the user how successful the equating process has been. However, one preliminary test that is necessary *before* equating the data is a test for *equal reliabilities*, since it is known that observed-score equating of two test scores X, X^* obeying model (10) is *impossible* unless these scores have equal reliabilities. This aspect of Rock's methodology should be seriously considered even by psychometricians who intend to use alternative equating methods.

These remarks are based upon the following theorem.

Theorem 1.   *Two test scores X and X^* obeying a single-factor model* (10) *are observed-score equatable if and only if they have equal reliabilities* $R(X)$, $R(X^*)$.

Theorem 1 appears to be well known. For example, in a survey of true-score equating methods, Peterson (1975) distinguishes between methods appropriate for equally reliable tests and methods for unequally reliable tests. In so doing, she notes that the former type of equating method achieves observed-score equating, while the latter type of method can yield only true-score equated test scores.

If X and X^* obey model (10), the quantity F serves as the "true score." The scores X and X^* can be (linearly) *true-score equated* if there exist constants a and b such that

$$aX^* + b = v + \Lambda F + ae^*;\tag{12}$$

that is, if

$$a\Lambda^* = \Lambda, \qquad av^* + b = v.\tag{13}$$

It is clear from (13) that as long as $\Lambda^* \neq 0$, true-score equating of X and X^* is always possible and is uniquely achieved by $a = \Lambda(\Lambda^*)^{-1} = 1/\Lambda^*, b = v - v^*(\Lambda^*)^{-1}$. On the other hand, X and X^* can be (linearly) *observed-score equated* if there exist constants a and b such that (12) holds and also ae^* has the same distribution as e. It follows from (10), (13), and (9) that X and X^* can be observed-score equated if and only if

$$R(X) = \phi/(\phi + \psi) = (\Lambda^*)^2\phi/[(\Lambda^*)^2\phi + \psi^*] = R(X^*).\tag{14}$$

This argument provides a short proof of Theorem 1.

Rock (this volume), in his first example, suggests a clever way of utilizing COFAMM to test hypothesis (14), using the easily derived fact that (14) is equivalent to

$$(\Lambda^*)^2\psi = \psi^*.\tag{15}$$

His method involves using e^* as a second factor for X^*, while requiring the error (residual) variance of X^* to be set to zero. (In Rock's example, $x_2 = X$, $x_2' = X^*$, and his x_3, x_4 are irrelevant to the procedure.) An anchor test (x_1, x_1') is required for the method to be valid. However, Rock's test of (14) is a large-sample test, and the literature on factor analysis is not clear as to how accurately the level of significance of such tests can be fixed in samples of practical size (particularly considering the unusual nature of the restriction imposed on the error variance of X^*), or whether such a test has power against reasonable alternatives to (14). Again, as in all of Rock's methodology, setting the level of significance at conventional values (0.05 or 0.01) will bias conclusions in favor of the null hypothesis of equal reliabilities, and thus can lead to false conclusions of observed-score equatability. Consequently, at the very least, users of this test should be prepared to use larger levels of significance (0.10, 0.25) so as to assure that unequal reliability situations of practical importance can be detected.

Suppose that the null hypothesis of equal reliabilities is rejected. Rock's (this volume) suggestion in such cases is to proceed with equating, accepting the fact that only true-score equating is achievable. However, Potthoff (this volume) notes that individuals with high ability would not be indifferent between true-score equated tests that are unequal in accuracy (do not have equal reliabilities). From this perspective, Rock's suggestion could result in a mental testing strategy that is not "fair" to all individuals (in the sense that it would matter to an individual which test of a true-score equated pair of tests was actually given). The goal of achieving testing strategies that are regarded as "fair" thus argues in favor of a different approach than that advocated by Rock. In Section 3 it is argued that by using weighted sums of subsection scores to summarize tests, it may be possible to observed-score equate tests even when the subsection scores for these tests are not equally reliable.

3.   Weighting Subsection Scores

Suppose that an old test τ_0 with m subsection scores $Z_1, Z_2, \ldots, Z_m$, and summary score $Z = \sum_{i=1}^m Z_i$ is to be replaced by a new test τ with m subsection scores $X_1, X_2, \ldots, X_m$. It is assumed that if both tests (τ_0 and τ) were given to the same randomly chosen individual, then Z and $X_1, X_2, \ldots, X_m$ would obey a single-factor model

$$Z = v_0 + F + e_0, \qquad X_i = v_i + \Lambda_i F + e_i, \qquad 1 \le i \le m, \qquad (16)$$

where the v_i's, $0 \leq i \leq m$, and Λ_j's, $1 \leq j \leq m$, are constants, $F, e_0, e_1, \ldots, e_m$ are mutually statistically independent, $F \sim N(0, \phi)$, and

$$e_i \sim N(0, \psi_i), \qquad 0 \leq i \leq m.$$

Typically, the summary score $X = \sum_{i=1}^{m} X_i$ is used for the test τ, and an attempt is made to linearly observed-score equate Z and X. As has been seen in Section 2, this is possible if and only if X and Z are equally reliable. On the other hand, Rock (this volume) suggests designs and a methodology in which each subsection score X_i is to be linearly observed-score equated to the corresponding subsection score Z_i, $1 \leq i \leq m$. Again, for this to be possible, X_i and Z_i must be equally reliable, $1 \leq i \leq m$. Futher, if the linear transformation of X_i that observed-score equates X_i to Z_i is $a_i X_i + b_i$, then the resulting summary score for the test τ that is linearly observed-score equated with Z is

$$X^* = \sum_{i=1}^{m} b_i + \sum_{i=1}^{m} a_i X_i = b + \sum_{i=1}^{m} a_i X_i. \tag{17}$$

Except for the constant b, this summary score is a *weighted sum* of the subsection scores X_i, $1 \leq i \leq m$. The requirement that $R(Z_i) = R(X_i)$, $1 \leq i \leq m$, where $R(W)$ is the reliability of W, appears to be even more difficult to achieve than the requirement $R(Z) = R(X)$ needed to observed-score equate Z and X. Instead, if the weights a_i are chosen in (17) to achieve $R(X^*) = R(Z)$, then X^* and Z can be linearly observed-score equated.

The fact that it is often possible to choose $a_1, a_2, \ldots, a_m$ so as to make $R(Z) = R(X^*)$ is a consequence of the following theorem.

Theorem 2. *Let $X_1, X_2, \ldots, X_m$ obey model (16), and let*

$$R_i = \frac{\Lambda_i^2 \phi}{\Lambda_i^2 \phi + \psi_i} \tag{18}$$

be the reliabilities of X_i, $1 \leq i \leq m$. Let

$$R(\mathbf{a}) = R\left(\sum_{i=1}^{m} a_i X_i\right) \tag{19}$$

be the reliability of $\sum_{i=1}^{m} a_i X_i$, for any vector $\mathbf{a} = (a_1, \ldots, a_m)$ of nonnegative weights $a_1, a_2, \ldots, a_m$; $a_i \geq 0$, $1 \leq i \leq m$. Then

$$\min_{1 \leq i \leq m} R_i \leq R(\mathbf{a}) \leq \frac{\sum_{i=1}^{m} R_i(1 - R_i)^{-1}}{1 + \sum_{i=1}^{m} R_i(1 - R_i)^{-1}} \tag{20}$$

with the upper bound in (20) achieved by

$$\mathbf{a}^* = (a_1^*, \ldots, a_m^*), \quad a_i^* = R_i^{1/2}(1 - R_i)^{-1}(\mathrm{Var}\, X_i)^{-1/2}, \qquad 1 \leq i \leq m, \tag{21}$$

and the lower bound in (20) *achieved by*

$$\mathbf{a}^{**} = (a_1^{**}, a_2^{**}, \ldots, a_m^{**}), \qquad a_i^{**} = \begin{cases} 1 & if \quad i = i_0, \\ 0 & if \quad i \neq i_0, \end{cases} \qquad (22)$$

where i_0 is any index i satisfying $R_{i_0} = \min_{1 \leq i \leq m} R_i$. Further, for any number R between the upper and lower bounds in (20), *there exists* $\mathbf{a} = (a_1, \ldots, a_m)$, $a_i \geq 0$, $1 \leq i \leq m$, *for which $R(\mathbf{a}) = R$.*

Proof. It follows from (16) and (19) and the definition of reliability of a test score, that

$$\bar{R}(\mathbf{a}) = \mathrm{Var}(\sum_{i=1}^{m} a_i \Lambda_i F) / [\mathrm{Var}(\sum_{i=1}^{m} a_i \Lambda_i F) + \mathrm{Var}(\sum_{i=1}^{m} a_i e_i)]$$

$$= (\sum_{i=1}^{m} a_i \Lambda_i)^2 \phi / [(\sum_{i=1}^{m} a_i \Lambda_i)^2 \phi + \sum_{i=1}^{m} a_i^2 \psi_i]. \qquad (23)$$

Let $\Lambda = (\Lambda_1, \Lambda_2, \ldots, \Lambda_m)$,

$$D_\psi = \begin{pmatrix} \psi_1 & & & & 0 \\ & \psi_2 & & & \\ & & \ddots & & \\ & & & \ddots & \\ 0 & & & & \psi_m \end{pmatrix},$$

a diagonal $m \times m$ matrix. From (23),

$$R(\mathbf{a})/(1 - R(\mathbf{a})) = \mathbf{a}\Lambda'\Lambda\mathbf{a}'/\mathbf{a}D_\psi\mathbf{a} . \qquad (24)$$

Since $t/(1 - t)$ is nondecreasing in t, $0 \leq t \leq 1$, it follows that maxima (minima) of $R(\mathbf{a})$ can be found by finding maxima (minima) of (24). It is well known that for all vectors $\mathbf{a}$,

$$\frac{\mathbf{a}\Lambda'\Lambda\mathbf{a}'}{\mathbf{a}D_\psi\mathbf{a}'} \leq \Lambda D_\psi^{-1}\Lambda' = \sum_{i=1}^{m} \frac{\Lambda_i^2}{\psi_i} = \frac{1}{\phi} \sum_{i=1}^{m} \frac{R_i}{1 - R_i}, \qquad (25)$$

where the last equality follows from (18). The upper bound $\phi^{-1} \sum_{i=1}^{m} R_i \cdot (1 - R_i)^{-1}$ is achieved for

$$\mathbf{a} = \Lambda D_\psi^{-1} = \left(\frac{\Lambda_1}{\psi_1}, \ldots, \frac{\Lambda_m}{\psi_m}\right) = \left(\sqrt{\frac{R_1}{(1 - R_1)\psi_1}}, \ldots, \sqrt{\frac{R_m}{(1 - R_m)\psi_m}}\right) = \mathbf{a}^*,$$

since $\psi_i = (1 - R_i) \mathrm{Var}(X_i)$, $1 \leq i \leq m$. The elements of $\mathbf{a}^*$ are clearly non-negative. Also, the upper bound in (20) follows directly from (24) and (25).

To obtain the lower bound in (20), let

$$\mathbf{c} = (\mathbf{a}D_\psi\mathbf{a}')^{-1/2}\mathbf{a}D_\psi^{1/2} = (c_1, c_2, \ldots, c_m).$$

Then since $c_i \geq 0$, $1 \leq i \leq m$, and $\sum_{i=1}^{m} c_i^2 = \mathbf{cc}' = 1$, and from (18),

$$\frac{\mathbf{a}\Lambda'\Lambda\mathbf{a}'}{\mathbf{a}D_\psi\mathbf{a}'} = \mathbf{c}D_\psi^{-1/2}\Lambda'\Lambda D_\psi^{-1/2}\mathbf{c} = (\mathbf{c}D_\psi^{-1/2}\Lambda')^2$$

$$= \left(\sum_{i=1}^{m} c_i \frac{\Lambda_i}{\psi_i^{1/2}}\right)^2 \geq \left[\min_{1 \leq i \leq m}\left(\frac{\Lambda_i^2}{\psi_i}\right)\right]\left(\sum_{i=1}^{m} c_i\right)^2$$

$$\geq \left[\min_{1 \leq i \leq m}\left(\frac{R_i}{(1 - R_i)\phi}\right)\right]\sum_{i=1}^{m} c_i^2$$

$$= \frac{1}{\phi}\min_{1 \leq i \leq m}\frac{R_i}{1 - R_i}. \tag{26}$$

This lower bound is clearly achieved by $\mathbf{a}^{**}$, and the lower bound in (20) follows from (24) and (26).

Finally, as can be seen from (23) or (24), $R(\mathbf{a})$ is a continuous function over all $\mathbf{a}$ in the positive orthant of m-dimensional space, so that if

$$\min_{1 \leq i \leq m} R_i \leq R \leq \frac{\sum_{i=1}^{m} R_i(1 - R_i)^{-1}}{1 + \sum_{i=1}^{m} R_i(1 - R_i)^{-1}}, \tag{27}$$

there exists $\mathbf{a}$ in this orthant for which $R(\mathbf{a}) = R$. $\square$

Remark 1. The requirement that the elements of $\mathbf{a}$ be nonnegative follows from the desire not to weight any subsection score negatively. This desire logically follows from the assumption (see Section 2) that each subsection score is directly related ($\Lambda_i \geq 0$) to the common factor F.

Remark 2. Theorem 2 reveals how to maximize the reliability of the weighted summary score $X^* = \sum_{i=1}^{m} a_i X_i$, a desirable end in itself. However, the use of Theorem 2 advocated here, to set $R(X^*) = R(Z)$, will usually lead to a suboptimal weighting. Although suboptimal weighting may be difficult to defend from the standpoint of ideal test construction, the procedure advocated here does have the advantage of allowing us to observed-score equate X^* to Z even when $R(X_i) \neq R(Z_i)$, all i.

Remark 3. If the optimal reliability of a weighted sum of subtest scores from test τ is less than the reliability of the score Z of τ_0, the only choices seem to be either to redesign the test τ (if this is possible) or to be content with true-score equating. However, if the test τ is given together with a subsection (or subsections) from test τ_0, used as anchors, and if the subsection from τ_0 is more reliable than the corresponding subsection from τ, matching of optimal reliabilities may yet be achieved by adding the score from the subsection from test τ_0 into the summary score for test τ [what Peterson (1975) calls *included anchor scoring*]. In any case the use of weighted sums of subsection scores to summarize a test opens up a rich variety of strategies for test equating.

Of course, the reliabilities $R(Z)$, $R(X_1)$, $R(X_2)$, ..., $R(X_m)$ and the subsection score variances $\text{Var}(X_i)$, $1 \leq i \leq m$, must either be known or estimated from data. It is here that the designs and methodology of estimation mentioned by Rock (this volume) may prove useful.

REFERENCES

Bancroft, T. A. (1944). On biases in estimation due to the use of preliminary tests of significance. *Annals of Mathematical Statistics*, *15*, 190–204.

Bancroft, T. A. (1968). *Topics in Intermediate Statistical Methods* (Vol. 1). Ames: Iowa State University Press.

Bancroft, T. A. (1972). Some recent advances in inference procedures using preliminary tests of significance. In *Statistical Papers in Honor of George W. Snedecor* (T. A. Bancroft, ed.). Ames: Iowa State University Press, 19–30.

Bock, M. E., Yancy, T. A., and Judge, G. G. (1973). The statistical consequences of preliminary test estimators in regression. *Journal of the American Statistical Association*, *68*, 109–116.

Cohen, A. (1974). To pool or not to pool in hypothesis testing. *Journal of the American Statistical Association*, *69*, 721–725.

Gleser, L. J. (1975). On the distribution of the number of successes in independent trials. *Annals of Probability*, *3*, 182–186.

Kitagawa, T. (1963). Estimation after preliminary tests of significance. *University of California Publications in Statistics*, *3*, 147–186.

Peterson, M. H. F. (1975). *Statistical Methods for Establishing Equivalent Scores in Successive Generations of a Testing Program* (Tech. Rep. No. 34). Stanford, Calif.: Department of Statistics, Stanford University.

Sclove, S. L., Morris, C., and Radhakrishnan, R. (1973). Nonoptimality of preliminary test estimators for the multinormal mean. *Annals of Mathematical Statistics*, *43*, 1481–1490.

Section Pre-Equating: A Preliminary Investigation*

Paul W. Holland

Lawrence E. Wightman

1. Introduction

Section pre-equating (SPE) refers to a class of test-equating methods that may be used in certain circumstances where other methods are unsatisfactory. In this paper we will describe several SPE methods and will give some numerical results that illustrate the sampling variability of several of these methods. To our knowledge, this paper is the first to describe section pre-equating in full generality, but Wing (1980) discusses a closely related technique.

We shall only consider *observed-score* test-equating methods here and will follow the language, definitions, and notation of Braun and Holland (this volume) wherever possible. To make this paper as self-contained as possible,

* This research was supported, in part, by the Program Statistics Research Project of the ETS Research Statistics Group and by the Law School Admission Services. The College Board supplied the data used in our empirical analyses. We would like to acknowledge the help of several people—Dorothy T. Thayer for both computer programming and general research assistance; Franklin Evans, Jane Faggen, Miles McPeek, Donald B. Rubin, and E. Elizabeth Stewart for their helpful discussions throughout our research; and Ingram Olkin, I. R. Savage, and Ted. H. Szatrowski for their comments on an earlier draft.

271

we shall give a brief description of the elements that enter into any observed-score equating method. The reader is referred to Braun and Holland (this volume) and the references therein for more details and alternative points of view.

1.1. Observed-Score Test Equating

Consider two tests X and Y that are parallel in form and content and a population P of potential examinees. Associated with each examinee in P are two potential raw scores—one for test X and one for test Y. When the examinees in P are tested with either X or Y, we then observe one (but generally not both) of these two raw scores.

Let $F(x)$ be the cumulative distribution function of potential X-raw-scores over all of P and let $G(y)$ be the corresponding cumulative distribution function for the potential Y-raw-scores. Observed-score equating methods are concerned with estimating the equipercentile equating function,

$$e(y) = F^{-1}(G(y)), \tag{1}$$

which transforms the Y-raw-scores so that they have the same distribution over P as do the X-raw-scores. Linear equating corresponds to approximating the equipercentile equating function in (1) by a linear function. This approximation usually takes the form

$$e_{\text{LIN}}(y) = \mu_X + (\sigma_X/\sigma_Y)(y - \mu_Y), \tag{2}$$

where μ_X, σ_X^2, μ_Y, σ_Y^2 are the means and variances of X and Y over P, respectively. We shall discuss only linear equating methods in this paper. The linear equating function in (2) is estimated by replacing the population moments in (2) by their sample estimates. The linear equating function in (2) matches the means and variance of the transformed Y-raw-scores to the mean and variance of the X-raw-scores. [Note that the equipercentile equating function in (1) matches *all* of the moments of X and $e(Y)$.]

1.2. The Ideas behind Section Pre-Equating

With this amount of background we are now in a position to give an informal description of section pre-equating. In many testing programs at ETS, it is usually not possible to observe both the X- and the Y-raw-score for the *same examinee*. However, it is often possible to include a *small portion* of the Y-test in the X-test. This may be accomplished in a variety of ways, but the simplest is through the use of "variable sections." A test is usually made up of a number of separately timed sections. A variable section is one whose content varies from one test booklet to the next. Suppose, for example, that test X is made of four sections, X_1, X_2, X_3, and X_4, and that the raw score on the whole test is obtained by summing the section raw scores

$$X = X_1 + X_2 + X_3 + X_4. \tag{3}$$

A test booklet for test X could contain the four X-sections and, in addition, one or more variable sections. These variable sections can be used for a number of purposes, including item pretesting, experimental testing of new item types, equating sections for common-item equating, and section pre-equating. We shall refer to the sections that do not vary across the test booklets as the *operational sections*. In normal circumstances the operational section scores count toward the final score, whereas the score obtained on a variable section does not.

For concreteness, suppose that the test booklet for test X has five sections —one variable section, V, plus the four operational X-sections. The order of these sections in the test booklet might be

$$X_1, X_2, V, X_3, X_4. \tag{4}$$

Of course, the location of V could be anywhere in the test. We have merely put it in the third position for illustration. The use of V to equate Y to X via SPE would proceed as follows. Suppose that Y, too, is made up of four sections Y_1, Y_2, Y_3, Y_4 and that X_i and Y_i are parallel in form and content. Four distinct types of test booklets (or subforms) would be made containing the following test sections:

$$
\begin{array}{llccccc}
\text{Type 1} & X_1, & X_2, & Y_1, & X_3, & X_4 \\
\text{Type 2} & X_1, & X_2, & Y_2, & X_3, & X_4 \\
\text{Type 3} & X_1, & X_2, & Y_3, & X_3, & X_4 \\
\text{Type 4} & X_1, & X_2, & Y_4, & X_3, & X_4
\end{array} \tag{5}
$$

Each examinee would be tested with only one of these subforms containing all four operational sections of X and just one section from Y. The idea behind SPE is to administer the four types of test booklets to distinct random samples of examinees from P and to use the resulting combination of complete information on X and incomplete information on Y to estimate the parameters of the desired equating function. Since linear equating functions are both widely used and most easily estimated, we shall focus on them exclusively in this paper.

We wish to emphasize here that in the type of applications that we have in mind, the population P is concrete and easily described. P consists of all persons taking the test on a given administration date. For many ETS tests these populations consist of tens (or hundreds) of thousands of individuals.

We note that if our objective were merely to equate Y_i to X_i, i.e., subscore equating, then the information described in (5) is clearly sufficient. Both X_i and Y_i raw scores are observed on random samples from P and consequently we may use standard linear or equipercentile methods to equate Y to X (see Angoff, 1971). This is the problem discussed by Wing (1980). However, we are not interested in the sections per se, but, rather, in the total scores $X = X_1 + X_2 + X_3 + X_4$, and $Y = Y_1 + Y_2 + Y_3 + Y_4$. Since X is observed on all examinees, its distribution over P can be estimated directly. Y, on the other hand, is not observed *in toto* for any examinee, and its distribution must be estimated by indirect means. We shall discuss a variety of indirect methods for estimating parameters of the distribution of Y in this next section.

1.3. The Psychological Assumptions of SPE

The basic psychological assumption underlying SPE is that the task of taking a single section Y_i embedded in the operational sections of X is the same task as taking Y_i as an operational section along with the other operational sections of Y. Clearly, these are not the same physical tasks and there are reasons why they may not be the same psychological tasks as well.

One effect might be labeled "differential motivation." This would arise when the examinee realized that Y_i was in a variable section and, therefore, not likely to count toward the final score. In this situation the examinee might not work as hard on Y_i when it appeared in a variable section as when it appeared in an operational section. This results in Y_i appearing to be a harder test section when it is in the variable section than when it is in an operational section.

A second effect is often termed a "practice effect." In (5) we see that forms of type 2 have section X_2 immediately followed by a similar section Y_2. This may make Y_2 appear *easier* when it is in the variable section than when it is not. Actually, practice effects can be quite complicated because they can affect both the partial test Y_i and the sections of the operational test X_i. For example, in forms of type 3, X_3 follows Y_3 and this could make X_3 easier, because of practice, than it would be if Y_3 followed X_3.

There is no way to be sure that practice effects and differential motivation are *not* operating when variable sections are used. It is important to be aware of them and to understand how large these effects might be in a given application. From a statistical point of view these effects, should they exist, would create *bias* in the estimates of the equating functions.

In this paper we shall ignore most effects that bias the estimates and will concentrate primarily on sampling variability in the equating functions. The study of differential motivation, practice, and other effects that can bias

SPE results in a large topic worthy of systematic investigation in each specific application and is beyond the scope of this paper.

1.4. Why Is It Pre-Equating?

The prefix "pre" is used in section pre-equating because the objective is to estimate an equating function for equating Y to X on P *before* Y is ever given as an intact test. Normally, observed-score equating functions are not estimated until both X and Y have been administered as intact tests to random samples from some population(s) of examinees.

2. Section Pre-Equating (SPE)

In this section we shall discuss several methods that may be used to estimate the distribution of Y when only one or two sections of Y have been taken by each examinee. Since linear equating only requires estimates of the means and variances of X and Y raw scores, we shall only discuss methods for estimating these parameters.

2.1. The Structure of the Data in SPE

In general, we shall have two tests X and Y that consist of m sections $X_1, \ldots, X_m$ and $Y_1, \ldots, Y_m$. We shall assume that the raw scores are computed by adding up the raw scores for each section, i.e.,

$$X = \sum_{i=1}^{m} X_i, \qquad Y = \sum_{i=1}^{m} Y_i, \tag{6}$$

and we assume that X_i and Y_i are parallel in form and content. We will consider two types of tests—one-variable-section tests (1VSTs) and two-variable-section tests (2VSTs). A one-variable-section test is a simple generalization of the example in Section 1 to m operational sections. In a 1VST there are $m + 1$ sections in total in the test booklet and there are m versions or subforms of the test booklet. A two-variable-section test has two, rather than one, variable sections. In a 2VST there are $m + 2$ sections in total in the test booklet and there are, at least,[1] $m(m - 1)/2$ subforms of the test booklet.

The data obtained from a 1VST may be represented as $m + 1$ section scores.

$$(X_1, \ldots, X_m, V), \tag{7}$$

---

[1] We shall ignore the order of sections in the test booklet in this paper. If order is taken into consideration then there may be up to $m(m - 1)$ subforms.

where $V = Y_1$, or $Y_2, \ldots$, or Y . The data obtained from a 2VST may be represented as $m + 2$ section scores.

$$(X_1, \ldots, X_m, V_1, V_2), \tag{8}$$

where $(V_1, V_2) = (Y_1, Y_2)$ or $(Y_1, Y_3), \ldots$ or (Y_{m-1}, Y_m). In a 2VST all *pairs* of Y-sections appear in the combinations of the two variable sections.

We may envision a hypothetical vector of length $2m$ associated with each examinee in P. This vector is

$$(X_1, \ldots, X_m, Y_1, \ldots, Y_m). \tag{9}$$

The data obtained in SPE consist of partial information on the vector in (9). We shall assume throughout that the subforms are administered to distinct random samples of examinees in P. This can be achieved by the technique of "spiraling" described in several sources, including Braun and Holland (this volume). Table 1 illustrates a schematic representation of the data (and the missing data) that arise in both a 1VST and a 2VST.

In summary, in SPE we regard the data that are obtained as partial information about the hypothetical $(2m)$-dimension vector in (9). The objec-

TABLE 1

Schematic Representation of the Data in 1VST and 2VST

Samples	X_1	X_2	$\cdots$	X_{m-1}	X_m	Y_1	Y_2	$\cdots$	Y_{m-1}	Y_m
				(a) Data obtained in a 1VST[a]						
P_1	$\checkmark$	$\checkmark$	$\cdots$	$\checkmark$	$\checkmark$	$\checkmark$				
P_2	$\checkmark$	$\checkmark$	$\cdots$	$\checkmark$	$\checkmark$		$\checkmark$			
$\vdots$	$\vdots$	$\vdots$	$\vdots$	$\vdots$	$\vdots$			$\vdots$		
P_{m-1}	$\checkmark$	$\checkmark$	$\cdots$	$\checkmark$	$\checkmark$				$\checkmark$	
P_m	$\checkmark$	$\checkmark$	$\cdots$	$\checkmark$	$\checkmark$					$\checkmark$

Samples	X_1	X_2	$\cdots$	X_{m-1}	X_m	Y_1	Y_2	Y_3	$\cdots$	Y_{m-1}	Y_m
				(b) Data obtained in a 2VST[b]							
P_{12}	$\checkmark$	$\checkmark$	$\cdots$	$\checkmark$	$\checkmark$	$\checkmark$	$\checkmark$				
P_{13}	$\checkmark$	$\checkmark$	$\cdots$	$\checkmark$	$\checkmark$	$\checkmark$		$\checkmark$			
$\vdots$	$\vdots$	$\vdots$	$\vdots$	$\vdots$	$\vdots$				$\vdots$		
$P_{m-1,m}$	$\checkmark$	$\checkmark$	$\cdots$	$\checkmark$	$\checkmark$					$\checkmark$	$\checkmark$

[a] P_i are random samples from P; $\checkmark$ = data are observed (otherwise, data are not observed).
[b] P_{ij} are random samples from P; $\checkmark$ = data are observed (otherwise, data are not observed).

tive is to use this partial information to estimate the parameters of interest in linear equating. These parameters and several techniques for estimating them are described in the remainder of this section.

2.2. The Linear Equating Parameters

The linear equating function in (2) requires estimates of four parameters: μ_X, σ_X, μ_Y, and σ_Y. Of these, μ_X and σ_X are easily estimated directly by the sample mean and standard deviation of X-raw-scores obtained by combining all the random samples, P_i or P_{ij} from P. (P_i and P_{ij} are defined in Table 1.) To estimate μ_Y and σ_Y we must make use of the equations in (6) and their consequences. From the usual rules for computing the mean and variance of a sum we have

$$\mu_Y = \sum_{i=1}^{m} \mu_{Y_i} \tag{10}$$

and

$$\sigma_Y^2 = \sum_{i=1}^{m} \sigma_{Y_i}^2 + 2 \sum_{i<j}^{m} \mathrm{Cov}(Y_i, Y_j), \tag{11}$$

where

$$\mathrm{Cov}(Y_i, Y_j) = \rho_{Y_iY_j}\sigma_{Y_i}\sigma_{Y_j}. \tag{12}$$

Thus the device we use to estimate μ_Y and σ_Y *indirectly* is to estimate each of the parameters μ_{Y_i}, $\sigma_{Y_i}^2$, $\rho_{Y_iY_j}$ and combine them via (10) and (11) into estimates of the parameters of interest. Our strategy for estimating these subscore parameters is to estimate the mean vector,

$$\boldsymbol{\mu} = (\mu_{X_1}, \ldots, \mu_{X_m}, \mu_{Y_1}, \ldots, \mu_{Y_m}) \tag{13}$$

and the covariance matrix,

$$\Sigma = \left[\begin{array}{cc|cc} \sigma_{X_1}^2 & \rho_{X_iX_j}\sigma_{X_i}\sigma_{X_j} & & \\ & & & \rho_{X_iY_j}\sigma_{X_i}\sigma_{Y_j} \\ - & \sigma_{X_m}^2 & & \\ \hline & & \sigma_{Y_1}^2 & \rho_{Y_iY_j}\sigma_{Y_i}\sigma_{Yj} \\ & & & \\ & - & & \sigma_{Y_m}^2 \end{array} \right]. \tag{14}$$

The parameters $\boldsymbol{\mu}$ and Σ are, respectively, the mean and covariance matrix of the $(2m)$-dimensional vector in (9).

There are various ways to estimate $\boldsymbol{\mu}$ and Σ and we discuss them in the remainder of this section.

2.3. Estimating $\boldsymbol{\mu}$ and Σ from a Test with Two Variable Sections

We consider the case of a 2VST first because it is the easier of the two. A test with two variable sections provides us with data in the form of Table 1b. The random sample P_{ij} is tested with sections Y_i and Y_j. We may pool all the examinees who were tested with Y_i and the resulting sample mean and variance of Y_i are unbiased estimates of the population mean and variance of Y_i. To estimate the correlation $\rho_{Y_iY_j}$ we use the data from sample P_{ij} to estimate the population cross-moment,

$$E(Y_iY_j). \tag{15}$$

This can be combined with the estimates of μ_{Y_i}, μ_{Y_j}, $\sigma_{Y_i}^2$, and $\sigma_{Y_j}^2$ to obtain an estimate $\rho_{Y_iY_j}$. All of these estimates are *consistent* in the sense that as the size of the samples P_{ij} increases, these estimates converge to the corresponding population values. The simple approach just described is called the "pairwise present" method for estimating $\boldsymbol{\mu}$ and Σ. It is the method that many computer programs use to "handle" missing data when computing means, variances, and covariances.

An alternative to the "pairwise present" method of estimation is to use maximum-likelihood estimates of $\boldsymbol{\mu}$ and Σ under the assumption that the $(2m)$-dimensional vector in (9) has a multivariate normal distribution over the population P. This approach must use iterative procedures to maximize the likelihood since the pattern of missing data displayed in Table 1b does not lead to a simple factorization of the likelihood function (see Rubin, 1974). We used a version of the "EM algorithm" to estimate $\boldsymbol{\mu}$ and Σ by maximum likelihood (Dempster *et al.*, 1977). One reason for using the maximum-likelihood method, even though the simple pairwise present method is easier to use, is the expectation that it will generally produce estimates of $\boldsymbol{\mu}$ and Σ that are closer to the population values since it uses *all* the information in the $(m + 2)$-dimensional vector in (8). When the P_{ij} are of modest size, this could be an important consideration. Our numerical results in Section 4 bear on this issue.

2.4. Estimating $\boldsymbol{\mu}$ and Σ from a Test with One Variable Section

When the test has only *one* variable section, the structure of the data is like that given in Table 1a. Since P_i is a random sample from P_i, the sample mean and variance of Y_i provide unbiased estimates of μ_{Y_i} and $\sigma_{Y_i}^2$. However, there is no simple way to obtain an estimate of the correlation between Y_i and Y_j from a 1VST because no one takes both Y_i and Y_j in a 1VST. However, a number of possibilities exist for imputing reasonable values for these correlations. We summarize three of these below.

2.4.1. BORROWING CORRELATIONS FROM ANOTHER POPULATION

In actual use of SPE with a 1VST, X would be the operational form and Y would be a new form to be pre-equated on P. However, at some future date Y would be the operational form on a new population Q. At that time we would obtain $\rho_{Y_iY_j}$ on Q. If we assume that

$$\rho_{Y_iY_j(Q)} = \rho_{Y_iY_j(P)}, \tag{16}$$

then we have sufficient information to equate Y to X on P (but not on Q) when Y is finally used operationally. Since this method can not be used until Y has been administered as an intact test, it does not, strictly speaking, *pre*-equate Y to X.

2.4.2. BORROWING CORRELATIONS FROM ANOTHER TEST

If X_i and Y_i are parallel in form and content, then in some circumstances it may be reasonable to assume that

$$\rho_{X_iX_j} = \rho_{Y_iY_j}. \tag{17}$$

If (17) were true, then we could use the correlations among the X-sections of the test as estimates of the corresponding correlations among the Y-sections of the test, and pre-equate Y to X on P.

2.4.3. IMPUTING THE UNKNOWN PARTIAL CORRELATIONS

Even though no one takes both Y_i and Y_j in a 1VST, it is not precisely correct to say that the correlation between these two sections is completely unknown. Since X_i and Y_i (and X_j and Y_j) will be strongly and positively correlated, by design, and since the correlation between X_i and X_j on P is known, it follows that limits can be put on the possible value of $\rho_{Y_iY_j}$. The higher the correlations $\rho_{X_iY_i}$ and $\rho_{X_jY_j}$, the tighter the limits are. Rubin and Thayer (1978) discuss this fact in a related context and illustrate a method for displaying these limits. Strictly speaking, it is the *partial* correlation between Y_i and Y_j after partialing out $X_1, \ldots, X_m$ that is inestimable with the data from a 1VST. Rubin and Thayer proposed to study the sensitivity of the estimates of Σ by varying the unknown partial correlations over a plausible range and combining these values with estimates of the estimable parameters in Σ. This results in a family of estimates of Σ. We used the same approach and studied the sensitivity of the resulting estimated equating lines to the values of the inestimable partial correlations.

Since X_i and Y_i are designed to be parallel, and therefore strongly correlated, we would expect the correlation of Y_i and Y_j partialing out $X_1, \ldots, X_m$ to be small but positive. Thus, in all of the sensitivity analyses that we will discuss in Section 4, we assumed that the partial correlation matrix of

$$\mathbf{Y} = (Y_1, \ldots, Y_m) \tag{18}$$

after partialing out

$$\mathbf{X} = (X_1, \ldots, X_m) \tag{19}$$

has the simple form

$$\begin{bmatrix} 1 & \rho & \cdots & \rho \\ \rho & & & \\ \vdots & & \ddots & \\ \rho & & \cdots & 1 \end{bmatrix}, \tag{20}$$

where ρ is a small value.

3. A Preliminary Study of Section Pre-Equating

In order to test the feasibility of section pre-equating (SPE) a variety of studies need to be made. The study that we report here is merely a first step in that process. It is concerned, primarily, with the statistical assumptions that need to be made in applying the versions of SPE described in Section 2. Our major goal was to obtain estimates of the accuracy of estimated linear equating functions for various sample sizes using test data that approximated the characteristics of testing programs which might need to use SPE.

This study was carried out using data from the May 1979 administration of the Scholastic Aptitude Test (SAT) and Test of Standard Written English (TSWE). These data were selected because they were conveniently available and had all the structure that we needed for an initial test of the feasibility of SPE. Of course, they were not originally collected with the intention of testing section pre-equating. Thus there are some minor ways in which these data are not exactly suited to our needs, but generally speaking we are reasonably satisfied that they gave us useful information about the variants of SPE described in Section 2.

In this section we describe the data used in the study, our method of using it to simulate several SPE options, and the criteria we used to make our comparisons of various SPE methods. Section 4 summarizes the results of the study.

3.1. The Data

The complete test given at the May 1979 SAT administration had the structure shown in Table 2, where M1 and M2 are two math sections, V1 and V2 are two verbal sections, and T1, T2, T3 and T4 are four sections containing TSWE items. The operational test consisted of V1, M1, T1, and V2 and M2. T2, T3, and T4 were three sets of TSWE items that were being pretested at this administration. In fact, there were seven other sets of items that could appear in the variable section of the test at this administration. These 10 subforms (one for each pretest item set) were spiraled[2] so that approximate random samples of one-tenth of the total administration received each subform. This resulted in approximately 24,000 people who took each test of the form

$$V1 \ M1 \ T1 \ V2 \ M2 \ Ti, \qquad i = 2, 3, 4.$$

All of the sections except T2, T3, and T4 were operational quality test sections, which means that they had gone through a long process of revision and refinement. The pretest item sets T2, T3, T4, had not gone through this process and so the item statistics for these three pretests were compared with the operational TSWE section T1. This examination led us to believe that T2 was the closest to being of operational quality of the three pretest sections and that T3 was the next best. In order to give a partial replication of our analyses on a different group of people and on a slightly different test, we selected the group who took (V1, M1, T1, V2, M2, T2) (i.e., the T2 group) and the group who took (V1, M1, T1, V2, M2, T3) (i.e., the T3 group) as the basic populations to be used in the study.

Furthermore, because we wanted to have two different populations of examinees who took the same test, we further subdivided these groups into

TABLE 2

	Content	No. of items	Time limit (min)
Operational sections	V1	45	30
	M1	25	30
	T1	50	30
	V2	40	30
	M2	35	30
Variable section	T2	50	30
	or T3	50	30
	or T4	50	30

[2] See Braun and Holland (this volume) for a description of "spiraling."

TABLE 3

	T2 Group	T3 Group	Total	%
Male	T2M⌋	T3M⌋		
	11,517	11,360	22,877	47.7
Female	T2F⌋	T3F⌋		
	12,761	12,357	25,118	52.3
Total	24,278	23,717	47,995	100.0

males and females. The population sizes for these four groups are given in Table 3.

We labled these four groups T2M, T2F, T3M, and T3F in the obvious way. These four groups formed the populations of this study.

The overall statistics by section for the T2 and T3 groups are given in Table 4.

TABLE 4

Summary Statistics

	(a) The T2 Group Number-right raw scores				Formula raw scores[a]	
	$\bar{X}$	SD	min	max	$\bar{X}$	SD
V1	22.8	7.6	1	45	18.8	8.9
V2	21.0	6.2	1	40	17.3	7.3
M1	12.2	5.1	0	25	9.9	6.2
M2	18.6	6.1	0	35	14.6	7.5
T1	34.4	8.9	3	50	31.0	10.7
T2	34.6	8.2	0	50	31.1	9.9

	(b) The T3 Group Number-right raw scores				Formula raw scores[a]	
	$\bar{X}$	SD	min	max	$\bar{X}$	SD
V1	22.9	7.6	0	45	18.9	8.9
V2	21.0	6.2	0	40	17.3	7.3
M1	12.3	5.2	0	25	9.9	6.2
M2	18.7	6.1	0	35	14.6	7.6
T1	34.5	8.9	3	50	31.1	10.7
T3	35.9	7.9	0	50	32.6	9.6

[a] Formula raw scores are obtained by subtracting a fraction (usually one-fourth) of the wrong answers (ignoring omits) from the number right. Negative formula scores were retained.

TABLE 5

Means for Formula Raw Scores for the Four Populations

			(a) T2 Groups			
	V1	M1	T1	V2	M2	T2
T2M	19.29	11.30	30.20	17.43	15.92	30.12
T2F	18.38	8.58	31.68	17.11	13.37	31.90
			(b) T3 Groups			
	V1	M1	T1	V2	M2	T3
T3M	19.29	11.30	30.16	17.43	16.00	31.59
T3F	18.47	8.67	31.92	17.24	13.36	33.57

There is enough consistency in Table 4 to make us believe that the spiraled samples were reasonably random and consequently that T3 is slightly easier than T2 and that T2 is more comparable to T1 than is T3. We came to the same conclusion from our examination of the item statistics of the three experimental TSWE sections. To simplify this report we shall only discuss results for formula raw scores[3]; the results for number right raw scores are similar.

The mean formula scores for the various section are given in Table 5 for the four groups T2M, T2F, etc. Again we get similar results across the two spiraled samples. The sex differences are consistent with previous experience with SAT data. These differences are large enough to make our comparisons of equating on the two populations interesting. Table 6 gives the variances, covariances, and correlations between the section formula raw scores for the four groups. There is a good deal of similarity among the correlations in the four groups. Ingram Olkin has pointed out (private communication) that there is a great deal of pattern in these correlation matrices. If we rearrange them so that the parallel sections are adjacent, the four correlation matrices all look very much like the one given in Table 7. Except for a small but consistent difference between males and females in the correlation between M1 and M2 (i.e., 0.84 versus 0.81), the corresponding entries in these four correlation matrices are very similar. This is quite important because it gives credence to the methods of "borrowing correlations from other populations or other tests" discussed in Sections 2.4.1 and 2.4.2.

3.2. The "True" Equating Functions

Before describing how the data shown in 3.1 were used to test various SPE methods, we will describe our criterion for judging the success or failure of

---

[3] Formula raw scores are the number of correct responses minus a fraction (usually one-fourth) of the incorrect responses, ignoring omits. Negative formula scores were retained.

TABLE 6

Variances, Covariances, and Correlations for Section, Formula Raw Scores
for the Four Populations[a]

	V1	M1	T1	V2	M2	T2
			(a) T2M Group			
V1	84.06	0.643	0.771	0.841	0.636	0.750
M1	37.52	40.50	0.630	0.648	0.839	0.618
T1	76.82	43.58	118.15	0.771	0.629	0.859
V2	56.78	30.37	61.71	54.26	0.639	0.740
M2	44.52	40.78	52.21	35.93	58.32	0.625
T2	70.17	40.15	95.32	55.65	48.71	104.29
			(b) T2F Group			
	V1	M1	T1	V2	M2	T2
V1	75.85	0.638	0.765	0.838	0.626	0.737
M1	31.66	32.47	0.627	0.640	0.813	0.611
T1	69.41	37.25	108.63	0.769	0.630	0.861
V2	53.41	26.67	58.62	53.53	0.629	0.739
M2	39.23	33.37	47.26	33.12	51.85	0.620
T2	61.50	33.35	86.07	51.83	42.78	91.90
			(c) T3M Group			
	V1	M1	T1	V2	M2	T3
V1	81.69	0.650	0.768	0.839	0.630	0.742
M1	37.58	40.90	0.638	0.647	0.838	0.611
T1	75.18	44.19	117.26	0.764	0.624	0.853
V2	54.87	29.95	59.88	52.35	0.634	0.743
M2	43.43	40.91	51.59	35.03	58.26	0.600
T3	66.73	38.85	91.87	53.44	45.53	98.94
			(d) T3F Group			
	V1	M1	T1	V2	M2	T3
V1	76.74	0.640	0.772	0.841	0.628	0.742
M1	32.49	33.59	0.632	0.649	0.814	0.601
T1	70.46	38.17	108.53	0.772	0.637	0.859
V2	54.39	27.76	59.37	54.48	0.639	0.749
M2	40.33	34.58	48.62	34.54	53.70	0.607
T3	59.36	31.81	81.79	50.55	40.63	83.49

[a] Variances on main diagonal, covariances below the diagonal, correlations above the
diagonal.

these methods. We took as the basic problem the estimation of the linear
equating function for equating

$$Y = V2 + M2 + T2 \tag{21}$$

to

$$X = V1 + M1 + T1 \tag{22}$$

TABLE 7

The Pattern of Correlations Exhibited in Tables 4a, b, c, and d

	V1	V2	M1	M2	T1	T2(T3)
V1	1	0.84	0.64	0.63	0.77	0.74
V2		1	0.65	0.63	0.77	0.74
M1			1	0.84 M[a]		
M1				0.81 F[a]	0.63	0.61
M2				1	0.63	0.61
T1					1	0.85
T2(T3)						1

[a] Separate correlations given here for males and females.

on the population T2M. As a replication we estimated the linear equating function for equating

$$Z = V2 + M2 + T3 \qquad (23)$$

to X on the population T3M. We selected X, Y, Z [defined in (21), (22), (23)] for study because these three test scores[4] are the sum of nonparallel section scores for which the corresponding summands are nearly parallel. These two conditions were desirable for a preliminary study of SPE because

(a)   heterogeneous composite scores are a common type of test score and one for which other methods of pre-equating (notably item response theory methods) may not be successful; and

(b)   in most potential applications of SPE the corresponding sections of the test forms will be parallel in form and content.

We could have used the raw-score points of X as the metric in which we expressed the equating results, but in order to put everything into units that were more like those of many ETS tests X was transformed so that its transformed-score distribution on the T2M group had a mean of 500 and a standard deviation of 100. This was accomplished by using the linear scaling function

$$h(X) = (4.2314)X + 242.76. \qquad (24)$$

The scaling function h is applied to X-raw-scores or to the transformed raw

---

[4] We emphasize here that the "test scores" X, Y, and Z defined in (21), (22), and (23) are not *real* test scores in the sense that no reported score has ever been based on them. However, they are composed of various section scores that have been used as the basis of reported scores. Our use of X, Y, and Z is simply to *simulate* test scores having certain characteristics.

scores from Y and Z which have been equated to X. We note, in passing, that even though h gives the raw scores from X a mean of 500 and a standard deviation of 100, the largest scaled score possible on X is achieved for $X = 120$ and this maximum scaled score is 750.5. Hence, even though our objective was to give X an appropriate 200 to 800 range of scaled scores, this was not quite achieved.

Table 8 gives the population raw means and standard deviation of X, Y, and Z in the appropriate populations T2M, T2F, T3M, and T3F. From this information the "true" linear equating function (i.e.. the population value) for equating Y to X on T2M can be obtained via Eq. (2). This results in the equating function

$$e_M(Y) = (1.0596)Y - 6.4630 \qquad (25)$$

To put $e_M(Y)$ into the scaled score units we form the *conversion* function by composing $h(\cdot)$ from (24) with $e_M(\cdot)$ from (25). The resulting conversion function is

$$C_M(Y) = h(e_M(Y)) = (4.4836)Y + 215.41. \qquad (26)$$

The objective of SPE methods is to come close to the conversion function in (26) when we must estimate the parameters from smaller *samples* of incomplete data from the population. How close is "close"? There are various ways of judging "close" and we offer the following observation as a possible guideline. *All* observed-score equating methods are, in principle, population dependent. Hence, we suggest using the variation that is due to the choice of plausible populations as a reference for evaluating the magnitude of the discrepancies between the "true" linear equating function and estimates of it using SPE methods (or any other method for that matter). To see what this amounts to in our data let us compare the conversion function obtained

TABLE 8

Means and Standard Deviations for X, Y, Z[a] over the Four Populations

Group	μ_X	σ_X	μ_Y	σ_Y
T2M	60.794	23.633	63.472	22.303
T2F	58.636	22.217	62.379	21.277
	μ_X	σ_X	μ_Z	σ_Z
T3M	60.741	23.532	65.019	21.853
T3F	59.055	22.385	64.175	21.050

[a] $X = V1 + M1 + T1$, $Y = V2 + M2 + T2$, $Z = V2 + M2 + T3$.

by equating Y to X on T2F with the one given in (26). The equating function of T2F is

$$e_F(Y) = (1.0442)Y - 6.5000 \tag{27}$$

and the corresponding conversion function is

$$C_F(Y) = h(e_F(Y)) = (4.4184)Y + 215.26. \tag{28}$$

The maximum difference between C_M and C_F occurs at the upper end of the raw score scale for Y. The maximum value of Y is 125 so that the maximum difference $|C_F - C_M|$ is 8.3 scaled score points. We may do the same computation to compare the conversion functions for Z on T3M and T3F. This yields the equating functions

$$e_M(Z) = (1.0768)Z - 9.2713 \tag{29}$$

and

$$e_F(Z) = (1.0634)Z - 9.1885 \tag{30}$$

and the corresponding conversion functions

$$C_M(Z) = (4.5564)Z + 203.53 \tag{31}$$

and

$$C_F(Z) = (4.4997)Z + 203.88. \tag{32}$$

The maximum difference $|C_M - C_F|$ occurs at the upper end of the raw-score scale again and is 6.7 scaled score points.

The average of these two maximum differences in conversion functions for the T2 and T3 groups is 7.5 scaled score points. In view of this value, we may regard discrepancies between estimated and population conversion functions of 5 to 10 points as of the same order of magnitude as the discrepancies that are due to the choice of population used for the equating. This comparison must be interpreted carefully since it only applies to the populations and test scores used in this study. We do not know how widely such a comparison can be generalized to other tests and tested populations.

3.3. Using the Data to Simulate SPE

We used the data described in 3.1 to simulate the type of data that arises in one-variable-section tests (1VSTs) and in two-variable-section tests (2VSTs). We did this by dividing the population, T2M or T3M, into three essentially random groups and then *ignoring* the appropriate values of V2, M2, or T2 (T3) in each of these groups. To simulate a 1VST in which the sections of Y appear in the variable section: we ignored M2 and T2 in P_1 (the first random third of T2M); we ignored V2 and T2 in P_2 (the second random third of T2M); and we ignored V2 and M2 in P_3 (the last random third of T2M).

To simulate a 2VST in which the pairs of sections of Y appear in the variable sections: we ignored T2 in P_1; we ignored M2 in P_2; and we ignored V2 in P_3. This method of simulating data from a 1VST or a 2VST results in data like those displayed schematically in Table 1a and b.

One of the basic questions concerns the sizes that the P_i must be in order to obtain reasonable estimates of the true linear equating function (as defined in Section 3.2). Thus we varied the sizes of the P_i over the values of 500, 1500, and 3000. The results of applying various methods of SPE to the resulting data sets are described in Section 4.

4. Results

We shall first discuss the results that bear on SPE methods for 2VSTs (Section 4.1) and then discuss results that apply to 1VSTs (Section 4.2). Our primary criterion for comparing estimates of equating functions with the corresponding population value was the maximum deviation between the two corresponding conversion functions. In all cases the conversion functions were obtained by composing the equating functions with the scaling function $h(X)$ given in (24). The maximum deviation between the conversion functions was taken over the formula raw-score range of 0 to 125. A formula raw score of zero represents "chance" performance while 125 is the largest possible value of Y and Z.

Three separate data sets form the basis of the results reported here. Two of these data sets come from T2M while one comes from T3M. These three data sets will be denoted T2M(1), T2M(2), and T3M(3), respectively. From each data set we extracted three samples which played the role of the SPE samples P_i or P_{ij} (see Table 1). The original T2M group was divided into spaced thirds to simulate spiraling of three test booklets. The samples of 500, 1500, and 3000 for data set T2M(1) were constructed by taking the first n records ($n = 500$, 1500, or 3000) of each third and thus these three samples are not independent. This was intended to simulate analyses on "early returns," for example. For data set T2M(2), the samples of 500, 1500, and 3000 were obtained by selecting every seventh case, every second case, and deleting every fifth case, respectively, up to the sample size desired for each third. This was intended to simulate more "random" (and less dependent) samples of the entire population. Thus the only difference between T2M(1) analysis groups and the T2M(2) analysis groups is the method of sampling. The T3M(3) data set was constructed from T3M in the same manner as the T2M(1) data set was from T2M. Thus there was a maximum of nine samples for each analysis.

4.1. Results for Tests with Two Variable Sections

For either the "pairwise present" method or the maximum-likelihood (MLE) method of estimating the parameters of a linear equating function, there are two possible overall approaches. The first is to use only the data from examinees who take the operational test X and two sections of the to-be-equated test Y. In terms of Table 1, this corresponds to using only the data from the samples P_{12}, P_{13}, and P_{23}. There is, however, a second approach that will generally be possible in the actual use of SPE. This stems from the fact that if SPE was to be used for ETS tests, there would also be examinees who took the operational test X but did not have sections of test Y in the variable sections of their test booklets. Instead they might have pretest sections for a test that is entirely different from Y. However, their X-test scores might still be used in equating Y to X. We considered two such uses in our examination of 2VST methods. For the pairwise present method this simply consisted of using population values of μ_X and σ_X in place of the sample values estimated from the combined samples P_{12}, P_{13}, and P_{23}. For the maximum-likelihood estimates (MLEs) this consists of adding a fourth pattern of missing data—namely, all X-sections observed but *no* Y-sections observed—to the estimation process. We shall refer to these two procedures that use the additional X-data as "population-adjusted" pairwise present or MLEs, respectively. We first discuss results for the simple pairwise present and MLE methods and then consider their "population-adjusted" versions.

Table 9 displays the maximum absolute differences[5] (over the range $Y = 0$ to 125) between the estimated and population conversion functions that result when Y (or Z) is equated to X. Results for the simple pairwise present and MLE methods are given in Table 7 for the three sizes of P_{ij} (500, 1500, and 3000) for each of the three basic data sets—T2M(1), T2M(2), and T3M(3).

In Table 7 we observe that all of the maximum discrepancies, columns 1 and 2, are less than 8.0 scaled score points, and only four of the 18 values exceed 3.0. Thus only the discrepancies for sample sizes of 500 are of the same order of magnitude as the discrepancies due to population differences (i.e., 7.5 scaled score points).

As expected, the discrepancies generally get smaller as the sample size increases, except for $n = 3000$ in the T3M(3) data set. The maximum discrepancies for the MLEs are smaller than those for the corresponding pairwise present estimates in six of the nine comparisons—i.e., column labeled "(1)–(2)" However, for the sample of size 500 for T3M(3) the MLE yields the

---

[5] These maximum differences must occur at the ends of the raw-score range. Some occur at $Y = 125$ while others occur at $Y = 0$.

TABLE 9

Maximum Differences between the Estimated and True Conversion Functions
for the Two Simple Methods of Estimation, Three Sample Sizes,
and Three Basic Data Sets

Sample size	Data set	(1) Pairwise present	(2) Maximum likelihood	(1)–(2)
500	T2M(1)	2.6	2.2	0.4
	T2M(2)	5.7	5.8	−0.1
	T3M(3)	6.3	7.9	−1.6
1500	T2M(1)	2.2	1.7	0.5
	T2M(2)	1.0	1.2	−0.2
	T3M(3)	2.2	1.0	1.2
3000	T2M(1)	0.7	0.0	0.7
	T2M(2)	0.5	0.1	0.4
	T3M(3)	2.8	2.1	0.7

largest maximum discrepancy in the entire table. In these data it seems that
the MLEs yield the most consistent improvements where they are needed
the least, i.e., for sample size 3000. In general, the improvements due to
MLEs are not very large.

In Table 10 we give the corresponding results for the "population-adjusted" versions of the pairwise present and MLE methods of estimation.
Since it seems intuitively obvious that using *more* data ought to improve our

TABLE 10

Maximum Differences between the Estimated
and True Conversion Functions for the Two Population-Adjusted
Methods of Estimation

Sample size	Data set	Pairwise (adj)	MLE (adj)
500	T2M(1)	14.6	3.9
	T2M(2)	8.0	4.2
	T3M(3)	5.3	7.2
1500	T2M(1)	10.3	2.8
	T2M(2)	9.3	0.7
	T3M(3)	8.2	2.0
3000	T2M(1)	5.5	0.7
	T2M(2)	3.6	0.4
	T3M(3)	6.0	2.5

estimates, the results in Table 10 are, at first, quite surprising. The population-adjusted pairwise present method gives results that are worse, in every case except one, than the corresponding "unadjusted" pairwise present estimates in Table 9. The population-adjusted MLEs are worse than their unadjusted versions in six of the nine cases. The population-adjusted MLEs improve upon the population-adjusted pairwise present estimates in eight of the nine cases, and they are better than the unadjusted pairwise present estimates in five of the nine cases. These results suggest that the sampling variability of these various estimation procedures does not follow simple intuitive ideas.

In order to provide some information that may be useful in understanding these results we give the conversion parameters for the estimates in Table 11 and the underlying estimates of μ_X, μ_Y, σ_X, and σ_Y in Table 12. In Table 12 we see that while the population-adjusted MLE of μ_Y is more accurate than either the simple MLE or the simple pairwise estimates, the corresponding estimate of σ_Y is not uniformly better than the other two for $n = 500$. Since $\hat{\sigma}_Y$ enters critically into estimates of formula (2), this may be part of the explanation of why the use of the population values of X to adjust the MLE did not make an improvement in the estimated conversion function.

TABLE 11

Estimates of the Slope and Intercept of the Conversion Function for Y and Z, for the Four Estimation Methods, for Three Sample Sizes in Each of the Three Data Sets

Sample size	Slopes				Intercepts			
	Pairwise	MLE	Pairwise (adj)	MLE	Pairwise	MLE	Pairwise (adj)	MLE (adj)
				T2M(1)—Data set				
500	4.521	4.518	4.687	4.545	213.35	213.36	200.84	211.54
1500	4.508	4.507	4.591	4.521	213.21	213.70	205.10	212.64
3000	4.491	4.484	4.540	4.492	214.68	215.42	209.93	214.75
All T2M[a]	4.484	4.484	4.484	4.484	215.41	215.41	215.41	215.41
				T2M(2)—Data set				
500	4.571	4.564	4.442	4.544	210.21	211.18	212.69	212.08
1500	4.487	4.470	4.584	4.487	214.42	216.01	206.16	214.75
3000	4.490	4.485	4.525	4.490	214.89	215.43	211.84	215.00
All T2M[a]	4.484	4.484	4.484	4.484	215.41	215.41	215.41	215.41
				T3M(3)—Data set				
500	4.457	4.430	4.512	4.441	209.73	211.39	203.72	210.71
1500	4.586	4.568	4.652	4.579	201.33	202.50	195.29	201.55
3000	4.588	4.576	4.610	4.581	200.69	201.40	197.48	200.99
All T3M[a]	4.556	4.556	4.556	4.556	203.52	203.52	203.52	203.52

[a] Values given in these rows are the population values.

TABLE 12

Estimates of μ_X, μ_Y, σ_X, σ_Y for the Three Methods and for
the Three Sample Sizes in Each of the Three Data Sets

Sample size	μ_X	σ_X	μ_Y			σ_Y		
			Pairwise	MLE	MLE (adj)	Pairwise	MLE	MLE (adj)
			T2M(1)—Data set					
500	61.25	22.80	63.83	63.88	63.47	21.34	21.35	22.00
1500	61.46	23.21	64.24	64.15	63.56	21.78	21.79	22.12
3000	61.18	23.38	63.90	63.82	63.50	22.03	22.06	22.26
All T2M[a]	60.79	23.63	63.47	63.47	63.47	22.30	22.30	22.30
			T2M(2)—Data set					
500	62.19	24.32	64.68	64.58	63.36	22.51	22.55	22.01
1500	61.28	23.14	64.10	64.00	63.58	21.82	21.90	22.29
3000	60.98	23.45	63.68	63.63	63.47	22.10	21.12	22.27
All T2M[a]	60.79	23.63	63.47	63.47	63.47	22.30	22.30	22.30
			T3M(3)—Data set[b]					
500	61.30	23.24	65.61	65.62	65.10	22.07	22.20	22.42
1500	61.15	23.20	65.46	65.46	65.13	21.41	21.49	21.75
3000	61.17	23.42	65.58	65.60	65.23	21.60	21.66	21.74
All T3M[a]	60.74	23.53	65.02	65.02	65.02	21.85	21.85	21.85

[a] Values given in these rows are the true population values.
[b] Means and standard deviations in the μ_Y and σ_Y columns for T3M(3)—Data set are those of Z.

4.2. Results for Tests with One Variable Section

In Section 2.4 we described three possible methods that may be used to impute reasonable values for the "unknown" correlations that arise from the use of a one-variable-section test (1VST) for section pre-equating. We examined versions of each of these three methods with our three data sets T2M(1), T2M(2), and T3M(3). Only one sample size was used, i.e., $n = 500$. This choice was partly due to the fact that when a 1VST is used, there are fewer variable sections available for all uses and therefore smaller sample sizes are the result because of the needs of the entire system of item pretesting, experimentation, and pre-equating. Furthermore, the results for 2VSTs suggest that sampling variability is substantial only for the smallest sample size used in this study.

To simulate "borrowing correlations from another population" we used T2F (and T3F) as the "lending population." To simulate "borrowing correlations from another test" we used the correlation matrix of (V1, M1, T1) over T2M (and T3M). The samples from T2M(1) or T2M(2) used T2F or T2M as their "lender" in these two methods, whereas T3M(3) borrowed from T3F or T3M respectively.

Table 13 summarizes the results of borrowing correlations for the three data sets T2M(1), T2M(2), and T3M(3) as well as for the entire populations of T2M and T3M. Borrowing the correlations from another population taking the same test was better than borrowing them from a slightly different test on the same population in two of the three comparisons with $n = 500$. When the entire populations were used instead of the samples of 500, the method of borrowing from a different population was better in both comparisons. This suggests that borrowing correlations from the same test on a subsequent test population may be preferable to using the correlations from another test.

To test the sensitivity of the estimated equating parameters to the values of the three unknown partial correlations (i.e., the partial correlations among V2, M2, and T2 after partialing out V1, M1, and T1) we used the method described in Rubin and Thayer (1978) and assumed all three partial correlations were equal to ρ. Here we report results for $\rho = 0.0$ and $\rho = 0.1$. The population values on T2M for these partial correlations were

$$\rho_{V_2 M_2 \cdot V_1 M_1 T_1} = 0.057, \quad \rho_{V_2 T_2 \cdot V_1 M_1 T_1} = 0.079, \quad \text{and} \quad \rho_{M_2 T_2 \cdot V_1 M_1 T_1} = 0.083.$$

In Table 14 we give the maximum difference between the conversion lines that are estimated under the assumption that the partial correlations all equal 0.0 and 0.1. These maximum differences are, on the average, half as large as those obtained in Table 13 for the method of borrowing correlations from another population. Neither the value of $\rho = 0.0$ or 0.1 seems to stand out as superior in Table 14. However the discrepancies in Table 14 are all

TABLE 13

Maximum Differences between the Estimated
and the True Conversion Functions for the Two Methods
of Borrowing Correlations

Data set	Borrowing from the females	Borrowing from test X
T2M(1)[a]	2.9	5.1
T2M(2)[a]	10.1	7.8
T3M(3)[a]	8.8	11.1
All of T2M	0.6	1.7
All of T3M	0.6	2.9

[a] $n = 500$.

TABLE 14

Maximum Differences between the Estimated
and True Conversion Function for the Two Values
of the Imputed Partial Correlations[a]

Data set	$\rho = 0.0$	$\rho = 0.1$
T2M(1)	2.9	3.0
T2M(2)	6.1	3.3
T3M(3)	1.9	4.7

[a] $n = 500$.

quite similar to the values obtained for sample size 500 in Table 9 for 2VSTs. This result suggests that the use of methods that impute reasonable values for the inestimable parameters in a 1VST may be comparable in accuracy with 2VST methods that do not need to impute these partial correlations. The results of Tables 13 and 14 suggest that imputing values for the partial correlation may be more accurate than methods that borrow correlations from other sources.

5. Discussion and Conclusions

This section is divided into two parts. In the first, we make some additional comments on the use of section pre-equating; in the second, we draw together the conclusions we have reached based on the data analyzed in Section 4.

5.1. General Comments on Section Pre-Equating

Section pre-equating (SPE) has one important advantage over other observed-score equating methods, namely, that it allows users to estimate a conversion function for a test long before the *intact* test is ever given. For the standard observed-score equating methods, as discussed in this volume, for example, by Braun and Holland and by Angoff, the traditional process of test equating must be accomplished between the time that the new test is administered and when the scores earned on this test are sent to the examinees. This time interval is often only a matter of a few weeks and the process of equating the new test is crammed into a complex and very busy score-reporting schedule. Thus SPE could be used to help relieve the need to produce a good conversion function in a short period of time.

Section pre-equating may be used to equate tests in an environment in which tests can not be reused. Such environments arise from several sources—

e.g., situations in which it is difficult to maintain the security of intact forms once they have been used, and recent legal requirements for public disclosure of test questions once they have been used operationally. Furthermore, the use of SPE does not require the tests to measure a single attribute or trait. This means that the sections that comprise the test can cover diverse material. Test parallelism *as a whole* across the different forms of the test is all that is required.

There are definite constraints that SPE imposes on the form and content of tests. First of all, in the versions described in Section 2, it is difficult to see how the sections could have differing time limits. However, in Holland and Thayer (1981) SPE is examined in a situation where the test sections must be split in half to fit into the time slot allowed for the variable section. This approach may prove to be a way of obtaining some flexibility in the time limits of sections in SPE. In any event, the effect of time limits can be profound and very serious consideration must be given to their determination for tests used in conjunction with SPE.

A second constraint that SPE imposes concerns the fact that the test sections that are being pre-equated must be given the same serious consideration by examinees as they will be given when they are used operationally. Whereas the placement of a variable section at the end of the test may be satisfactory for item-pretesting purposes, they may not be so for pre-equating purposes. This raises the question of where to place the variable section(s) in each test form. Generally speaking, the printing costs will make *variable-position* variable section(s) impractical at a given test administration. Hence the variable section may need to be shifted from one position (or pair of positions) to another from administration to administration. This will be necessary if it is important to insure that it is difficult for examinees to identify the nonoperational test section(s).

As we mentioned in Section 1.3, SPE assumes small or negligible practice effects. This is related to the problem of the position of the variable sections because these positions determine which sections are "practiced" and which are not. The need for negligible practice effects when SPE is to be used may require changing some types of test items to those that are not as susceptible to the effects of practice during a testing session. Such a change of item types may be desirable for other reasons as well.

In the versions of SPE that we anticipate, the sections of the yet-to-be-equated test Y that are embedded in the operational section(s) of test X are of *operational* quality. This means that they have undergone reviews as well as pretesting on real examinee populations. Thus SPE requires exposure of whole sections of operational quality tests prior to their operational use. This prior exposure will need to be balanced against the needs of test security in specific applications.

If printing or other types of errors occur so that the sections of Y as they appear in the variable sections of X are not identical to those of Y when they are used operationally, then additional noise is added to the system and some accommodations will need to be made for it. This problem arises whenever test questions are "reprinted" and used as if the tests were identical on two tested populations. Since SPE estimates the parameters associated with entire sets of questions, it is likely that the small corrections needed to accommodate printing errors that involve only one or two items can be developed either from test theory considerations or through empirical experience.

5.2. Conclusions and Recommendations

In general, we think that the results of this study are very encouraging for the potential usefulness of SPE. Of course, a number of new questions are raised by this research. For a 2VST, the maximum-likelihood method of estimating the equating parameters is slightly better than the simpler "pairwise present" method. However, we were surprised to find that this improvement was largest for the larger sample sizes and negligible for $n = 500$. Furthermore, our attempts to increase the accuracy of the estimates by using the available population values were not successful. This is a problem that needs to be investigated through more theoretical and empirical research.

For 1VSTs, in which the values of the intercorrelations must be "borrowed" from other sources, or otherwise "imputed," the results of this study are very encouraging. In particular, the method that simply imputes a small positive value to the inestimable *partial* correlations among the Y-sections gave quite satisfactory results for samples of size 500. This method appears to be preferable to "borrowing" correlations from either another population of examinees or from another, parallel, test. This too should be investigated more carefully in subsequent research. We were generally surprised to find that 1VST and 2VST results were comparable for $n = 500$. One must remember that the sample sizes for a 1VST and a 2VST are not strictly comparable since there are more data per observation in a 2VST. Although there were only 500 observations from which to estimate the mean and variance of Y_i in the 1VSTs with ns of 500, there were 1000 observations for the mean and variance of Y_i in the 2VSTs with ns of 500. This suggests that it may be profitable to consider using 1VST methods even in the case of 2VSTs. The advisability of such an approach will depend on the accuracy with which the the correlations are estimated. If the sample sizes are sufficiently small, it may be advisable to impute values for the partial correlations rather than to estimate them from small samples. Such an approach should be investigated in future research.

There are various limitations of the present study that should be remembered in interpreting the results. As we mentioned in Section 1.3 no attempt has been made here to deal with the problem of "practice" and other types of effects that interfere with using the incomplete information assembled in SPE to estimate the relevant parameters of the distributions of Y-test scores. The biases that practice effects create could, in some applications, dominate the sampling variability that we have examined in this paper. In addition, we should emphasize that the "tests" that were examined here were not *real* but merely constructed in specific ways from the data of real SAT examinees. Consequently the two tests were not strictly parallel in form and content and were not constructed with our use of them in mind. However, even with these limitations we believe that this study indicates that one- and two-variable section SPE is a promising test-equating technique that should be investigated further.

REFERENCES

Angoff, W. H. (1971). Scales, norms, and equivalent scores. In R. Thorndike (Ed.), *Educational Measurement* (2nd ed.). Washington, D. C.: American Council on Education, 508–600.

Dempster, A. P., Laird, N. M., and Rubin, D. B. (1977). Maximum likelihood from incomplete data via the EM algorithm. *Journal of the Royal Statistical Society*, B, *39*(1), 1–38.

Holland, P. W., and Thayer, D. T. (1981). *Section Pre-Equating the Graduate Record Examination* (Program Stat. Res. Tech. Rep. No. 81-13). Princeton, N. J.: Educational Testing Service.

Rubin, D. B. (1974). Characterizing the estimation of parameters in incomplete-data problems. *Journal of the American Statistical Association*, *69*, 467–474.

Rubin, D. B., and Thayer, D. T. (1978). Relating tests given to different samples. *Psychometrika*, *43*, No. 1, 3–10.

Wing, H. (1980). Practice effects with traditional mental test items. *Applied Psychological Measurement*, *4*, 141–155.

Discussion of "Section Pre-Equating: A Preliminary Investigation"

I. Richard Savage

In looking at this paper, I am struck by the fact that the "imputations" are merely replacements for a Bayesian concept. A Bayesian framework would be useful just to carry out the sensitivity studies that are of interest. The book by Box and Tiao (1973) contains good examples of sensitivity analyses done from the Bayesian viewpoint. Even if one is not a Bayesian, their approach is useful.

For the procedures discussed in this paper, analytical work could give interesting answers because if one has analytical formulas it is much easier to map out the sensitivity functions. One should rather easily be able to obtain the asymptotic means and variances of things of interest by using the delta method. One would only need to have the fourth moments of the things of interest, and these could be obtained either from the sampled data or from theoretical models.

The analyses done here are really a case study and are quite limited in scope. Either one has to generate a large number of samples via a Monte Carlo study or do the analytical work. Monte Carlo might be feasible for studying smaller sample sizes where the use of the delta method might not be appropriate. But for the large-sample problems discussed in this paper, the delta method appears to me to be the best choice.

299

Discussion of "Section Pre-Equating: A Preliminary Investigation"

Donald B. Rubin

Ted H. Szatrowski

Some recent research on patterned covariance matrices and missing data may be helpful for the section pre-equating problem. The potential relevance of patterned covariance matrices was pointed out at the conference by Ingram Olkin. The point is that unique maximum-likelihood estimates of the covariance matrix in the one-variable-section test can be obtained under assumptions about patterns in the covariances or correlations. Also, hypothesis tests can be made to determine, to some extent, whether a proposed pattern is reasonable.

In the one-variable-section test, assuming only a general structure for the correlation matrix, one cannot uniquely estimate the correlations between the Y_1, Y_2, and Y_3 variables (using Holland and Wightman's notation). However, if the variables are reordered for the T2-M sample in Table 4a, the correlation matrix becomes

$$
\begin{array}{c}
\begin{array}{cccccc}
\ \ V_1 & V_2 & M_1 & M_2 & T_1 & T_2
\end{array}\\
\begin{array}{c}
V_1\\ V_2\\ M_1\\ M_2\\ T_1\\ T_2
\end{array}
\left[
\begin{array}{cccccc}
1 & 0.841 & 0.643 & 0.636 & 0.771 & 0.750\\
0.841 & 1 & 0.648 & 0.639 & 0.771 & 0.740\\
0.643 & 0.648 & 1 & 0.839 & 0.630 & 0.618\\
0.636 & 0.639 & 0.839 & 1 & 0.629 & 0.625\\
0.771 & 0.771 & 0.630 & 0.629 & 1 & 0.859\\
0.750 & 0.740 & 0.618 & 0.625 & 0.859 & 1
\end{array}
\right].
\end{array}
\tag{1}
$$

This correlation matrix seems to have the pattern represented by

$$
\begin{array}{c}
\begin{array}{cccccc}
V_1 & V_2 & M_1 & M_2 & T_1 & T_2
\end{array}\\
\begin{array}{c}
V_1\\ V_2\\ M_1\\ M_2\\ T_1\\ T_2
\end{array}
\left[
\begin{array}{cccccc}
1 & b & c & c & d & d\\
b & 1 & c & c & d & d\\
c & c & 1 & b & c & c\\
c & c & b & 1 & c & c\\
d & d & c & c & 1 & b\\
d & d & c & c & b & 1
\end{array}
\right].
\end{array}
\tag{2}
$$

The other correlation matrices in Table 4 also seem to exhibit this pattern.

If these patterns in the correlation matrix hold, then maximum-likelihood estimates (MLEs) assuming this patterned structure would be more efficient than MLEs based on general structure. Because the correlations are so close to 1, the gain of efficiency if we had complete data on all variables would not be that great, at least using the squared error efficiency measure used in Szatrowski (1980b). However, with data from a one-variable-section test, the assumption of this structure allows the calculation of unique MLEs for correlations among the Y tests by avoiding ridges in the likelihood function that exist with the general correlation structure.

1.   Using the EM Algorithm for Patterned Covariances with a Fully Observed Data Matrix

Even though section pre-equating leads to partially observed variables, assume for the moment a fully observed data matrix. For covariance patterns and a fully observed data matrix, Szatrowski (1980a) has characterized the conditions for explicit, noniterative MLEs. This work shows that many interesting patterns do not have explicit noniterative solutions.

A large class of patterned covariance matrices can be generated by integrating over hypothetical fully missing variables in a multivariate normal. The advantage of conceptualizing patterned covariance structures this way is that, in many cases, a general missing data algorithm, the EM algorithm (Dempster, Laird, and Rubin, 1977), can be used to find maximum-likelihood estimates using only the usual regression operators (Rubin and Szatrowski, 1981).

Let "complete-data" refer to all variables, missing and observed, and "incomplete-data" refer to just the observed variables. The patterned covariance matrices most appropriate for attack by EM are those for which the complete-data problem has an explicit MLE because then the M-step of the EM algorithm has an explicit form; the E-step for these problems always has a simple explicit form because of the normality.

For example, a stationary covariance matrix (a pattern without explicit solutions) is a hypothetical fully missing data version of the circular symmetry pattern. As an illustration, consider the 3×3 stationary covariance pattern of the form

$$\Sigma_{ST} = \begin{bmatrix} a & b & c \\ b & a & b \\ c & b & a \end{bmatrix},$$

which does not have an explicit MLE, and the 4×4 circular symmetry pattern

$$\Sigma_{CIR} = \left[\begin{array}{ccc|c} a & b & c & b \\ b & a & b & c \\ c & b & a & b \\ \hline b & c & b & a \end{array}\right],$$

which does have an explicit MLE (e.g., Olkin and Press, 1969). Note if $\text{Cov}(\mathbf{X}) = \Sigma_{CIR}$, then with $\mathbf{X}' = (\mathbf{X}_1' : \mathbf{X}_2')$, $\text{Cov}(\mathbf{X}_1) = \Sigma_{ST}$.

More explicitly, suppose we have a random sample of size n from a multivariate normal distribution, $N_3(\mathbf{0}, \Sigma_{ST})$. These observations could be considered as the first three components out of four from a random sample of size n, $\mathbf{x}_1, \mathbf{x}_2, \ldots, \mathbf{x}_n$ (4×1 column vectors), from a multivariate normal distribution, $N_4(\mathbf{0}, \Sigma_{CIR})$. The first three components of each $\mathbf{x}_i$ are observed, and the last component is missing. Let $\mathbf{C} = (\sum_{i=1}^n \mathbf{x}_i \mathbf{x}_i')/n$ and partition $\mathbf{C}$ and $\Sigma = \Sigma_{CIR}$ by

$$\mathbf{C} = \begin{bmatrix} \mathbf{C}_{11} & \mathbf{C}_{12} \\ \mathbf{C}_{21} & \mathbf{C}_{22} \end{bmatrix}, \qquad \Sigma = \begin{bmatrix} \Sigma_{11} & \Sigma_{12} \\ \Sigma_{21} & \Sigma_{22} \end{bmatrix}, \qquad \Sigma_{11} \text{ and } \mathbf{C}_{11} \text{ (3 × 3) matrices.}$$

The matrix $\mathbf{C}$ is the complete-data sufficient statistic and $\mathbf{C}_{11}$ is the observed sufficient statistic, $\mathbf{C}_{12}$ and $\mathbf{C}_{22}$ being missing or unobserved. Then the E-step involves calculating the expected value of $\mathbf{C}$ given $\mathbf{C}_{11}$ and the current estimate of $\mathbf{\Sigma}$:

$$E(\mathbf{C}|\mathbf{C}_{11}, \mathbf{\Sigma}) = \begin{bmatrix} \mathbf{C}_{11} & \mathbf{C}_{11}\mathbf{\Sigma}_{11}^{-1}\mathbf{\Sigma}_{12} \\ \mathbf{\Sigma}_{21}\mathbf{\Sigma}_{11}^{-1}\mathbf{C}_{11} & \{\mathbf{\Sigma}_{22} - \mathbf{\Sigma}_{21}(\mathbf{\Sigma}_{11}^{-1} - \mathbf{\Sigma}_{11}^{-1}\mathbf{C}_{11}\mathbf{\Sigma}_{11}^{-1})\mathbf{\Sigma}_{12}\} \end{bmatrix}.$$

The explicit MLE of $\mathbf{\Sigma} = \mathbf{\Sigma}_{\text{CIR}}$ is then obtained for the complete-data case by simple averaging (Szatrowski, 1978). Thus the M-step is given by

$$\hat{a} = (\sum_1^4 C_{ii})/4, \qquad \hat{b} = (\sum_1^3 C_{i,i+1} + C_{14})/4, \qquad \hat{c} = (C_{13} + C_{24})/2.$$

These values of $\hat{a}$, $\hat{b}$, and $\hat{c}$ yield a new value of $\hat{\mathbf{\Sigma}}$ which is used in the next E-step, and so on. Including a nonzero mean is a trivial extension.

We are working on characterizing the class of linear patterned covariance matrices that are the marginal covariance matrices of multivariate normal covariance matrices with an explicit MLE.

2. Patterned Covariances with Data Matrices Arising from Section Pre-Equating

Two ideas have been presented: (a) using EM to handle missing data such as in section pre-equating, and (b) using EM to find MLEs in patterned covariance matrices without explicit MLEs by creating hypothetical fully missing variances. These two ideas can be combined to find unique MLEs for section pre-equating data assuming a patterned covariance matrix. Using the correlation pattern in (2) suggested by the Holland–Wightman data with a common variance among the variables yields the covariance pattern given by the upper left 6×6 matrix, $\mathbf{\Sigma}_{11}$ of

$$\mathbf{\Sigma} = \begin{bmatrix} \mathbf{\Sigma}_{11} & \mathbf{\Sigma}_{12} \\ \mathbf{\Sigma}_{21} & \mathbf{\Sigma}_{22} \end{bmatrix} = \begin{matrix} & \begin{matrix} V_1 & V_2 & M_1 & M_2 & T_1 & T_2 & Z_1 & Z_2 \end{matrix} \\ \begin{bmatrix} a & b & c & c & d & d & c & c \\ b & a & c & c & d & d & c & c \\ c & c & a & b & c & c & d & d \\ c & c & b & a & c & c & d & d \\ d & d & c & c & a & b & c & c \\ d & d & c & c & b & a & c & c \\ c & c & d & d & c & c & a & b \\ c & c & d & d & c & c & b & a \end{bmatrix} \end{matrix}.$$

The patterned covariance Σ_{11} does not have explicit MLE whereas the pattern Σ does have explicit MLE. Suppose that sections V_1, M_1, and T_1 are always observed, whereas V_2 is observed only in the first sample of n_1 observations, M_2 in the second sample of n_2 observations, and T_2 in the third sample of n_3 observations. Variables Z_1 and Z_2 are never observed but have been introduced so that the covariance matrix Σ has an explicit MLE. If we knew the sample covariance matrix $C = (\sum_{i=1}^{n} x_i x_i')/n$ for complete data based on $n = n_1 + n_2 + n_3$ observations, averaging elements of C with the same element in Σ yields the MLE for that element in Σ (cf. Szatrowski, 1978):

$$\hat{a} = (\operatorname{tr} C)/8,$$

$$\hat{b} = (C_{12} + C_{34} + C_{56} + C_{78})/4,$$

$$\hat{c} = (C_{13} + C_{14} + C_{23} + C_{24} + C_{17} + C_{18} + C_{27} + C_{78}$$
$$+ C_{35} + C_{36} + C_{45} + C_{46} + C_{57} + C_{58} + C_{67} + C_{68})/16,$$

$$\hat{d} = (C_{15} + C_{16} + C_{25} + C_{26} + C_{37} + C_{38} + C_{47} + C_{48})/8.$$

These estimates of $\hat{a}$, $\hat{b}$, $\hat{c}$, and $\hat{d}$ create the MLE of Σ.

The EM algorithm begins with an initial estimate of Σ, $\Sigma^{(0)}$. The E-step calculates the expected value of C given $\Sigma^{(0)}$ and the observed data. Using the above expressions for $\hat{a}$, $\hat{b}$, $\hat{c}$, and $\hat{d}$, the M-step calculates the next iterate of Σ, $\Sigma^{(1)}$, as if the expected value of C were the observed value of C, and so on. The E-step is carried out using the usual regression operators, just as with the general EM algorithm with missing data from the multivariate normal. For example, in order to find the expectation of the cross products for the n_1 individuals with V_1, M_1, T_1, V_2 observed and M_2, T_2, Z_1, Z_2 missing, sweep V_1, M_1, T_1, V_2 from $\Sigma^{(0)}$ in order to find the conditional distribution of missing variables given the observed variables. This conditional distribution and the observed value of the cross products for the observed variables yield the expected value of the cross products for the unobserved variables (conditional variance plus square of conditional means).

Szatrowski (1981a,b) provides details of the precise form of the EM algorithm for more general patterned means, covariances, and correlations. He also discusses the likelihood ratio tests that can be used to test the appropriateness of special covariance patterns that must be assumed in order to analyze the one-variable test, and presents the standard asymptotic chi-squared distribution of the likelihood ratio test under the null hypothesis as well as the asymptotic normal distribution under the alternative hypothesis. We can also test whether this structure is appropriate by gathering data with at least two variables in the sections to be equated so that all correlations can be estimated directly. If the special structure hypothesis is appropriate, then one can in principle estimate all parameters using only a one-variable-section test.

If these special correlation patterns are found to be appropriate, additional work is needed to determine sample sizes required for accurate estimates of the equating parameters and to determine the forms of the confidence intervals for the estimated equating lines. The delta method can be used for large sample results, and Monte Carlo methods can be used to check the accuracy of large sample results for small samples.

Further research also needs to be done on the application of the EM algorithm to structured covariance and correlation matrix estimation problems. The EM algorithm has the good property of guaranteed convergence to a maximum for a wide class of problems [Dempster, Laird, and Rubin (1977) provide theory]. Based on several working examples, it appears that the EM algorithm takes more iterations than the method of scoring for these missing data problems. The EM iterations are faster, however, and the algorithm is quite simple to implement.

ACKNOWLEDGMENTS

Donald B. Rubin was supported by the Program Statistics Research Project of the Educational Testing Service Research Statistics Group. Ted H. Szatrowski received partial support from the Program Statistics Research Project and from National Science Foundation Grant MCS77-28184.

REFERENCES

Dempster, A. P., Laird, N. M., and Rubin, D. B. (1977). Maximum likelihood from incomplete data via the EM algorithm. *Journal of the Royal Statistical Society, B, 39,* 1–38.

Olkin, I., and Press, S. J. (1969). Testing and estimation for a circular stationary model. *Annals of Mathematical Statistics, 40,* 1358–1373.

Rubin, D. B., and Szatrowski, T. H. (1981). Finding maximum likelihood estimates of patterned covariance matrices of the EM algorithm. To be published in *Biometrica.*

Szatrowski, T. H. (1978). Explicit solutions, one iteration convergence and averaging in the multivariate normal estimation problem for patterned means and covariances. *Annals of Statistical Mathematics, 30,* Part A, 81–88.

Szatrowski, T. H. (1980a). Necessary and sufficient conditions for explicit solutions in the multivariate normal estimation problem for patterned means and covariances. *Annals of Statistics, 8* (4), 802–810.

Szatrowski, T. H. (1980b). Efficiency of estimates using patterned covariances or correlations in the multivariate normal estimation problem. (Working paper.)

Szatrowski, T. H. (1981a). Missing data in the multivariate normal patterned mean and patterned covariance matrix testing and estimation problem. (Working paper.)

Szatrowski, T. H. (1981b). Missing data in the multivariate normal patterned mean and correlation matrix testing and estimation problem. (Working paper.)

Part IV

RELATED TOPICS AND DISCUSSION

Item and Score Conversion by Pooled Judgment

Robert L. Thorndike

With the current spate of "truth in testing" laws and the pressure to release the answers to examinees and to the public, it may become necessary to prepare new test forms at more frequent intervals in the future. With this it may become increasingly difficult to carry out advance tryouts of test items so as to have an empirical basis for selecting those items that have the desired ability to discriminate levels of ability and the desired levels of difficulty to permit maintaining comparability of successive test forms. We may then be thrown back upon reliance on the judgments of presumably knowledgeable persons to maintain comparability of our instruments from one testing to the next.

The problem is not a new one. Keys to civil service examinations in New York City for such positions as firefighter and police officer have always been published. Examinees have had the right to challenge the keying of any item, and items have been deleted or the keying changed in response to such challenges. This has always required the preparation of a new examination each time one is scheduled, and the circumstances have made it virtually impossible to try out test items in advance. Naturally, the exam-makers have wanted to maintain as much comparability as possible from one testing to the next, so that "passing" scores expressed as a percent of items gotten right have maintained the same standard.

309

The problem led my late colleague, Irving Lorge, together with some of his students (Lorge and Kruglov, 1952, 1953; Tinkelman, 1947) 30 years ago to explore the possibility of using groups of judges to estimate the difficulty of test items. If judges were able to make judgments of item difficulty with sufficient accuracy, these judgments could then be used as the means of controlling the difficulty of new test forms so that they would be comparable to earlier forms and/or so that they would be appropriate in difficulty for the groups to be tested and the selection decisions to be made.

I have made a somewhat cursory survey of the literature, and find little or nothing since that early work, so the state of our knowledge seems to be about where it was in 1953. Basically, what Lorge and the others found was that judges showed moderately good agreement in appraising the *relative* difficulty of test exercises, but that they differed widely in the absolute difficulty level that they assigned to the items in a set. It appeared that the pooling of the results from several judges would yield a rather stable ranking of which were the more difficult and which the less difficult items. However, it was not clear that one could depend on the judgments to provide a sound estimate of the actual proportion of examinees that would get the items right, or of the average score level that a group of examinees would achieve.

The natural next step was to provide judges with a certain number of anchor items, the difficulty of which was known from their use in an earlier test. The items were displayed, together with their difficulty indices, and this information was available as the judges evaluated a new set of items. Not surprisingly, the agreement among judges as to the absolute level of difficulty improved.

There are, of course, various ways in which one can use anchoring items. One is, as above, to identify them as anchors and to have them available for reference, separate from the items to be judged. An alternate plan would be to embed the anchor items in the new set of items without identifying them in any way, and to judge them along with the new items. The judged scale values of the anchor items could then be plotted against their empirical difficulty scale values, and a line of relationship established by which judged difficulty values of the other items could be converted into estimates of empirical scale equivalents.

I must admit to having always had an aesthetic preference for this second procedure, and I was moved by the occasion of this conference to accumulate some data to see how the procedure worked in practice. It is these data that I shall share with you now.

I happened to have in my files some rather solid data on the empirical difficulty of the items in the current form of our Cognitive Ability Tests, so I used three of the subtests of the battery as the items to be rated for difficulty. Ratings were obtained from 20 graduate students at Teachers College,

Columbia University, drawn mostly from among the majors in measurement and evaluation. They were presumably interested in measurement problems, but in most cases had had little practical experience in item writing or item evaluation. Thus they might be considered a group of interested amateurs.

The item types that were used for the ratings were of three sorts: verbal analogies, quantitative relations, and figure analogies. Raters were asked to rate the items on a 9-point scale, each step of which was described in terms of a grade group's typical performance. The extreme scale points were "Would be passed by not more than 30% of twelfth graders," designated 9, and "Would be passed by 75% or more of third-graders," designated 1. Empirical difficulty (and discrimination) data were available for groups of not less than 2000 in each grade from 3 through 12. The test is in multilevel format, with different grade groups starting and stopping at different points in the set of items. Thus any item is taken by several grade groups, and the overlapping permits putting all the items on a single continuous difficulty scale. Normal deviate values were used to estimate the change in ability level from one grade to the next, and thus to place all the items on a common scale, in which the units were at least approximately equal.

We ask first how much agreement there was among judges in their relative difficulty estimates. Analysis-of-variance procedures were used to estimate the between-items variance and the interaction variance between judges and items. This latter was used as an estimate of error variance, and the correlation was computed (1) between the sum of the 20 judges and a hypothetical other set of 20 judges; and (2) between one judge and one other judge, on the average. The value was also estimated for a set of four judges, on the assumption that this might be a reasonable number to obtain in a practical scaling problem. The results are shown in Table 1. The agreement for the whole set of 20 judges is, of course, very high. For research purposes, one can get quite stable relative difficulty estimates for an item set. The agreement for single judges is fairly good, running about 0.60 for verbal and

TABLE 1

Between-Judge Agreement for Difficulty Estimates

	Total of 20 judges	Single judge	Estimated for 4 judges
Verbal analogies	0.967	0.59	0.85
Quantitative relations	0.971	0.62	0.87
Figure analogies	0.947	0.47	0.78

quantitative items and about 0.50 for the less familiar figural items. However, one would certainly need to pool the results from several judges to feel comfortable using the results in practical work. The correlations for the set of four judges indicate that this number might be satisfactory, though in careful work one might want a larger group. It should perhaps be noted that these items did have a pretty wide spread of empirical difficulty, ranging from easy third-grade to difficult twelfth-grade material. In that sense, the judges were being asked to make fairly gross distinctions, and the results may present them in an unduly favorable light.

We ask next how much variation there was among judges in the absolute level of the ratings assigned. Table 2 shows the range of average ratings, the standard deviation of the average values, and the size of the between-judges variance in relation to the total variance that includes variance between items and the interaction of judges with items. Clearly, the range of average ratings for these judges is substantial. Referring to the steps on the nine-point scale that was described earlier, the verbal items varied in average judged difficulty from "about 50% of fourth-grade students" to slightly above "about 50% of eighth-grade students." In the case of the figure analogies items, the range is from "about average for fourth-grade students" to "about average for eleventh-grade students." The percent of variance that is between-raters variance ranges from 28 in the quantitative relations test to 64 for figure analogies.

It is not surprising that the greatest variation among judges occurs for the figure analogies test, where there is no familiar referent as there is in the test dealing with numerical matters that have a fairly clear relation to school instruction. The lack of a frame of reference appears to have lowered the judges' consistency in ordering the items and at the same time to have made them more variable in their absolute standard.

In general, our results confirm earlier studies in indicating that satisfactory agreement can be obtained in ranking items for relative difficulty, but that

TABLE 2

Variability of Judges in Severity of Rating Standard[a]

	Range of average rating	SD of average rating	Percent of variance between-rater variance
Verbal analogies	3.1–6.2	0.77	39
Quantitative relations	3.5–5.7	0.58	28
Figure analogies	3.0 7.4	0.98	64

[a] Ratings on a scale extending from 1 to 9.

a single judge—or perhaps a team of judges—cannot be depended on to be accurate in assessing the absolute difficulty of an item set. The need for some type of anchoring procedure is confirmed.

Basically more crucial than the question of reliability of the ratings for difficulty is that of their validity, here represented by the correlation of the ratings, singly or pooled, with the empirically determined difficulty scale values. Do the mean ratings converge on these values, or are there systematic displacements in the ratings so that some items are judged to be significantly harder or significantly easier than their true position in the set of items? This is the factor that really sets the limit on using anchor items, because if the judged difficulty both of the anchors and of the new items does not correlate closely with their true empirical difficulty, controlling the judged difficulty of the new set of items will be relatively ineffective in controlling their true empirical difficulty.

The relevant information is summarized in Table 3. Obtained correlations between judged and empirical difficulty are shown, and also estimates corrected for attenuation due to the unreliability of the ratings. There may be some unreliability in the empirical difficulty scale values, but the reliability should be high in view of the large samples upon which the scale values are based.

Though the correlations between pooled judgment based on 20 raters and empirical difficulty based on several thousand cases are substantial, they fall far short of perfection. Using the somewhat more realistic and less-comforting squared correlation, we see that proportions of empirical difficulty variance that cannot be accounted for by even perfectly reliable ratings for difficulty range from 29% on the verbal items to 45% on the figural items. Clearly, there is room for sizable discrepancies in empirical difficulty even when items have been matched for difficulty as judged by a team of raters. Of course, these were untrained raters with no particular experience with items of the type that they were judging, and the validities might improve with experience with the items or with practice in judging them. However,

TABLE 3

Correlation of Pooled Rating (for 20 Raters)
with Empirically Determined Difficulty

	Obtained	Corrected for attenuation
Verbal analogies	0.828	0.842
Quantitative relations	0.742	0.753
Figure analogies	0.721	0.741

it seems clear that one cannot count on accurate determination of the difficulty of single items, even if the judgments of a number of raters are pooled. Of course, we would use ratings primarily to estimate the average difficulty and the spread of difficulties in a set of items, and errors in the estimation of single items would to some extent cancel out. We may inquire now how much variation we may expect to encounter when a set of items is rated.

I have suggested using a number of items whose empirical difficulty is known as anchors to control the difficulty of a new form of a test. The procedure that I have been interested in investigating is one in which the anchor items are included with the potential new items, without being identified as in any way separate from the set of new items, and each item in the complete set is rated for difficulty. The simplest and perhaps the most satisfactory model to use would be to select as anchor items a subset from the prior test that directly represented the complete test in average difficulty and in dispersion of difficulties. The mean and dispersion of rated difficulties for this subset should then provide an unbiased estimate of the mean and dispersion of rated difficulties that would have been obtained for the complete prior test. The strategy would then be to select from the pool of new items a set whose mean and dispersion of rated difficulties matched those of the set of anchor items. This set, it is hoped, would reproduce in the new test the average and spread of empirical difficulties found in the prior test.

However, we can expect slippage at both ends of this equating. Since the correlation of empirical with rated difficulty appears to fall well short of unity, even when allowance is made for unreliability in the pooled ratings and in the empirical indices, there will not be a one-to-one correspondence between empirical and rated difficulty, and different subsets that might be chosen to represent the empirical parameters of the prior test will vary somewhat in rated difficulty. By the same token, different sets of items that might be chosen because they match the anchor items in rated difficulty can be expected to vary somewhat in empirical difficulty when this is subsequently determined. As I see it, the correlation between the actual empirical difficulties of a set of items from the prior test and a set of new items that were chosen because each one individually matched one of the items in the prior set in rated difficulty would be the square of the correlation between empirical and rated difficulty—a value that may be distressingly low. The standard error for estimating the empirical difficulty of a single new item would then be given by

$$\sigma_{\text{Est}} = \sigma(1 - r_{\text{ER}}^4)^{1/2},$$

where σ is the standard deviation of the pool of empirical difficulty values, and r_{ER} is the correlation between empirical and rated difficulty. For single items, this standard error of estimation is discouragingly large. If the correlation is 0.8, which more or less matches the values in Table 3, the standard error is 77% of the original standard deviation.

Of course, one is not picking a single item, but rather a set of items. Since the match of the set depends on two stages of matching, it is a function of

(1)   how adequately the empirical-to-rated difficulty bridge provided by the anchor items represents the total prior test, and

(2)   how adequately the rated-to-empirical bridge for the selected items represents the correspondence between rated and empirical validity for the total item pool.

The adequacy of each bridge will, of course, depend on the number of items upon which the bridge is based, and we can expect the precision to be approximately in proportion to the square root of that number. The number constituting the bridge to the new test is fixed by the length of that test, so the only element that we are free to manipulate is the number of anchor items. The limiting factor at this end is the length of the prior test and tolerance and resistance to fatigue on the part of the raters.

To provide an illustration, and to give me (and you) some sense of the precision that can be expected, I divided each of the three types of items, of which there were 59 or 60, into six subsets that were closely matched in mean and standard deviation of rated difficulty. These could be thought of as replicated samples from the prior test, or as replicated samples of a set of items to comprise the new test. Various analyses were made of the comparability of empirical difficulty in these matched sets.

First, we can look at the differences between subsets of 10 taken in all possible combinations. These are summarized, so far as mean difficulty is concerned in the top section of Table 4. The differences for pairs range from near zero to an amount that can be almost as much as a full standard deviation of the distribution of empirical difficulties of the items in the pool. The root-mean-square difference is about one third of the standard deviation of item difficulties. Another frame of reference for interpreting the differences is that their average size is from one to one and a half times the difference between the average performance in adjacent school grades. That is, if you started out to match a test that was of average difficulty for sixth-graders, two-thirds of your products would fall within a range from about fifth- to seventh-grade difficulty.

The lower parts of the table show the level of precision that is achieved when the 10 anchor items are compared with a full length test of 50 items, that is, with the sum of all the remaining sets of 10 items. Finally, 20 items are used as anchor items, and are compared with a 40-item test.

TABLE 4

Variation in Empirical Difficulty for Sets Matched in Rated Difficulty

		Verbal analogies	Quantitative relations	Figure analogies
Two subsets of 10 items	Root-mean-square difference	4.7	8.9	5.1
	Range of difference	0.2–9.6	0.3–18.4	0.3–10.1
10-anchor vs. 50-item test	Root-mean-square difference	3.8	6.8	4.0
	Range of difference	1.2–7.5	0.3–13.5	2.0–6.5
20-anchor vs. 40-item test	Root-mean-square difference	3.0	5.4	3.1
	Range of difference	0.2–5.7	0.6–8.9	0.0–5.5
SD of test item pool		18.7	21.4	17.1
Average difference between adjacent grades		4	5	3

It should be pointed out that the several samples matched for rated difficulty differ not only in mean empirical difficulty but also in dispersion of difficulty values. Thus the tests that they serve to represent would vary not only in mean score but also in standard deviation, and the discrepancies are such that the matching is a good deal shakier for extreme scores than it is at the means of the several tests.

All in all, these results are not too encouraging to a person who is interested in equating a new form of a test to an existing form through the medium of pooled difficulty judgments. What might be done to improve the situation?

Clearly, the crux of the matter is the validity of the pooled judgments, and any improvement in accuracy depends on improving that validity. Three suggestions can be offered.

(1) *Select better judges.* Judges in the present little experiment had no extended practical experience in item writing or editing, and no particular expertise in the type of material being judged. It is at least possible that using judges with either a broader background of test-making experience or, for example, contact with the school arithmetic curriculum could have resulted in judgments that corresponded more closely to reality.

(2) *Train the judges.* It seems likely, though I have no evidence on the matter, that a few periods of training, with feedback on errors made in judging, could sensitize judges to aspects of item difficulty to which ours were currently inadequately responsive. For example, naive judges may pay too much attention to the stem of an item and not enough to the seductiveness (or lack of it) in the response options.

(3) *Differentiate the anchor items from those to be judged.* It may have been a mistake to embed the anchor items in the total set being judged. Perhaps one of the two bridges would be more clearly and more validly established if the anchor items were separated out, the actual difficulty of each indicated, and the judges required to place each newly judged item in relation to the specified scale. This requires a good deal of referring back and forth between the anchor scale and the items to be judged, and seems a little awkward. However, if it results in only one locus for error rather than two, it may prove to be the preferred procedure. The items chosen for anchor items should probably be ones on which judged and empirical difficulty are in close agreement, so that the scale gradations will be apparent to judges using those items as anchors, as well as empirically true.

I have found this little exercise in judging items instructive. It has caused me to revise my preconceptions as to the best strategy for assembling judgments. If the need to equate in the absence of tryout becomes a pressing one, perhaps others will be moved to explore further to determine optimal strategies for using judgments as a tool to control test parameters.

REFERENCES

Lorge, I., and Kruglov, L. (1952). A suggested technique for the improvement of difficulty prediction of test items. *Educational and Psychological Measurement, 12,* 554–561.
Lorge, I., and Kruglov, L. (1953). The improvement of estimates of test difficulty. *Educational and Psychological Measurement, 13,* 34–46.
Tinkelman, S. (1947). *Difficulty Prediction of Test Items.* New York: Bureau of Publications, Teachers College, Columbia University.

Discussion of "Item and Score Conversion by Pooled Judgment"*

David Rogosa

Thorndike's paper opens up a novel and potentially important area for research on test-equating methods—namely, the use of experts' judgments to estimate item difficulties and to equate tests. The use of expert judgment to equate tests might appear to be "test equating without any data," especially when it is compared with procedures described throughout this volume that entail extensive and sophisticated data collections. But like examinee data, ratings of item difficulties are information about the test items, albeit "subjective" information. In fact, informal test equating through judgment is common in educational institutions, and is familiar to any instructor put in the position of having to construct a make-up exam.

Throughout this discussion I refer to the use of "experts" and "expert judgment" in test equating. Here, expertise implies the experience and skills of those professionals who write test items and construct tests for an organization such as ETS. Obviously, Thorndike's group of 20 graduate students, referred to as "interested amateurs," falls short of expert qualifications. Therefore, assessments of the ability of these students to estimate item

* Preparation of the comment was supported by a National Academy of Education Spencer Fellowship.

319

difficulties should not be taken to represent the ability of experts to do so. Still, further examination of these data and, in particular, consideration of relevant statistical issues and literature should help clarify some of the issues for test equating and generate interest in this topic. Two areas in which I focus my comments are described by the headings: "Methods for assessing judgments" and "Procedures for combining individual judges."

1. Methods for Assessing Judgments

In order to review what Thorndike did to assess his raters and to examine alternative procedures it is necessary to introduce some notation to describe Thorndike's data and analyses. The rated difficulty of item i by rater j on test k is $d_{ij}^{(k)}$, which is on a 9-point scale of ratings such as "Would be passed by not more than 30% of twelfth-graders." Recall that there are three tests— verbal analogies, quantitative relations, and figure analogies—each with 60 items and each item was rated by each of the 20 graduate students; $k = 1, 2, 3$, $i = 1, \ldots, 60$, and $j = 1, \ldots, 20$. The "actual" difficulty of item i on test k is $\delta_i^{(k)}$. Here, the $\delta_i^{(k)}$ range from 0 to 10. These values were constructed by Thorndike through a nonlinear transformation of the empirical proportion correct (for examinees at multiple grades) on each item; I divided Thorndike's values by 10 to give some numerical correspondence with the $d_{ij}^{(k)}$.

The d and δ scales could have been made equivalent by using the student responses at each grade level to provide empirical answers to the same questions that were used in the ratings. Surely the relation between the d and δ scales is monotone, but it is not necessarily linear, and certainly the natural reference to the 45 degree line in the plot of d and δ cannot be used. The lack of equivalence between the scales for the raters' responses and the empirical difficulties limits the information contained in the discrepancy $d_{ij}^{(k)} - \delta_i^{(k)}$ and thus limits the applicability of loss functions to measure the success of the raters.

1.1. Individual Raters

One basic description of an individual rater's performance is the scatterplot of $d_{ij}^{(k)}$ versus $\delta_i^{(k)}$ for rater j and test k. Figure 1 presents this scatterplot for rater 10 on the 60 items of the verbal analogies test ($k = 1$). The validity for rater 10, which is the correlation between $d_{i,10}^{(1)}$ and $\delta_i^{(1)}$ is 0.71 (one of the highest individual rater correlations). By itself, this correlation might be taken to indicate the rater 10 is a useful and dependable judge of item difficulties. However, the scatterplot reveals that in the middle of the $\delta^{(1)}$

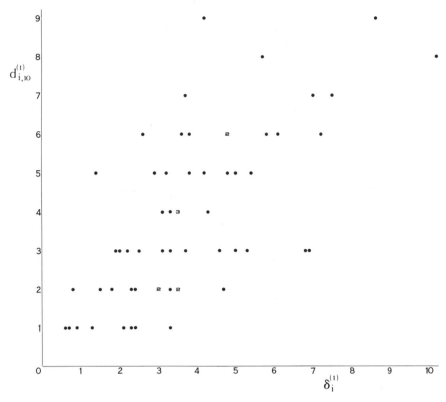

Fig. 1 Plot of ratings versus empirical difficulties for an individual judge on the 60 items from the verbal analogies test.

distribution (say $2.5 < \delta_i^{(1)} < 5.0$) the $d_{i,10}^{(1)}$ corresponding to a narrow slice of $\delta_i^{(1)}$ are spread over nearly the entire range of the ratings. That is, the validity does not reveal that even the better raters are poor except when rating the least difficult and most difficult items.

1.2. Comparing Raters

Many diagnostic questions about performance and possible remedial training for raters can be addressed through various summaries and plots of the $d_{ij}^{(k)}$. Are raters exchangeable? Are some raters clearly superior to others? Do different raters have different kinds of shortcomings? What items are the most difficult to rate? Thorndike presents two analyses directed toward the consistency among individual raters. Thorndike's Table 2 examines consistency among individual raters on "absolute" difficulty. Thorndike computes the average difficulty for each rater, $\bar{d}_{\cdot j}^{(k)}$, and reports the range of these averages for each of the three tests. A five-number summary for the $\bar{d}^{(1)}_{.j}$ is 3.1 [4.0, 4.4, 4.9] 6.2. The middle 10 raters are not greatly discrepant; the large range is created by one high rater and two low raters. In addition to $\bar{d}^{(k)}_{.j}$, which are the mean values of the ordinate in scatterplots like Fig. 1, the slope of the regression in this scatterplot is important in examining systematic and nonsystematic differences among raters.

Thorndike's other comparison among raters in his Table 1 follows the psychometric tradition of studies of interrater reliability. These psychometric procedures are based on agreement between the ratings of individual judges or between groups of judges. Scatterplots of individual ratings for two different raters (and also of sets of means of four raters) reveal agreement for the high and low ratings but large disagreement for middle values.

1.3. Average Rating

Pooled judgments can, of course, be displayed in the same way as judgments of individual raters. In Thorndike's study pooled judgment is represented solely by the arithmetic mean of the individual ratings, written as $\bar{d}^{(k)}_{i.}$. Figure 2 presents a scatterplot of $\bar{d}^{(1)}_{i.}$ versus $\delta^{(1)}_{i}$. The correlation for this scatterplot is 0.83; this correlation and the correlation for the other two tests are reported in Thorndike's Table 3. Again, although the correlation is high, the scatterplot shows that in the middle of the $\delta^{(1)}$ distribution $(2.5 < \delta^{(1)}_{i} < 4.5)$ for a specific value of $\delta^{(1)}_{i}$ (e.g., $\delta^{(1)}_{i} \simeq 3.5$) there is considerable dispersion in the $\bar{d}^{(1)}_{i.}$. That is, for items having (nearly) identical empirical difficulties, even difficulty ratings averaged over all 20 raters differ considerably.

The poor performance of the individual and average ratings should not generate despair about the usefulness of experts' ratings. Studies contrasting expert and novice problem-solving skills have shown large performance differences and also qualitative differences in the processing and use of knowledge (see Larkin *et al.*, 1980). For example, in rating item difficulties experts may be far more sensitive than novices to the influence of distractors on the difficulty of the item. Among the studies of experts' judgments, a modest success story familiar to statisticians is the study by Relles and Rogers (1977) in which experienced statisticians estimated location parameters. The statisticians performed far better than least squares but were not as good as some popular robust estimators of location.

1.4. Quantifying Rater Performance

In the statistical literature, a number of procedures incorporating various penalty functions and scoring rules have been used to assess judgments. Much of this literature is based on developments in the theory of personal

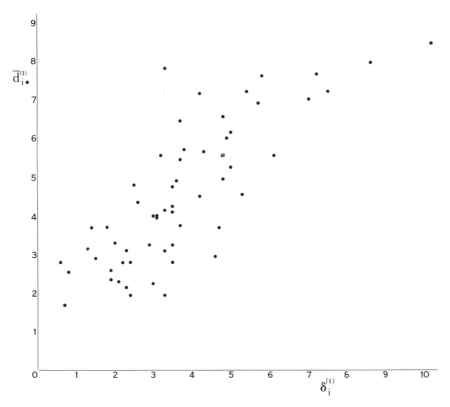

Fig. 2 Plot of average ratings versus empirical difficulties on the 60 items from the verbal analogies test.

probability (see Hogarth, 1975; Winkler, 1967). An active and relevant area of application is the assessment of weather forecasts, especially precipitation probability forecasts. In these forecasts the forecasters assign probabilities (e.g., 5, 10, 20, ..., 100%) to the event "measureable precipitation." The forecasts can be compared with the observed relative frequency of precipitation on days for which a specific probability of precipitation was forecast. The Brier score (a proper quadratic scoring rule) is often used by meteorologists. This and other scoring rules are examined in a survey of the calibration of probabilities by Lichtenstein *et al.* (1977). Additional statistical procedures for evaluating the performance of individual forecasters are presented by Murphy and Winkler (1977). Their "reliability diagrams" consist of plots of observed relative frequency versus forecast probability and correspond to the plots of the δ_i versus d_{ij} or $\bar{d}_i$ in Figs. 1 and 2. In this literature three attributes of the forecaster—reliability, accuracy, and skill—are defined and computed.

2. Combining Judgments of Individual Raters

In Thorndike's study the term "pooled judgment" is synonyomous with the arithmetic mean of the individual ratings. However, pooled judgments can be constructed in many ways, and several procedures have been studied in the statistical literature. Some of these methods appear promising for rating item difficulties. The broadest division of methods for combining judgments is on the basis of whether there exists interaction among the judges—specifically, feedback about the other judges and judgments. Distinctions among different types of feedback are used to categorize procedures further.

2.1. Combining Judgments without Feedback

Pooled judgments consist of a statistical summary of the individual ratings. The arithmetic mean is an obvious, but not the only, mechanism for summarizing the individual ratings. Trimmed means of the individual ratings might provide more useful judgments of the item difficulties. Probably the most familiar application of trimmed means for combining judgments is in Olympic diving competition. When a 20% trimmed mean is used in place of $\bar{d}_{i.}^{(1)}$ in Fig. 2, the vertical range of pooled judgments for a specific $\delta_i^{(1)}$ value is greatly decreased and the validity is increased. Of course, the median of the individual ratings is included here as a limiting case of the trimmed mean.

Additional procedures for summarizing the individual ratings can be constructed by using a weighted average of the individual ratings, $\sum_{j=1}^{20} w_j d_{ij}^{(k)}$. The arithmetic mean and the trimmed mean are obtained from specific forms of the w_j. Other useful values for the w_j could be obtained by asking each rater to indicate confidence or sureness for each rating on a numerical scale and using those data to form the w_j. This weighted average is an example of an "opinion pool" studied by Stone (1961).

2.2. Combining Judgments with Feedback

Pooled judgments consist of a group consensus or, failing that, a small collection of subgroup judgments. The interaction among the judges can be characterized by whether the feedback is quantitative or qualitative. The best-known procedure for group decision making is the Delphi method, which employs quantitative feedback and no face-to-face meeting. That is, after each round of responses each of the judges receive a quantitative summary of all responses (e.g., the median or a five-number summary). Details of implementations of the Delphi and criticisms of the method can be found in the references in Hogarth (1975) and Press (1978). A different quantitative feedback procedure is the Shang inquirer described in Press (1978).

Qualitative feedback procedures permit a richer interaction among the judges. In the qualitative controlled-feedback procedure examined by Press (1978) no quantitative feedback is provided; instead each judge is given a composite list of reasons elicited from each judge underlying the individual response. DeGroot (1974) proposed an interesting model for reaching a group consensus that involves each judge constructing revised opinions (for a point estimate or a probability distribution) by weighting the response of each of the judges by an individual assessment of that judge's importance or expertise. Chatterjee and Seneta (1977) investigated the convergence of this process when the weights change over time.

3.   Test-Equating Procedures

The following test-equating situation is constructed in Thorndike's analysis. A set of anchor items is identified, for which the empirical difficulties are known. These anchor items may be considered a sample of items from an existing test. Ratings are obtained for the anchor items and for a pool of new items, from which an equated test is to be formed. In terms of this study, both $\bar{d}_{i\cdot}^{(k)}$ and $\delta_i^{(k)}$ are available for anchor items, and $\bar{d}_{i\cdot}^{(k)}$ are available for the new items. Thorndike discusses both linear adjustment and matching of individual items as methods for constructing an equated test with the correct mean and variance. The analysis reported in the study (Table 4) involves the matching of anchor and new items on values of $\bar{d}_{i\cdot}^{(k)}$. The equating is evaluated through comparison of the distributions of the $\delta_i^{(k)}$ for the anchor items and the $\delta_i^{(k)}$ for the matched set of new items. Given the properties exhibited by the $\bar{d}_{i\cdot}^{(1)}$ in Fig. 2, it is not surprising that this equating is not successful.

What use can be made of expert judgment in practical test-equating procedures? One naive application would be an adaptation of basic observed-score test-equating techniques. Ratings of item difficulties for tests A and B for specific examinee populations could be used in place of examinee data; for example, linear equating could be carried out using means and variances for the total test score calculated from the sum of the individual item parameter ratings. But even with expert raters and sophisticated construction of pooled judgments, there is no guarantee of success. However, this is an empirical question and may be worthwhile to pursue.

A more promising strategy is the use of judgments in conjunction with examinee data. A general question is: How can expert judgments be combined with data? Of special interest is the use of judgments to provide information that is difficult or impossible to obtain from the test-equating design. For example, in the common-item equating (CIE) procedures described by Braun and Holland (elsewhere, in this volume) judgment data may be used to specify the unknown conditional distributions of the test scores or particularly the unknown conditional moments in the "Tucker equating" procedure.

REFERENCES

Chatterjee, S., and Seneta, E. (1977). Towards consensus: Some convergence theorems on repeated averaging. *Journal of Applied Probability*, *14*, 89–97.
DeGroot, M. H. (1974). Reaching a consensus. *Journal of the American Statistical Association*, *69*, 118–121.
Hogarth, R. M. (1975). Cognitive processes and the assessment of subjective probability distributions. *Journal of the American Statistical Association, 70*, 271–294.
Larkin, J., McDermott, J., Simon, D., and Simon, H. (1980). Expert and novice performance in solving physics problems. *Science, 208*, 1135–1142.
Lichtenstein, S., Fischhoff, B., and Phillips, L. D. (1977). Calibration of probabilities: The state of the art. In H. Jungermann and G. deZeeuw (Eds.), *Decision Making and Change in Human Affairs*. Dordrecht, Netherlands: Reidel Publ., 275–324.
Murphy, A. H., and Winkler, R. L. (1977). Reliability of subjective probability forecasts of precipitation and temperature. *Applied Statistics, 26*, 41–47.
Press, S. J. (1978). Qualitative controlled feedback for forming group judgments and making decisions. *Journal of the American Statistical Association, 73*, 526–535.
Relles, D. A., and Rogers, W. H. (1977). Statisticians are fairly robust estimators of location. *Journal of the American Statistical Association, 72*, 107–111.
Stone, M. (1961). The opinion pool. *Annals of Mathematical Statistics, 32*, 1339–1342.
Winkler, R. L. (1967). The quantification of judgment: Some methodological suggestions. *Journal of the American Statistical Association, 62*, 1105–1120.

Partial Orders and Partial Exchangeability in Test Theory

Joseph Deken

1. Introduction

The point of view I will take in this paper is that the problem of scoring (and ultimately of equating) tests is the problem of extending a natural partial order that is always present in a collection of test outcomes, beyond the limited number of test-takers it usually covers. This natural order is simply that we judge outcome A to be preferable to outcome B if for every item i in the test the response of A to item i is as good as the response of B, and sometimes better. (If item responses are classified as "right" and "wrong," then A gets every item right that B does and some that B does not.) In most of the discussion we will assume that the goal is to determine the relative ranks of all pairs of individuals, i.e., to produce a total ordering. In other cases, it may only be necessary to select a subset of the test-takers (the "high" or "accepted" group), and we only judge A to rank above B if A is selected and B is not. Here we would not be concerned with comparisons within the selected or unselected groups, so that the extension of the natural partial order will itself be only a partial order.

Whenever any partial ordering is extended, discrepancies may arise between the extended ordering and the original data. At the most fundamental level, we may observe that for some pair of test-takers A and B, our extended

327

ordering ranks A above B, and yet for some item of the test, B's answer is preferred to A's. I have denoted such an occurrence, involving only one pair of test-takers (outcomes) and one item as a fundamental discrepancy. Other terms, such as "atomic reversal," might be more descriptive in that these discrepancies represent the smallest (indecomposable) ways in which the data can go against an imposed extended ordering. I contend that any evaluation of a test ranking (or scoring or equating) procedure must ultimately be based on its performance in minimizing its lack of fit, as a function of the elemental discrepancies, to the data. The definition of the "correct" procedure (i.e., extended ordering) for the population may well depend on exactly how the lack of fit is defined, but no definition of fit that does not address the elemental discrepancies will be considered fundamental. The most convincing argument for this approach is its success (and unquestioned acceptance) in dealing with regression problems. In such contexts the elemental discrepancy is the individual residual or difference of one observation from its fitted value. Although hundreds of variants of regression procedures are in practical use, they invariably are motivated by the goal of making the residuals small in some sense.

In taking the point of view outlined above, I do not wish to minimize the usefulness of previous approaches to the problem, such as the response-prediction of "reproducibility" methods of Guttman (1944, 1947), factor-analytic techniques (i.e., Lazarsfeld, 1950), or other methods gathered under the umbrella of "scaling" (Torgerson, 1958). By setting a framework in terms of elemental discrepancies and extended (partial) orders, however, I believe we can evaluate many approaches to scaling in a consistent fashion. At the same time, I believe (taking the regression situation as an indication) that within this framework there can be developed new procedures that are both computationally feasible and potentially more informative than methods currently in use.

One might argue convincingly that little progress will be made from the "extend the partial order with minimum lack of fit" approach unless further assumptions are introduced. The equivalence of certain (potential) types of items in a test seems to be a likely simplification, especially as it enters in the definition of parallel tests. When this equivalence is assumed (or operationally guaranteed), the probabilistic idea of partial exchangeability becomes relevant. Some suggestions are offered here for models that reduce the complexity of the ranking problem, starting from the assumption of partial exchangeability.

Since the quantification issue of how to rank individuals in a population, given all information possibly observable, logically precedes the statistical issue of how to best approximate this ranking based on the limited information present from a test, sampling issues will be discussed only as a secondary issue here.

2. Tests as Instruments for Comparison

For the purpose of this discussion, I will suppose that the purpose of the tests we are discussing is to compare individual test-takers, i.e., to decide for every pair of test-takers A and B whether to rank A above B or vice versa. I do not take this point of view in order to ignore the uses of tests for evaluation and guidance. Although ranking is of primary interest for selection, I believe that this formalism can be adapted to evaluation and guidance as well. (Specifically, for guidance or evaluation, we may be interested in comparing A to a fictitious B, which may represent a previous A or a standard.) Similarly, the issue of aggregation is also not addressed here (how to rank one group of test-takers above or below another), as it is felt to be a separate topic. I believe that a discussion in terms of individual comparisons will delineate the basic issues.

In the context of individual comparisons, our problem is to take the record on test T of an individual A, and the record on test T' (possibly different, but parallel to T) of an individual B, plus information about other individuals and parallel tests, and to decide whether to rank A above B or vice versa. Equivalently, we may speak of attempting to rank records (possible test outcomes), rather than individuals, since individuals with identical outcomes will be treated identically.

3. The Natural Partial Order for Tests

If a test of N items is given, where each item may be answered correctly or incorrectly, there is a natural ordering of all the possible test outcomes. The raw data available consist of right–wrong vectors for each individual, where the vector for individual A has a "1" in position i if A gets item i right, and an "0" if A gets item i wrong, so that the number of components of the vector is the number of items in the test. The natural ordering of the individuals, based on this data, is to say that A ranks above B if B misses every item that A does and at least one item that A did not. While this ordering is natural in the sense that we would be unhappy with any scoring or equating procedure that went against it, it is severely limited. In equating situations, individuals are divided into (at least) two groups, and individuals within the same group have taken the same test, but different groups have taken different (but parallel in some sense) tests. Not only does the natural ordering not extend between groups (the equating problem), but in any realistic situation it will probably only cover a small portion of the pairs of individuals A and B even within the same group. That is, with high probability there will be both an item that A gets right and B gets wrong and an item that B gets right and A gets wrong.

The point of view I will take here is that the problem of scoring (and ultimately of equating) tests is the problem of extending the natural partial order beyond the limited number of test-takers it covers. In most of the discussion we will assume that the goal is to determine the relative ranks of all pairs of individuals, i.e., to produce a total ordering. In other cases, for example, if it is only necessary to select a subset of the test-takers (the "high" or "accepted" group), we only judge A to rank above B if A is selected and B is not, and would not be concerned with comparisons within the selected or unselected groups, so that the extension of the natural partial order will itself be only a partial order. The following discussion, in particular the definition of an elemental discrepancy, would carry over unchanged.

4.   Goodness of Fit and Elemental Discrepancies

Whenever any partial ordering is extended to a total ordering, there will be some lack of fit to the underlying data. The definition of the "correct" total ordering for the population intuitively should be the ordering that minimizes this lack of fit. Different definitions of "fit" may give rise to different "correct" orderings.

It may be useful here to recall the usual regression situation: In that case, our total population would consist of a large number of (x, y) pairs (a joint distribution of x and y if you like), and a regression of y on x is just a function $f(x)$. In order to choose the "correct" f, we choose f to minimize the lack of fit of the set of pairs $(y, f(x))$ to the set of pairs (y, x). Whatever the notion of lack of fit, it is usually built up from the set of elemental discrepancies, or residuals, $(y - f(x))$ for (x, y) in the population. (Well-known examples include the sum of the squares or absolute values of the elemental discrepancies.)

In extending the natural partial ordering based on items $I_1 I_2, \ldots, I_s$ and examinees $A_1 A_2, \ldots, A_N$ to a total ordering " $>$," we will consider the elemental discrepancies as triples (I_i, A_j, A_k) or (i, j, k) for short such that $A_j > A_k$ but A_j gets I_i wrong and A_k gets item I right.

These elemental discrepancies are as fundamental to the notion of fit for a total ordering as the differences $(y - f(x))$ are in the regression context. Note that if A ranks above B in the natural partial ordering, then there are no elemental discrepancies involving A and B. We may also consider cases where responses are not dichotomous. At this finer level, an elemental discrepancy occurs when $A > B$ but on item i, B chooses response r, which is better than A's response s. The discrepancy is then specified by the five-tuple, (i, r, s, A, B).

The notation for discrepancies used here is not meant to suggest that these discrepancies must be treated in isolation from each other. To pursue the regression analogy, we must think of $f(x)$ in general as a vector, in which case our notion of fit might combine discrepancies related to the same individual (or item or response) differently than discrepancies related to different individuals or items or responses. (For example, if we consider the sum of squared Euclidean distances between observation vectors y and fitted vectors $f(x)$, the components of the residual vectors $(y - f(x))$ are in effect treated separately, whereas if we sum up some other norm of the vector differences, they might not be.)

We might take our measure of goodness of fit of a total ordering to be just the number of elemental discrepancies. In this case, the optimal ranking is that based on giving every individual a score equal to the total number of items correct, then ranking by this score. In fact, a more general result can be stated:

Scoring Theorem.   *Let the lack of fit L of a total ordering ">" be defined as the sum over all items of the weight of the item times the number of elemental discrepancies involving that item. Then the ordering that minimizes the lack of fit is that based on assigning a score to each individual equal to the sum of the weights of that individual's correct items.*

Proof.   For any individuals A and B, we can compare the contribution to L of weighted elemental discrepancies involving A and B when (i) A is ranked above B or (ii) B is ranked above A, and conclude that this contribution can be minimized by ranking A above B just when the weighted sum of A's correct items is greater than the weighted sum of B's correct items. The conclusion holds for every pair of individuals, so the result follows.

In the context of equating, the novelty is that we do not assume that two tests must first be scored and then these scores equated, but ask, "For any new ordering of all individuals in the combined groups of test-takers, how well does this ordering agree with the available test data"? The answer must be based, in the same way that regression solutions are based on the analysis of residuals, on the elemental discrepancies between our new ordering and the data. Current methods of scoring and equating must be considered as the standard to improve, as well as the status quo, but they do not represent a fundamental or logical starting point.

5. Is the Statistical Problem of Equating Well Defined?

It is a serious mistake to launch into a discussion of test equating without resolving the basic, nonstatistical issue of what it is we are trying to reconstruct from our typical sample of data; i.e., what is the "population ranking." After all, the New Jersey lottery does not have an equating problem, because it is not trying to estimate anything. Tests could always be equated by assigning scores at random. The question to be answered is: "How would we rank the test-takers if we had all the data we could want?"

A familiar context in which to treat the preceding question is the case where we envision a (finite but usually very large) universe of possible tests which are considered to be equivalent. In this context, knowing "all the data" would mean knowing

for every individual A in the population,
for every class T of equivalent tests,
for every possible 0–1 vector of scores x,

the probability that individual A will obtain score vector x on a test t from T. (The definition of equivalence is that this probability is the same for all t in the same equivalence class T.) Supposedly, if we have all this information, we can look at individuals A and B and decide whether to rank A above or below B. I am not proposing certainly that we will ever collect all the data (we could then fire the statisticians), but the point is that if we could not rank A and B given all the data, we have no hope of saying anything sensible about test equating, which boils down to ranking A and B given only part of the data.

For example, suppose we are giving a test of mental arithmetic, with strata corresponding to addition, multiplication, subtraction, and division. Suppose we know, for every possible parallel test (five items each from each stratum, say only problems of two-digit numbers), and every score vector of 20 0's and 1's, the probability that an individual A and an individual B will obtain that score. What is the rule for ranking A above B or vice versa?

Once we have a procedure for inducing a ranking in a population based on complete test information, we may always (as in the case of "generic true score" to be discussed below) take the ranking induced by the test as defining the construct of interest. One then evaluates the ranking procedure in terms of the psychological meaningfulness of the resulting construct. Alternatively, an external criterion may be present, in which case the population ranking procedure must be evaluated in terms of the validity of the test. Only when these conceptual issues have been resolved can one embark on a clear discussion of reconstructing the population ordering from samples. Since we take the fundamental point of view that we are trying to extend available partial orderings in such a way as to minimize elemental discrepancies (residuals), the first place in which to apply this idea is in the population of test items and test-takers.

6. Some Possible Extensions to Total Ordering

Two examples will be given here, representing two different possible ways of extending the natural partial order. The first example is well known, and merely serves to show that the usual notions fit in this scheme. The second example is a new suggestion that I believe has much intuitive appeal. Other approaches can be built up without assuming that the final ranking is compatible with any score function that is a sum of item weights.

We will simplify matters by supposing that for any item on any test and any test-taker, the probability that the test-taker gets the item right is either 0 or 1. (You might imagine a collection of geography questions: What is the capital of the United States? What is the capital of South Dakota? of Nepal? etc.)

6.1. Generic True Score

A "generic" true score, as defined by Lord and Novick (1968), is based on a collection of equivalent tests as described earlier. Assuming a scoring function that assigns a number to the outcome of every possible test, a person's generic true score is defined as the expected value on a test chosen at random from among all the equivalent tests. Any reasonable scoring function will have the property that if A gets every question right that B does and some that B does not, then A's score is higher than B's. Thus the ranking implied by subjects' generic true scores (A ranks above B if A's score is higher than B's) extends the partial ordering present in the "full data" or "population." A typical scoring function is to assign to A a score equal to the number of items correct, or if items have been assigned weights, A's score is the sum of the weights of the items A gets correct.

The preceding definition of generic true score may be criticized if the equivalence implied between sets of items that have the same sum of weights is not psychologically reasonable. For example, in the mental arithmetic context, your generic mental arithmetic true score is higher if you get one-third of all problems right than if you get all division problems right and don't know anything else. (One usually supposes, under the rubric of single-factor tests or similar assumptions, that such difficulties do not arise. The assumptions may or may not be checked.)

If the scoring method is a weighted sum of items correct, then the scoring theorem applies to show that the ranking produced minimizes the weighted number of discrepancies involving each item. If the weights themselves are not based on discrepancies or equally fundamental considerations, the motivation for the scoring method is consequently weak. An example of such a nonfundamental consideration is the assignment of weights based on factor-analytic methods, which produce a maximum-variance ranking. Why maximize variance, unless this maximization produces other more interpretable consequences? Current scoring methods, from which generic true score may be defined, are typically not justified in terms of minimizing discrepancies, even under strong models of items, traits, and populations. Would we be satisfied with a theory of regression that was not based on, and indeed seldom made explicit contact with, the idea of minimizing residuals?

6.2. Discrepancy-Weighted Total Orders

By allowing the weights for items to depend on the elemental discrepancies involving the item, we suggest here an ordering extension method that is directly analogous to iteratively reweighted regression procedures. It should be emphasized that this method, and especially the use of a particular weight function g, is meant to be only one possible example of the use of the elemental discrepancies directly, without appeal to other assumptions. Knowledgeable readers will observe that for some choices of the function g, the ranking obtained, and even the final weights, may be the same as those obtained by the factor-analytic approach of Guttman (1944, 1947). While giving some plausibility to this approach, this fact also serves to put it in perspective—as a special case of one specific methodology of fitting. (There are two possible motivations given for Guttman scaling. The first posits that it is desirable to maximize the variance between subjects as a fraction of the total variance, the assumptions being primarily that a ratio of this sort is a good thing to maximize, and that sum-of-squares is the right way to measure variability. Buying the first assumption does not require buying the second. A second rationale is based on the assumption that test outcomes are well quantified as vectors in euclidean space, and that arbitrary directions in this space are meaningful. A particular one of these directions—a factor—is used to define item weights, and we are further required to agree that it is the sum of squared norms of the deviations of outcomes from their projections in this direction that justify the direction as best. When these or more questionable assumptions are the basis of a methodology, the method ought to be examined in the light of more fundamental concepts, and with a view toward viable alternatives.)

Let g be a nondecreasing function such that for $x < 0$, $g(x) = 0$, and let the lack of fit L of the ranking based on scores $s(A)$, $s(B)$, . . . be defined as the sum over all elemental discrepancies (i, j, k) of $w(i)$, where $w(i)$ is a weight assigned to item i as the solution of the following two equations:

$s(A)$ = sum of weights $w(i)$ of items correctly answered by A, for all individuals A.

$w(i) = 1/(1 +$ sum over elemental discrepancies (i, j, k) of $g(s(A_j) - s(A_k)))$.

The effect of the preceding definition is to rank individuals according to a weighted sum of items, where items involved in a large number of elemental discrepancies are downweighted. This procedure is appealing in that it relates sensibly to the elemental discrepancies in the same way that iteratively reweighted regression procedures relate to the deviations $(y - f(x))$ and is guaranteed to preserve the natural partial order. An example will illustrate the use of this procedure and the role of the weight function g:

Consider the case of a population of ten items and five individuals. The right–wrong vectors for the individuals are

$$
\begin{array}{cccccccccc}
1 & 1 & 1 & 1 & 1 & 1 & 1 & 1 & 1 & 1 \\
1 & 1 & 1 & 1 & 1 & 0 & 0 & 0 & 0 & 0 \\
1 & 1 & 1 & 1 & 1 & 0 & 0 & 0 & 0 & 0 \\
1 & 0 & 0 & 0 & 0 & 0 & 0 & 0 & 0 & 0 \\
0 & 0 & 0 & 0 & 0 & 0 & 0 & 0 & 1 & 1 \\
\end{array}
$$

That is, the first individual gets all items correct, the second two get the first five items correct, etc. Using "total correct" as a score, the ranking of the individuals would be $1 > 2 > 3 > 5 > 4$.

If we choose the function $g(x) = (1 + \text{signum}(x))/2$, we obtain a solution that assigns weight 1 to items 1–8, and weight $\frac{1}{4}$ to items 9 and 10, with the resulting ranking $1 > 2 > 3 > 4 > 5$ and corresponding scores 8.5, 5, 5, 1, 0.5. The value of L obtained is 1.5. (Another solution exists for the ranking $1 > 2 > 3 > 5 > 4$, with weights $\frac{1}{2}$ for item 1, 1 for items 2–8, and $\frac{1}{3}$ for items 9 and 10. The scores are 8.17, 4.5, 4.5, 0.5, 0.67. The value of L is $11/6 > 1.5$, so the first solution is optimal.)

If we choose the function $g(x) = (x + \text{abs}(x))/2$, we obtain weights of 1 for items 1–8, $1 - \text{sqrt}(30)/6 = 0.087$ for items 9 and 10 and ranking of $1 > 2 > 3 > 4 > 5$, with scores 8.17, 5, 5, 1, 0.17.

The purpose of a small example is for simplicity of presentation. It should be computationally feasible to draw moderate-sized samples from a larger population in order to check stability of the results. Further development of theory and practice will tell whether such an approach is manageable and illuminating.

7. Reducing the Complexity of the Full Data

In any realistic situation the number of potential items in the population will be enormous. (For the arithmetic example, assuming two-digit arguments, there are 40,000 items. This is nearly the smallest nontrivial example imaginable.) In order to develop procedures of any practical use, simplifying assumptions and/or models become necessary. The basic simplifying idea is the equivalence of items in some sense to certain other items. Depending on the mathematical form given to this idea of equivalence, different procedures can be developed. In any case, the form should always be such that the model can be potentially verified. Although many notions of equivalence have been proposed, and embodied in mathematical definitions of traits and assumptions of local independence, we will argue here that the basic mathematical form of equivalence is exchangeability, which may then be strenghtened in one of several directions to give practical procedures.

As is often done in practice, we may define strata for items and declare items within a given stratum to be equivalent. Such strata are used in some definitions of parallel tests. For example, "L-parallel tests" are tests constructed by random sampling a fixed number of items from each of a given set of strata. The notions of partial order given above carry through, taking the partial order now as $A > B$ just when A gets at least as many items correct in every stratum as B does. Computations and numerical procedures are vastly simplified; e.g., weights are given or found for strata rather than for items.

When items are stratified, it is natural to assume that the right–wrong vector relating to items within the same stratum is exchangeable. We will call this equivalence "partial exchangeability" to emphasize the fact that it does not extend between strata. The naturalness of this assumption is that any information about order between the items is discarded in practice. That is, we might agree to use, instead of the right–wrong vector, a random permutation of this vector. Exchangeability is then not an assumption; it is built in. But if our procedure uses only the number correct within a stratum (or some function of this number), then we can agree to random permutation because it leaves the procedure invariant. The point is that if the data are analyzed at a level that seems manageable (using functions of the total number correct within strata), exchangeability is built in.

From the assumption of partial exchangeability, we obtain from (the finite form of) de Finetti's theorem that the "full data" can be specified by giving a vector parameter P having one coordinate for each stratum, and specifying a distribution for this parameter with all coordinates lying between 0 and 1. When the coordinate parameter P for a given stratum is known, the right–wrong vector within this stratum consists of independent Bernouilli trials with success probability p (Diaconis and Freedman, 1971). The specification of the joint distribution of P may itself be unmanageable. Further simplifications have been or could be assumed: We might specify that all the coordinates of P are obtained by monotone nondecreasing transformations of a single random variable, as is familiar from latent trait theory. Since most realistic tests are not actually "single factor," other (log linear?) models for the coordinates of P could be investigated. Individuals A and B are then described with respect to the test by their respective distributions F and G of the vector parameter P. The basic quantification issue of comparing individual A to individual B amounts to comparing F to G, perhaps under some model for the form of these distributions. The development of such a model might be similar to previous approaches using independence as a starting point (Bahadur, 1961; Lazarsfeld, 1961), or perhaps log linear models utilizing information about the strata would be useful. In any case, the adequacy of any description would ultimately be judged by the magnitude of the resulting elemental discrepancies.

8. The Statistical Problem of Equating

A final (solitary) statistical issue needs to be raised with regard to equating. Admittedly, scores are time-honored methods of expressing test results, and if we have ranked a single group of subjects, we can (and might as well) express the results of our ranking by giving scores to the subjects. But to conclude from this that the test results are coextensive with the scores, even for practical purposes, is unwarranted.

Provided that a population ranking (or score) is well defined, and I have argued that it must be before equating can sensibly be discussed, the actual result of a test administration is a reduction of our uncertainty about where various subjects fit into this ranking. Specifically, we may have confidence intervals, likelihood functions, or posterior distributions for each tested individual, depending on our statistical persuasion. If this information is available for two groups tested with different tests, then we may have some disagreement as to how to rank the combined groups, but reasonable procedures will be making use of all the information available. It will also be clear that if we have less information about one group than the other, there is no possible way to remedy this. (If one insists on "fairness," there is no recourse but to throw away information about the more precisely measured group. Barring this, subjects will prefer one test to the other.) If we throw away all the information about subjects except the final ranking within each group (scores), and then try to combine rankings or "equate," the task is made much more difficult. It is even difficult to state what the criteria for a good solution are. (In contrast, a Bayesian who keeps posterior distributions could simply decide to rank A above B if the posterior probability that A's full data score is higher than B's is $> \frac{1}{2}$.)

Despite precedent, I think there are viable alternatives to simply (recording and) reporting scores, certainly within the framework of L-parallel tests. Whether these alternatives should be as sophisticated as posterior distributions or as simple as confidence intervals and whether the test construction process can be systematized and formalized in such a way as to make the L-parallel test a believable model are questions worth answering. Perhaps a hypothetical instrument similar to the Test of English as a Foreign Language might provide a starting point. Could a fixed corpus of written English be delineated and strata for items be based on well-defined and distinct sampling/construction procedures for items based on this corpus? Such a formal framework for test construction will make explicit our uncertainty above interpreting test results, even under optimal conditions, because of the inability of any short test to map accurately the large and complex structure from which it is extracted.

REFERENCES

Bahadur, R. R. (1961). A representation of the joint distribution of responses to n dichotomous items. In H. Solomon (Ed.), *Studies in Item Analysis and Prediction*. Stanford, Calif.: Stanford University Press, Chap. 9.

Diaconis, P., and Freedman, D. (1980). De Finetti's generalizations of exchangeability. In R. Carnap (Ed.), *Studies in Inductive Logic and Probability*, Vol. 2. Berkeley: University of California Press, 233–249.

Guttman, L. (1944). A basis for scaling qualitative data. *American Sociological Review, 9*, 139–150.

Guttman, L. (1947). The Cornell technique for scale and intensity analysis. *Educational and Psychological Measurement, 7*, 247–280.

Lazarsfeld, P. F. (1950). The logical and mathematical foundation of latent structure analysis. In S. A. Stouffer, L. Guttman, E. A. Suchman, P. S. Lazarsfeld, S. A. Star, J. A. Clausen (Eds.), *Measurement and Prediction*. Princeton, N. J.: Princeton University Press, Chap. 10.

Lazarsfeld, P. F. (1961). The algebra of dichotomous systems. In H. Solomon (Ed.), *Studies in Item Analysis and Prediction*. Stanford, Calif.: Stanford University Press, Chap. B.

Lord, F. M., and Novick, M. R. (1968). *Statistical Theories of Mental Test Scores*. Reading, Mass.: Addison-Wesley.

Torgerson, W. S. (1958). *Theory and Methods of Scaling*. New York: Wiley.

Discussion of "Partial Orders and Partial Exchangeability in Test Theory"

Donald B. Rubin

Joseph Deken's interesting paper focuses on the fundamental ideas underlying the use of test scores to rank individuals. His position apparently is that sorting out this aspect of testing necessarily comes before the problem of equating two different tests.

Without any external assumptions, the result of a test given to a group of individuals only yields a partial ordering of the individuals. That is, A is better than B if and only if (1) A missed no items that B got right and (2) A got at least one item right that B missed. If all items are to be treated as equivalent or exchangeable in some sense, then symmetry implies that "total number right" is the obvious measure of a test score and should be used to order the test-takers (I am assuming here that each item is scored as either right or wrong). If items are to be (or can be) treated differently, then a unique transformation from a partial ordering to a total ordering relies on some external assumptions or principles.

For an illustrative example, let us examine Deken's 10-item, 5-individual display summarized in Fig. 1. Assume that each row (individual) in fact represents many individuals (say 10,000) so that the display represents a population and in that population there are only four patterns of response instead of the 2^{10} that are possible with 10 items.

339

Number of individuals	Pattern
10,000	1111111111
20,000	1111100000
10,000	1000000000
10,000	0000000011

Fig. 1 Hypothetical right-wrong vectors for population of 50,000 individuals on 10-item test.

Note that item 1 is the easiest item because 40,000 out of 50,000 got it right and, except for items 6, 7, and 8 (which only the 10,000 top individuals got correct along with all other items), items 9 and 10 are the hardest items since only 20,000 got them right. Consequently, any rule that produces a total ordering must explicitly or implicitly decide whether it is better (1) to get two hard items right and miss everything else or (2) to get the easiest item right and miss all the other (harder) items. That is, minimally, the total ordering must decide whether (0, 0, 0, 0, 0, 0, 0, 0, 1, 1), is better or worse than (1, 0, 0, 0, 0, 0, 0, 0, 0, 0).

Such a decision is not easy because intuition can lead to either answer. For example, one story goes that getting two harder items right clearly demonstrates knowledge—missing the easier items suggests that the test is not unidimensional and that the testees with this pattern do not know certain fields commonly known by other individuals in the population; those getting only the easiest item right clearly have minimal knowledge. On the other hand, one could argue that missing the easy items suggests nearly no knowledge at all, and the fact that two hard items were correctly answered is the result of guessing and should be effectively discarded—if these individuals really knew the answers to those two hard questions, they would have known the answers to the easier questions. Consequently, the group with only the easiest item correct is superior. Of course, if the last two items were not easy to answer correctly without knowledge of the correct answer, then getting 2 items out of 10 right by guessing is not likely, and the first argument might be viewed as more compelling.

The logical way to resolve such conflicts in intuitive descriptions of test-taking behavior is to understand the formal reasoning behind the arguments and test the arguments using data. I believe that the best way to do this is within the framework of formal statistical models of test-taking behavior. In any case, I doubt that the choice of a particular scoring function is going to lend much insight into the solution of this problem, unless the form of the function is mathematically tied to psychological understanding about the nature of the test and the test-takers. When the scoring function and Deken's ideas of discrepancy are explicitly related to such factors, perhaps by being derived from particular models of test behavior, then these ideas may play an important role in scoring tests and thereby equating tests.

Another alternative is to finesse the problem by declaring that people who get hard items right and easy items wrong have *aberrant* test results and should be put aside for further consideration. Such an approach has been proposed by Levine and Rubin (1979) and seems to have some appeal in cases without good models or obvious choices for moving from a partial to a total ordering. This suggestion to discard those people who distort the natural partial ordering from a complete ordering is dual to and similar in spirit to Deken's suggestion that by weighting, items that destroy the partial ordering can effectively be discarded.

REFERENCE

Levine, M. V., and Rubin, D. B. (1979). Measuring the appropriateness of multiple-choice test scores. *Journal of Educational Statistics*, *4*(4), 269–290.

General Remarks

I. Richard Savage, Elizabeth Scott,

and Madeline M. Wallmark

I. Richard Savage

I am an outsider but I do have a special interest in this area. In the past, I have never had anything to do with aptitude testing. I have never worked on any of the related statistical problems. Now, I am serving on the Panel on Aptitude Testing of the Handicapped for the National Research Council, and therefore, have the needs of that panel very much in mind.

The present meeting has been highly technical. As far as I can tell, nearly all of the discussion and certainly all of the papers were directed at the technical problems of equating two tests. There was very little discussion of why tests should be equated and, more importantly, of how *well* tests need to be equated. It is unwise to have a statistical problem if you do not have some bounds on how well the job should be done. If there are no bounds, then one tends to use as much money as is available whether it is useful or not. It does seem to me that the equating of tests is not exactly what is needed. What we want to do is find out how well the tests are measuring something that we are interested in. How well they measure each other is not really

343

relevant. Unless we have some very strong assurance that the tests are doing an outstanding job, the matching of tests can be absolutely disastrous.

It is worthwhile in a problem of this kind, if you are going to equate tests, to design your procedures to equate two successive tests. But those procedures may be quite inadequate for equating tests that are far separated in time. There is likely to be drift that would be hard to detect. Tukey (1979) discusses the use of statistical procedures in the government and he considers this problem in the measurement of employment.

All of the discussion is concerned with very large samples. This is, in part, because you wish to obtain very good fits. But the motivation is obscure. This mentality prevents one from looking at the more interesting current problems. In particular, how can we relate ability tests given to the various groups of handicapped individuals or to people with different linguistic backgrounds? In the case of the handicapped, the 504 regulations put considerable pressure on ETS and others to attempt to make some comparisons. The effort should be going into how to equate tests that are in different forms and administered to different populations. The great difficulty here is that the populations themselves and the samples also are likely to be rather small by your standards. But are they large enough for doing what should be done?

I really don't like to use the phrase "test equating" since I am sure that the ideas about validation or calibration are much closer to what is needed.

The main point of this discussion is that the use of extremely large samples obscures the need for serious work on small samples. The heterogeneity of the handicapped population is only part of what causes trouble. Trouble comes from the necessity of giving a great variety of forms of the tests. It is my understanding that the validation should be on the predictive value of the test as administered. The handicapped condition of the individual could be used in the calibration effort. That is a subject open to discussion.

Elizabeth Scott

The papers we have heard on test equating are very interesting, but I think that they are not directed to the problems Educational Testing Service should be considering. The fundamental aim should be to produce more meaningful predictions of student performance, with better emphasis on producing fair predictors. At this time, Educational Testing Service has the opportunity, and even the responsibility, to rethink what it is trying to accomplish, what colleges and universities want to do, and what students want to do. One problem deserving special attention is the possibility of producing reliable alternative paths for admission to college that do not entail the standard test questions. These possibilities are of great importance to students who are culturally or physically handicapped as well as to students of outstanding but nonroutine abilities.

Madeline M. Wallmark

I feel that the success of the conference was somewhat marred by the failure to address the practical problems that would have been encountered if someone attempted to apply to real-life situations the theoretical methods discussed. Some of the attendees may have come away with the impression that ETS is slow to accept innovative ideas, but if they had been challenged to apply their theories to special situations, such impressions may have been tempered by a better understanding of how imperfect the equating world we live in really is.

REFERENCE

Tukey, J. W. (1979). Methodology, and the statistician's responsibility for BOTH accuracy AND relevancy. *Journal of the American Statistical Association, 74,* 786–793.

Bibliography on Test Equating and Related Topics

Abu-Sayf, F. K. (1977). A new formula score. *Educational and Psychological Measurement*, *37*, 853–862.

Andrulis, R. H., Starr, L. M., and Furst, L. M. (1978). The effects of repeaters on test equating. *Educational and Psychological Measurement*, *38*, 341–349.

Angoff, W. H. (1957). The "equating" of nonparallel tests. *Journal of Experimental Education*, *25*(3), 241–247.

Angoff, W. H. (1960). Measurement and scaling. In C. W. Harris (Ed.), *Encyclopedia of Educational Research* (3rd ed.). New York: Macmillan, 807 817.

Angoff, W. H. (1961). Basic equations in scaling and equating. In S. S. Wilks (Ed.), *Scaling and Equating College Board Tests*. Princeton, N.J.: Educational Testing Service, 120–129.

Angoff, W. H. (1961). Language training study: 1947. In S. S. Wilks (Ed.), *Scaling and Equating College Board Tests*. Princeton, N.J.: Educational Testing Service, 130–143.

Angoff, W. H. (1966). Can useful general-purpose equivalency tables be prepared for different college admissions tests? In A. Anastasi (Ed.), *Testing Problems in Perspective*. Washington, D.C.: American Council on Education, 251–264.

Angoff, W. H. (1968). How we calibrate College Board scores. *College Board Review*, *68*, 11–14.

Angoff, W. H. (Ed.) (1970). *The College Board Technical Manual: A Description of Research and Development for the College Board Scholastic Aptitude Test and Achievement Tests*. Princeton, N.J.: Educational Testing Service.

Angoff, W. H. (1971). Scales, norms, and equivalent scores. In R. L. Thorndike (Ed.), *Educational Measurement* (2nd ed.). Washington, D.C.: American Council on Education, 508–600.

Angoff, W. H. (1971). Test scores and norms. In L. C. Deighton (Ed.), *Encyclopedia of Education*. New York: Macmillan, 153–165.

Angoff, W. H. (1982). Norms and scales. In H. E. Mitzel (Ed.), *Encyclopedia of Educational Research*. New York: Macmillan, in press.

Angoff, W. H., and Dyer, H. S. (1971). The Admissions Testing Program. In W. H. Angoff (Ed.), *The College Board Admissions Testing Program: A Technical Report on Research and Development Activities Relating to the Scholastic Aptitude Test and Achievement Tests.* New York: College Entrance Examination Board, 1–13.

Angoff, W. H., and Modu, C. C. (1973). *Equating the Scales of the Spanish-Language Prueba de Aptitude Academic and the English-Language Scholastic Aptitude Test of the College Entrance Examination Board.* College Entrance Examination Board, New York, N.Y. (RB-73-4 and RDR-72-73-4). Princeton, N.J.: Educational Testing Service.

Angoff, W. H., and Stern, J. (1971). *The Equating of the Scales for the Canadian and American Scholastic Aptitude Tests.* Princeton, N.J.: Educational Testing Service.

Angoff, W. H., and Waite, A. (1961). Study of double part-score equating for the Scholastic Aptitude Test. In S. S. Wilks (Ed.), *Scaling and Equating College Board Tests.* Princeton, N.J.: Educational Testing Service, 73–85.

Bahadur, R. R. (1961). A representation of the joint distribution of responses to n dichotomous items. In H. Solomon (Ed.), *Studies in Item Analysis and Prediction.* Stanford, Calif.: Stanford University Press, Chap. 9.

Beaton, A., Hilton, T., and Schrader, W. (1977). *Changes in the Verbal Abilities of High School Seniors, College Entrants, and SAT Candidates between 1960 and 1972.* New York: College Entrance Examination Board.

Bianchini, J. C., and Loret, P. G. (1974). *Anchor Test Study* (Final report in 33 volumes). Berkeley, Calif.: Educational Testing Service.

Birnbaum, A. (1968). In F. M. Lord and M. R. Novick (Eds.), *Statistical Theories of Mental Test Scores.* Reading, Mass.: Addison-Wesley, Chaps. 17–20.

Boldt, R. F. (1972). *Anchored Scaling and Equating: Old Conceptual Problems and New Methods* (RB-72-28 and RDR-72-73-2). Princeton, N.J.: Educational Testing Service.

Boldt, R. F. (1974). Comparability of scores from different tests though on the same scale. *Educational and Psychological Measurement, 34,* 239–246.

Box, G. E. P., and Tiao, G. C. (1973). *Bayesian Influence in Statistical Analysis.* Addison-Wesley, Reading, Mass.

Braswell, J., and Petersen, N. (1977). *An Investigation of Item Obsolescence in the Scholastic Aptitude Test.* New York: College Entrance Examination Board.

Conrad, H. S. (1950). Norms. In W. S. Monroe (Ed.), *Encyclopedia of Educational Research* (Rev. ed.). New York: Macmillan, 795–802.

Cureton, E. E. (1941). Minimum requirements in establishing and reporting norms on educational tests. *Harvard Educational Review, 11,* 287–300.

Cureton, E. E., and Tukey, J. W. (1951). Smoothing frequency distributions, equating tests, and preparing norms. *American Psychologist, 6,* 404. (Abstract)

Dempster, A. P., Laird, N. M., and Rubin, D. B. (1977). Maximum likelihood from incomplete data via the EM algorithm. *Journal of the Royal Statistical Society, B, 39,* 1–38.

Echternacht, G. (1974). *Alternate Methods of Equating GRE Advanced Tests* [Project Report PR 71-17 (October 1971). GRE Board Professional Report (GREB No. 69-2p)]. Princeton, N.J.: Educational Testing Service.

Fan, C. T. (1957). On the applications of the method of absolute scaling. *Psychometrika, 22,* 175–183.

Flanagan, J. C. (1939). *The Cooperative Achievement Tests: A Bulletin Reporting the Basic Principles and Procedures used in the Development of their System of Scaled Scores.* New York: American Council on Education, Cooperative Test Service.

Flanagan, J. C. (1951). Units, scores, and norms. In E. F. Lindquist (Ed.), *Educational Measurement.* Washington, D.C.: American Council on Education, 695–763.

Flanagan, J. C. (1962). Symposium: Standard scores for achievement tests (Discussion). *Educational and Psychological Measurement*, 22, 35–39.

Gardner, E. F. (1966). The importance of reference groups in scaling procedure. In A. Anastasi (Ed.), *Testing Problems in Perspective*. Washington, D.C.: American Council on Education, 272–280.

Gleser, L., and Olkin, I. (1973). Multivariate statistical inference under marginal structure. *British Journal of Statistical Psychology*, 26, 98–123.

Gleser, L., and Olkin, I. (1975). Multivariate statistical inference under marginal structure, II. In J. N. Srivastava (Ed.), *A Survey of Statistical Design and Linear Models*. Amsterdam: North-Holland Publishing Co., 165–179.

Goulet, L. R., Linn, R. L., and Tatsuoka, M. M. (1975). *Vertically Equated Test Forms*. Chapter 5 in investigation of methodological problems in educational research-longitudinal methodology. Final Report, Project No. 4-1114. Urbana-Champaign, Illinois: University of Illinois.

Green, B. F. (1979). 3 decades of quantitative methods in psychology. *American Behavioral Scientist*, 23, 811–834.

Gulliksen, H. (1950). *Theory of Mental Tests*. New York: Wiley.

Gustafsson, J. E. (1979). Rasch Model in vertical equating of tests—critique of Slinde and Linn. *Journal of Educational Measurement*, 16, 153–158.

Guttman, L. (1941). The quantification of a class of attributes: A theory and method of scale construction. In P. Horst (Ed.), *The Prediction of Personal Adjustment*. New York: Social Science Research Council, 321–347.

Guttman, L. (1944). A basis for scaling qualitative data. *American Sociological Review*, 2, 139–150.

Guttman, L. (1947). The Cornell technique for scale and intensity analysis. *Educational and Psychological Measurement*, 2, 247–280.

Guttman, L. (1950). The basis for scalogram analysis. In S. A. Stouffer *et al.* (Eds.), *Studies in Social Psychology in World War II* (Vol. 4). Princeton, N.J.: Princeton University Press, 60–90.

Hambleton, R. K., and Cook, L. L. (1977). Latent trait models and their use in the analysis of educational test data. *Journal of Educational Measurement*, 14, 75–96.

Horst, P. (1931). Obtaining comparable scores from distributions of dissimilar shape. *Journal of the American Statistical Association*, 26, 455–460.

Horst, P. (1932). Comparable scores from skewed distributions. *Journal of Experimental Psychology*, 15, 465–468.

Huddleston, E. M. (1957). *Equating (/TDM-57-1)*. Princeton, N.J.: Educational Testing Service.

Hull, C. L. (1922). The conversion of test scores into series which shall have any assigned mean and degree of dispersion. *Journal of Applied Psychology*, 6, 298–300.

Jöreskog, K. G. (1971). Statistical analysis of sets of congeneric tests. *Psychometrika*, 36, 109–133.

Karon, B. P. (1955). *The Stability of Equated Test Scores* (RB-55-21). Princeton, N.J.: Educational Testing Service.

Karon, B. P., and Cliff, R. H. (1957). *The Cureton-Tukey Method of Equating Test Scores* (RB-57-6). Princeton, N.J.: Educational Testing Service.

Keats, J. A., and Lord, F. M. (1962). A theoretical distribution for mental test scores. *Psychometrika*, 27, 59–72.

Keats, J. B. (1970). A program for equating scores. *Educational and Psychological Measurement*, 30, 143–145.

Krus, D. J., and Krus, P. H. (1977). Lost: McCall's T Scores: Why? *Educational and Psychological Measurement, 37,* 257–261.

Lawley, D. N. (1942–1943). On problems connected with item selection and test construction. *Proceedings of the Royal Society of Edinburgh, 61* (Section A, Part III), 273–287.

Lazarsfeld, P. F. (1950). In S. A. Stouffer *et al.* (Eds.), *Studies in Social Psychology in World Wat II* (Vol. 4). Princeton, N.J.: Princeton University Press, Chaps. 10 and 11.

Lazarsfeld, P. F. (1961). The algebra of dichotomous systems. In H. Solomon (Ed.), *Studies in Item Analysis and Prediction.* Stanford, Calif.: Stanford University Press, Chap. B.

Lennon, R. T. (1964). Equating nonparallel tests. *Journal of Educational Measurement, 1,* 15–18.

Lennon, R. T. (1966). Norms: 1963. In A. Anastasi (Ed.), *Testing Problems in Perspective.* Washington, D.C.: American Council on Education, 1966, 243–250.

Levine, R. S. (1955). *Equating the Score Scales of Alternate Forms Administered to Samples of Different Ability* (RB-55-23). Princeton, N.J.: Educational Testing Service.

Lindquist, E. F. (1930). Factors determining reliability of test norms. *Journal of Educational Psychology, 21,* 512–520.

Lindquist, E. F. (1953). Selecting appropriate score scales for tests (Discussion). In *Proceedings of the 1952 Invitational Conference on Testing Problems.* Princeton, N.J.: Educational Testing Service, 34–40.

Lindquist, E. F. (1966). Norms by schools. In A. Anastasi (Ed.), *Testing Problems in Perspective.* Washington, D.C.: American Council on Education, 269–271.

Lindsay, C. A., and Prichard, M. A. (1971). An analytical procedure for the equipercentile method of equating tests. *Journal of Educational Measurement, 8,* 203–207.

Lord, F. M. (1950). *Notes on Comparable Scales for Test Scores* (RB-50-48). Princeton, N.J.: Educational Testing Service.

Lord, F. M. (1952). A theory of test scores. *Psychometric Monographs, 7.* University of Chicago Press, Chicago.

Lord, F. M. (1955). Equating test scores—a maximum likelihood solution. *Psychometrika, 20,* 193–200.

Lord, F. M. (1955). A survey of observed test-score distributions with respect to skewness and kurtosis. *Educational and Psychological Measurement, 15,* 383–389.

Lord, F. M. (1957). Do tests of the same length have the same standard errors of measurement? *Educational and Psychological Measurement, 17,* 510–521.

Lord, F. M. (1959). Test norms and sampling theory. *Journal of Experimental Education, 27,* 247–263.

Lord, F. M. (1962). Estimating norms by item-sampling. *Educational and Psychological Measurement, 22,* 259–267.

Lord, F. M. (1974). Estimation of latent ability and item parameters when there are omitted responses. *Psychometrika, 39,* 247–264.

Lord, F. M. (1975). *A Survey of Equating Methods Based on Item Characteristic Curve Theory* (RB-75-13). Princeton, N.J.: Educational Testing Service.

Lord, F. M. (1977). Practical applications of item characteristic curve theory. *Journal of Educational Measurement, 14*(2), 117–138.

Lord, F. M. (1980). *Applications of Item Response Theory to Practical Testing Problems.* Hillsdale, N.J.: Lawrence Erlbaum Associates.

Lord, F. M., and Novick, M. R. (1968). *Statistical Theories of Mental Test Scores.* Reading, Mass.: Addison-Wesley, Chaps. 17–20.

Marco, G. L. (1977). Item characteristic curve solutions to 3 intractable testing problems. *Journal of Educational Measurement, 14,* 139–160.

Marks, E., and Lindsay, C. A. (1972). Some results relating to test equating under relaxed test form equivalence. *Journal of Educational Measurement, 9*, 45–56.

Maxey, E. J., and Lenning, O. T. (1974). Another look at concordance tables between ACT and SAT. *Journal of College Student Personnel, 15*, 300–304.

McCall, W. A. (1939). *Measurement*. New York: Macmillan.

McGee, V. E. (1961). Towards a maximally efficient system of braiding for Scholastic Aptitude Test equating. In S. S. Wilks (Ed.), *Scaling and Equating College Board Tests*. Princeton, N.J.: Educational Testing Service, 86–96.

Modu, C. C., and Stern, J. (1977). *The Stability of the SAT-Verbal Score Scale*. New York: College Entrance Examination Board.

Myers, C. T. (1973). *Bias and Interpretation: Cases for Ordinal Measurement*. Paper presented at the Annual Meeting of the American Psychological Association (81st, Montreal, Quebec, Canada, August 27–31).

Myers, C. T. (1975). *Test fairness: A Comment on Fairness in Statistical Analysis* (ETS-RB-75-12). Princeton, N.J.: Educational Testing Service.

Peterson, M. H. F. (1975). *Statistical Methods for Establishing Equivalent Scores in Successive Generations of a Testing Program* (Tech. Rep. No. 39). Stanford, Calif.: Stanford University, Department of Statistics.

Potthoff, R. F. (1966). Equating of grades or scores on the basis of a common battery of measurements. In P. R. Krishnaiah (Ed.), *Multivariate Analysis*. New York: Academic Press, 541–559.

Rasch, G. (1960). *Probabilistic Models for Some Intelligence and Educational Tests*. Copenhagen, Denmark: Danish Institute for Educational Research.

Reilly, R. R. (1975). Empirical option weighting with a correction for guessing. *Educational and Psychological Measurement, 35*, 613–619.

Rentz, R. R., and Bashaw, W. L. (1975). *Equating Reading Tests with the Rasch Model* (Vols. 1 and 2). Athens: Educational Research Laboratory, College of Education, University of Georgia.

Rentz, R. R., and Bashaw, W. L. (1977). The National Reference Scale for reading: An application of the Rasch model. *Journal of Educational Measurement, 14*, 161–179.

Rubin, D. B., and Thayer, D. (1978). Relating tests given to different samples. *Psychometrika, 43*, 3–10.

Schrader, W. B. (1960). Norms. In C. W. Harris (Ed.), *Encyclopedia of Educational Research* (3rd ed.). New York: Macmillan, 922–927.

Schultz, M. K., and Angoff, W. H. (1956). The development of new scales for the Aptitude and Advanced Tests of the Graduate Record Examinations. *Journal of Educational Psychology, 47*, 285–294.

Slinde, J. A., and Linn, R. L. (1977). Vertically equated tests: Fact or phantom? *Journal of Educational Measurement, 14*, 23–32.

Slinde, J. A., and Linn, R. L. (1978). An exploration of the adequacy of the Rasch model for the problem of vertical equating. *Journal of Educational Measurement, 15*, 23–35.

Slinde, J. A., and Linn, R. L. (1979). Vertical equating via the Rasch model for groups of quite different ability and tests of quite different difficulty. *Journal of Educational Measurement, 16*, 159–165.

Stevens, S. S. (1951). Mathematics, measurements, and psychophysics. In S. S. Stevens (Ed.), *Handbook of Experimental Psychology*. New York: Wiley, 1–49.

Swineford, F., and Fan, C. T. (1957). A method of score conversion through item statistics. *Psychometrika, 22*, 185–188.

Szatrowski, T. H. (1978). Explicit solutions, one iteration convergence and averaging in the multivariate normal estimation problem for patterned means and covariances. *Annals of Statistical Mathematics, 30*, Part A, 81–88.

Szatrowski, T. H. (1980). Necessary and sufficient conditions for explicit solutions in the multivariate normal estimation problem for patterned means and covariances. *Annals of Statistics, 8*(4), 802–810.

Terman, L. M., and Merrill, M. A. (1937). *Measuring Intelligence.* New York: Houghton Mifflin.

Thorndike, E. L. (1922). On finding equivalent scores in tests of intelligence. *Journal of Applied Psychology, 6*, 29–33.

Thorndike, R. L. (1971). Concepts of culture—fairness. *Journal of Educational Measurement, 8*, 63–70.

Thorndike, E. L., Bregman, E. O., Cobb, M. V., and Woodyard, E. (1927). *The Measurement of Intelligence.* New York: Columbia University, Teachers College, Bureau of Publications.

Thurstone, L. L. (1925). A method of scaling psychological and educational tests. *Journal of Educational Psychology, 16*, 433–451.

Thurstone, L. L. (1927). A law of comparative judgment. *Psychological Review, 34*, 273–286.

Thurstone, L. L. (1947). The calibration of test items. *American Psychologist, 2*, 103–104.

Tinsley, H. E. A., and Dawis, R. V. (1977). Test-free person measurement with the Rasch simple logistic model. *Applied Psychological Measurement, 1*, 483–487.

Torgerson, W. S. (1958). *Theory and Methods of Scaling.* New York: Wiley.

Tucker, L. R (1953). Scales minimizing the importance of reference groups. In *Proceedings of the 1952 Invitational Conference on Testing Problems.* Princeton, N.J.: Educational Testing Service, 22–28.

Werts, E. E., Rock, D. A., Linn, R. L., and Jöreskog, K. G. (1977). Validating psychometric assumptions within and between several populations. *Educational and Psychological Measurement, 37*, 863–872.

Wilks, S. S. (1946). Sample criteria for testing equality of means, equality of variances, and equality of covariances in a normal multivariate distribution. *Annals of Mathematical Statistics, 17*, 257–281.

Wilks, S. S. (Ed.) (1961). *Scaling and Equating College Board Tests.* Princeton, N.J.: Educational Testing Service.

Wood, R. L., Wingersky, M. S., and Lord, F. M. (1976). *LOGIST: A Computer Program for Estimating Examinee Ability and Item Characteristic Curve Parameters* (RM-76-6). Princeton, N.J.: Educational Testing Service.

Wright, B. D. (1968). Sample-free test calibration and person measurement. In *Proceedings of the 1967 Invitational Conference on Testing Problems.* Princeton, N.J.: Educational Testing Service, 85–101.

Author Index

Numbers in italics refer to the page on which the complete reference is listed.

Subject Index